Communications
in Computer and Information Science 1371

More information about this series at http://www.springer.com/series/7899

Tzung-Pei Hong · Krystian Wojtkiewicz ·
Rathachai Chawuthai · Pawel Sitek (Eds.)

Recent Challenges in Intelligent Information and Database Systems

13th Asian Conference, ACIIDS 2021
Phuket, Thailand, April 7–10, 2021
Proceedings

Springer

Editors
Tzung-Pei Hong ⓘ
National University of Kaohsiung
Kaohsiung, Taiwan

Rathachai Chawuthai ⓘ
King Mongkut's Institute of Technology
Ladkrabang
Bangkok, Thailand

Krystian Wojtkiewicz ⓘ
Wrocław University of Science
and Technology
Wrocław, Poland

Pawel Sitek ⓘ
Kielce University of Technology
Kielce, Poland

ISSN 1865-0929   ISSN 1865-0937 (electronic)
Communications in Computer and Information Science
ISBN 978-981-16-1684-6   ISBN 978-981-16-1685-3   (eBook)
https://doi.org/10.1007/978-981-16-1685-3

This Springer imprint is published by the registered company Springer Nature Singapore Pte Ltd.
The registered company address is: 152 Beach Road, #21-01/04 Gateway East, Singapore 189721, Singapore

Preface

ACIIDS 2021 was the 13th edition of the Asian Conference on Intelligent Information and Database Systems. The aim of ACIIDS 2021 was to provide an international forum for research workers with scientific backgrounds on the technology of intelligent information and database systems and its various applications. The ACIIDS 2021 conference was co-organized by King Mongkut's Institute of Technology Ladkrabang (Thailand) and Wrocław University of Science and Technology (Poland) in cooperation with the IEEE SMC Technical Committee on Computational Collective Intelligence, the European Research Center for Information Systems (ERCIS), The University of Newcastle (Australia), Yeungnam University (South Korea), Leiden University (The Netherlands), Universiti Teknologi Malaysia (Malaysia), BINUS University (Indonesia), Quang Binh University (Vietnam), Nguyen Tat Thanh University (Vietnam), and the "Collective Intelligence" section of the Committee on Informatics of the Polish Academy of Sciences. ACIIDS 2021 was at first scheduled to be held in Phuket, Thailand during April 7–10, 2021. However, due to the COVID-19 pandemic, the conference was moved to the virtual space and conducted online using the ZOOM videoconferencing system.

The ACIIDS conference series is already well established. The first two events, ACIIDS 2009 and ACIIDS 2010, took place in Dong Hoi City and Hue City in Vietnam, respectively. The third event, ACIIDS 2011, took place in Daegu (South Korea), followed by the fourth event, ACIIDS 2012, in Kaohsiung (Taiwan). The fifth event, ACIIDS 2013, was held in Kuala Lumpur (Malaysia) while the sixth event, ACIIDS 2014, was held in Bangkok (Thailand). The seventh event, ACIIDS 2015, took place in Bali (Indonesia), followed by the eighth event, ACIIDS 2016, in Da Nang (Vietnam). The ninth event, ACIIDS 2017, was organized in Kanazawa (Japan). The 10th jubilee conference, ACIIDS 2018, was held in Dong Hoi City (Vietnam), followed by the 11th event, ACIIDS 2019, in Yogyakarta (Indonesia). The 12th and 13th events were initially planned to be held in Phuket (Thailand). However, the global pandemic relating to COVID-19 resulted in both editions of the conference being held online in virtual space.

This volume contains 35 peer-reviewed papers, selected for poster presentation from 291 submissions in total. Papers included in this volume cover the following topics: data mining and machine learning methods, advanced data mining techniques and applications, intelligent and contextual systems, natural language processing, network systems and applications, computational imaging and vision, decision support and control systems, and data modelling and processing for industry 4.0.

The accepted and presented papers focused on new trends and challenges facing the intelligent information and database systems community. The presenters showed how research work could stimulate novel and innovative applications. We hope that you found these results useful and inspiring for your future research work.

We would like to express our sincere thanks to the honorary chairs for their support: Prof. Suchatvee Suwansawat (President of King Mongkut's Institute of Technology Ladkrabang, Thailand), Prof. Arkadiusz Wójs (Rector of Wrocław University of Science and Technology, Poland), and Prof. Moonis Ali (President of the International Society of Applied Intelligence, USA). We would like to express our thanks to the keynote speakers for their world-class plenary speeches: Dr. Edwin Lughofer from Johannes Kepler University Linz (Austria), Prof. Manuel Núñez from Universidad Complutense de Madrid (Spain), and Prof. Agachai Sumalee from Chulalongkorn University (Thailand).

We cordially thank our main sponsors: King Mongkut's Institute of Technology Ladkrabang (Thailand) and Wrocław University of Science and Technology (Poland), as well as all aforementioned cooperating universities and organizations. Our special thanks are also due to Springer for publishing the proceedings and to all the other sponsors for their kind support.

Our special thanks go to the Program Chairs, Special Session Chairs, Organizing Chairs, Publicity Chairs, Liaison Chairs, and Local Organizing Committee for their work towards the conference. We sincerely thank all the members of the international Program Committee for their valuable efforts in the review process, which helped us to guarantee the highest quality of the selected papers for the conference. We cordially thank all the authors, for their valuable contributions, and the other participants of the conference. The conference would not have been possible without their support. Thanks are also due to the numerous experts who contributed to making the event a success.

April 2021

Tzung-Pei Hong
Krystian Wojtkiewicz
Rathachai Chawuthai
Paweł Sitek

Conference Organization

Honorary Chairs

Suchatvee Suwansawat	President of King Mongkut's Institute of Technology Ladkrabang, Thailand
Arkadiusz Wójs	Rector of Wrocław University of Science and Technology, Poland
Moonis Ali	President of the International Society of Applied Intelligence, USA

General Chairs

Ngoc Thanh Nguyen	Wrocław University of Science and Technology, Poland
Suphamit Chittayasothorn	King Mongkut's Institute of Technology Ladkrabang, Thailand

Program Chairs

Dusit Niyato	Nanyang Technological University, Singapore
Tzung-Pei Hong	National University of Kaohsiung, Taiwan
Edward Szczerbicki	University of Newcastle, Australia
Bogdan Trawiński	Wrocław University of Science and Technology, Poland

Steering Committee

Ngoc Thanh Nguyen (Chair)	Wrocław University of Science and Technology, Poland
Longbing Cao	University of Science and Technology Sydney, Australia
Suphamit Chittayasothorn	King Mongkut's Institute of Technology Ladkrabang, Thailand
Ford Lumban Gaol	Bina Nusantara University, Indonesia
Tzung-Pei Hong	National University of Kaohsiung, Taiwan
Dosam Hwang	Yeungnam University, South Korea
Bela Stantic	Griffith University, Australia
Geun-Sik Jo	Inha University, South Korea
Le Thi Hoai An	University of Lorraine, France
Toyoaki Nishida	Kyoto University, Japan
Leszek Rutkowski	Częstochowa University of Technology, Poland
Ali Selamat	Universiti Teknologi Malaysia, Malaysia

Special Session Chairs

Krystian Wojtkiewicz   Wrocław University of Science and Technology,
                              Poland
Rathachai Chawuthai   King Mongkut's Institute of Technology Ladkrabang,
                              Thailand

Liaison Chairs

Ford Lumban Gaol   Bina Nusantara University, Indonesia
Quang-Thuy Ha   VNU-University of Engineering and Technology,
                              Vietnam
Mong-Fong Horng   National Kaohsiung University of Applied Sciences,
                              Taiwan
Dosam Hwang   Yeungnam University, South Korea
Le Minh Nguyen   Japan Advanced Institute of Science and Technology,
                              Japan
Ali Selamat   Universiti Teknologi Malaysia, Malaysia
Paweł Sitek   Kielce University of Technology, Poland

Organizing Chairs

Wiboon Prompanich   King Mongkut's Institute of Technology Ladkrabang,
                              Thailand
Amnach Khawne   King Mongkut's Institute of Technology Ladkrabang,
                              Thailand
Adrianna Kozierkiewicz   Wrocław University of Science and Technology,
                              Poland
Bogumiła Hnatkowska   Wrocław University of Science and Technology,
                              Poland

Publicity Chairs

Natthapong Jungteerapani   King Mongkut's Institute of Technology Ladkrabang,
                              Thailand
Marek Kopel   Wrocław University of Science and Technology,
                              Poland
Marek Krótkiewicz   Wrocław University of Science and Technology,
                              Poland

Webmaster

Marek Kopel   Wrocław University of Science and Technology,
                              Poland

Local Organizing Committee

Pakorn Watanachaturaporn	King Mongkut's Institute of Technology Ladkrabang, Thailand
Sathaporn Promwong	King Mongkut's Institute of Technology Ladkrabang, Thailand
Putsadee Pornphol	Phuket Rajabhat University, Thailand
Maciej Huk	Wrocław University of Science and Technology, Poland
Marcin Jodłowiec	Wrocław University of Science and Technology, Poland
Marcin Pietranik	Wrocław University of Science and Technology, Poland

Keynote Speakers

Edwin Lughofer	Johannes Kepler University Linz, Austria
Manuel Núñez	Universidad Complutense de Madrid, Spain
Agachai Sumalee	Chulalongkorn University, Thailand

Special Sessions Organizers

1. ADMTA 2021: Special Session on Advanced Data Mining Techniques and Applications

Chun-Hao Chen	Tamkang University, Taiwan
Bay Vo	Ho Chi Minh City University of Technology, Vietnam
Tzung-Pei Hong	National University of Kaohsiung, Taiwan

2. CIV 2021: Special Session on Computational Imaging and Vision

Manish Khare	Dhirubhai Ambani Institute of Information and Communication Technology, India
Prashant Srivastava	NIIT University, India
Om Prakash	HNB Garwal University, India
Jeonghwan Gwak	Korea National University of Transportation, South Korea

3. CoSenseAI 2021: Special Session on Commonsense Knowledge, Reasoning and Programming in Artificial Intelligence

Pascal Bouvry	University of Luxembourg, Luxembourg
Matthias R. Brust	University of Luxembourg, Luxembourg
Grégoire Danoy	University of Luxembourg, Luxembourg
El-ghazil Talbi	University of Lille, France

4. DMPI-4.0 vol. 2 – 2021: Special Session on Data Modelling and Processing for Industry 4.0

Du Haizhou	Shanghai University of Electric Power, China
Wojciech Hunek	Opole University of Technology, Poland
Marek Krótkiewicz	Wrocław University of Science and Technology, Poland
Krystian Wojtkiewicz	Wrocław University of Science and Technology, Poland

5. ICxS 2021: Special Session on Intelligent and Contextual Systems

Maciej Huk	Wrocław University of Science and Technology, Poland
Keun Ho Ryu	Ton Duc Thang University, Vietnam
Goutam Chakraborty	Iwate Prefectural University, Japan
Qiangfu Zhao	University of Aizu, Japan
Chao-Chun Chen	National Cheng Kung University, Taiwan
Rashmi Dutta Baruah	Indian Institute of Technology Guwahati, India
Tetsuji Kuboyama	Gakushuin University, Japan

6. ISCEC 2021: Special Session on Intelligent Supply Chains and e-Commerce

Arkadiusz Kawa	Łukasiewicz Research Network – The Institute of Logistics and Warehousing, Poland
Justyna Światowiec-Szczepańska	Poznań University of Economics and Business, Poland
Bartłomiej Pierański	Poznań University of Economics and Business, Poland

7. MMAML 2021: Special Session on Multiple Model Approach to Machine Learning

Tomasz Kajdanowicz	Wrocław University of Science and Technology, Poland
Edwin Lughofer	Johannes Kepler University Linz, Austria
Bogdan Trawiński	Wrocław University of Science and Technology, Poland

Senior Program Committee

Ajith Abraham	Machine Intelligence Research Labs, USA
Jesus Alcala-Fdez	University of Granada, Spain
Lionel Amodeo	University of Technology of Troyes, France
Ahmad Taher Azar	Prince Sultan University, Saudi Arabia
Thomas Bäck	Leiden University, Netherlands
Costin Badica	University of Craiova, Romania
Ramazan Bayindir	Gazi University, Turkey
Abdelhamid Bouchachia	Bournemouth University, UK
David Camacho	Universidad Autonoma de Madrid, Spain

Leopoldo Eduardo Cardenas-Barron — Tecnologico de Monterrey, Mexico

Oscar Castillo — Tijuana Institute of Technology, Mexico

Nitesh Chawla — University of Notre Dame, USA

Rung-Ching Chen — Chaoyang University of Technology, Taiwan

Shyi-Ming Chen — National Taiwan University of Science and Technology, Taiwan

Simon Fong — University of Macau, Macau SAR

Hamido Fujita — Iwate Prefectural University, Japan

Mohamed Gaber — Birmingham City University, UK

Marina L. Gavrilova — University of Calgary, Canada

Daniela Godoy — ISISTAN Research Institute, Argentina

Fernando Gomide — University of Campinas, Brazil

Manuel Grana — University of the Basque Country, Spain

Claudio Gutierrez — Universidad de Chile, Chile

Francisco Herrera — University of Granada, Spain

Tzung-Pei Hong — National University of Kaohsiung, Taiwan

Dosam Hwang — Yeungnam University, South Korea

Mirjana Ivanovic — University of Novi Sad, Serbia

Janusz Jeżewski — Institute of Medical Technology and Equipment ITAM, Poland

Piotr Jedrzejowicz — Gdynia Maritime University, Poland

Kang-Hyun Jo — University of Ulsan, South Korea

Jason J. Jung — Chung-Ang University, South Korea

Janusz Kacprzyk — Systems Research Institute, Polish Academy of Sciences, Poland

Nikola Kasabov — Auckland University of Technology, New Zealand

Muhammad Khurram Khan — King Saud University, Saudi Arabia

Frank Klawonn — Ostfalia University of Applied Sciences, Germany

Joanna Kolodziej — Cracow University of Technology, Poland

Józef Korbicz — University of Zielona Gora, Poland

Ryszard Kowalczyk — Swinburne University of Technology, Australia

Bartosz Krawczyk — Virginia Commonwealth University, USA

Ondrej Krejcar — University of Hradec Králové, Czech Republic

Adam Krzyzak — Concordia University, Canada

Mark Last — Ben-Gurion University of the Negev, Israel

Le Thi Hoai An — University of Lorraine, France

Kun Chang Lee — Sungkyunkwan University, South Korea

Edwin Lughofer — Johannes Kepler University Linz, Austria

Nezam Mahdavi-Amiri — Sharif University of Technology, Iran

Yannis Manolopoulos — Open University of Cyprus, Cyprus

Klaus-Robert Müller — Technical University of Berlin, Germany

Saeid Nahavandi — Deakin University, Australia

Grzegorz J. Nalepa — AGH University of Science and Technology, Poland

Ngoc-Thanh Nguyen	Wrocław University of Science and Technology, Poland
Dusit Niyato	Nanyang Technological University, Singapore
Yusuke Nojima	Osaka Prefecture University, Japan
Manuel Núñez	Universidad Complutense de Madrid, Spain
Jeng-Shyang Pan	Fujian University of Technology, China
Marcin Paprzycki	Systems Research Institute, Polish Academy of Sciences, Poland
Bernhard Pfahringer	University of Waikato, New Zealand
Hoang Pham	Rutgers University, USA
Tao Pham Dinh	INSA Rouen, France
Radu-Emil Precup	Politehnica University of Timisoara, Romania
Leszek Rutkowski	Częstochowa University of Technology, Poland
Juergen Schmidhuber	Swiss AI Lab IDSIA, Switzerland
Björn Schuller	University of Passau, Germany
Ali Selamat	Universiti Teknologi Malaysia, Malaysia
Andrzej Skowron	Warsaw University, Poland
Jerzy Stefanowski	Poznań University of Technology, Poland
Agachai Sumalee	Chulalongkorn University, Thailand
Edward Szczerbicki	University of Newcastle, Australia
Ryszard Tadeusiewicz	AGH University of Science and Technology, Poland
Muhammad Atif Tahir	National University of Computing & Emerging Sciences, Pakistan
Bay Vo	Ho Chi Minh City University of Technology, Vietnam
Gottfried Vossen	University of Münster, Germany
Lipo Wang	Nanyang Technological University, Singapore
Junzo Watada	Waseda University, Japan
Michał Woźniak	Wrocław University of Science and Technology, Poland
Farouk Yalaoui	University of Technology of Troyes, France
Sławomir Zadrożny	Systems Research Institute, Polish Academy of Sciences, Poland
Zhi-Hua Zhou	Nanjing University, China

Program Committee

Muhammad Abulaish	South Asian University, India
Bashar Al-Shboul	University of Jordan, Jordan
Toni Anwar	Universiti Teknologi PETRONAS, Malaysia
Taha Arbaoui	University of Technology of Troyes, France
Mehmet Emin Aydin	University of the West of England, UK
Amelia Badica	University of Craiova, Romania
Kambiz Badie	ICT Research Institute, Iran
Hassan Badir	École Nationale des Sciences Appliquées de Tanger, Morocco
Dariusz Barbucha	Gdynia Maritime University, Poland

Paulo Batista	University of Évora, Portugal
Maumita Bhattacharya	Charles Sturt University, Australia
Bülent Bolat	Yildiz Technical University, Turkey
Mariusz Boryczka	University of Silesia, Poland
Urszula Boryczka	University of Silesia, Poland
Zouhaier Brahmia	University of Sfax, Tunisia
Stephane Bressan	National University of Singapore, Singapore
Peter Brida	University of Zilina, Slovakia
Andrej Brodnik	University of Ljubljana, Slovenia
Grażyna Brzykcy	Poznań University of Technology, Poland
Robert Burduk	Wrocław University of Science and Technology, Poland
Aleksander Byrski	AGH University of Science and Technology, Poland
Dariusz Ceglarek	WSB University in Poznań, Poland
Zenon Chaczko	University of Technology Sydney, Australia
Somchai Chatvichienchai	University of Nagasaki, Japan
Chun-Hao Chen	Tamkang University, Taiwan
Leszek J. Chmielewski	Warsaw University of Life Sciences, Poland
Kazimierz Choroś	Wrocław University of Science and Technology, Poland
Kun-Ta Chuang	National Cheng Kung University, Taiwan
Dorian Cojocaru	University of Craiova, Romania
Jose Alfredo Ferreira Costa	Federal University of Rio Grande do Norte (UFRN), Brazil
Ireneusz Czarnowski	Gdynia Maritime University, Poland
Piotr Czekalski	Silesian University of Technology, Poland
Theophile Dagba	University of Abomey-Calavi, Benin
Phuc Do	Vietnam National University, Ho Chi Minh City, Vietnam
Tien V. Do	Budapest University of Technology and Economics, Hungary
Rafal Doroz	University of Silesia, Poland
El-Sayed M. El-Alfy	King Fahd University of Petroleum and Minerals, Saudi Arabia
Keiichi Endo	Ehime University, Japan
Sebastian Ernst	AGH University of Science and Technology, Poland
Usef Faghihi	Université du Québec à Trois-Rivières, Canada
Rim Faiz	University of Carthage, Tunisia
Victor Felea	Alexandru Ioan Cuza University, Romania
Dariusz Frejlichowski	West Pomeranian University of Technology, Szczecin, Poland
Blanka Frydrychova Klimova	University of Hradec Kralove, Czech Republic
Janusz Getta	University of Wollongong, Australia
Gergő Gombos	Eötvös Loránd University, Hungary
Antonio Gonzalez-Pardo	Universidad Autónoma de Madrid, Spain

Quang-Thuy Ha	VNU-University of Engineering and Technology, Vietnam
Dawit Haile	Addis Ababa University, Ethiopia
Pei-Yi Hao	National Kaohsiung University of Applied Sciences, Taiwan
Marcin Hernes	Wrocław University of Economics and Business, Poland
Kouichi Hirata	Kyushu Institute of Technology, Japan
Bogumiła Hnatkowska	Wrocław University of Science and Technology, Poland
Huu Hanh Hoang	Posts and Telecommunications Institute of Technology, Vietnam
Quang Hoang	Hue University of Sciences, Vietnam
Van-Dung Hoang	Ho Chi Minh City University of Technology and Education, Vietnam
Jeongky Hong	Yeungnam University, South Korea
Maciej Huk	Wrocław University of Science and Technology, Poland
Zbigniew Huzar	Wrocław University of Science and Technology, Poland
Dmitry Ignatov	National Research University Higher School of Economics, Russia
Hazra Imran	University of British Columbia, Canada
Sanjay Jain	National University of Singapore, Singapore
Khalid Jebari	Abdelmalek Essaâdi University, Morocco
Joanna Jędrzejowicz	University of Gdansk, Poland
Przemysław Juszczuk	University of Economics in Katowice, Poland
Dariusz Kania	Silesian University of Technology, Poland
Mehmet Karaata	Kuwait University, Kuwait
Arkadiusz Kawa	Institute of Logistics and Warehousing, Poland
Zaheer Khan	University of the West of England, UK
Marek Kisiel-Dorohinicki	AGH University of Science and Technology, Poland
Attila Kiss	Eötvös Loránd University, Hungary
Jerzy Klamka	Silesian University of Technology, Poland
Shinya Kobayashi	Ehime University, Japan
Marek Kopel	Wrocław University of Science and Technology, Poland
Jan Kozak	University of Economics in Katowice, Poland
Adrianna Kozierkiewicz	Wrocław University of Science and Technology, Poland
Dalia Kriksciuniene	Vilnius University, Lithuania
Dariusz Król	Wrocław University of Science and Technology, Poland
Marek Krótkiewicz	Wrocław University of Science and Technology, Poland
Marzena Kryszkiewicz	Warsaw University of Technology, Poland

Jan Kubíček	VSB - Technical University of Ostrava, Czech Republic
Tetsuji Kuboyama	Gakushuin University, Japan
Elżbieta Kukla	Wrocław University of Science and Technology, Poland
Julita Kulbacka	Wrocław Medical University, Poland
Marek Kulbacki	Polish-Japanese Academy of Information Technology, Poland
Kazuhiro Kuwabara	Ritsumeikan University, Japan
Jong Wook Kwak	Yeungnam University, South Korea
Halina Kwaśnicka	Wrocław University of Science and Technology, Poland
Annabel Latham	Manchester Metropolitan University, UK
Yue-Shi Lee	Ming Chuan University, Taiwan
Florin Leon	Gheorghe Asachi Technical University of Iasi, Romania
Horst Lichter	RWTH Aachen University, Germany
Tony Lindgren	Stockholm University, Sweden
Sebastian Link	University of Auckland, New Zealand
Igor Litvinchev	Nuevo Leon State University, Mexico
Doina Logofătu	Frankfurt University of Applied Sciences, Germany
Lech Madeyski	Wrocław University of Science and Technology, Poland
Bernadetta Maleszka	Wrocław University of Science and Technology, Poland
Konstantinos Margaritis	University of Macedonia, Greece
Takashi Matsuhisa	Karelia Research Centre, Russian Academy of Science, Russia
Tamás Matuszka	Eötvös Loránd University, Hungary
Michael Mayo	University of Waikato, New Zealand
Vladimir Mazalov	Karelia Research Centre, Russian Academy of Science, Russia
Héctor Menéndez	University College London, UK
Mercedes Merayo	Universidad Complutense de Madrid, Spain
Jacek Mercik	WSB University in Wrocław, Poland
Radosław Michalski	Wrocław University of Science and Technology, Poland
Peter Mikulecky	University of Hradec Kralove, Czech Republic
Miroslava Mikusova	University of Zilina, Slovakia
Marek Milosz	Lublin University of Technology, Poland
Jolanta Mizera-Pietraszko	Wrocław University of Science and Technology, Poland
Dariusz Mrozek	Silesian University of Technology, Poland
Leo Mrsic	IN2data Ltd., Croatia
Agnieszka Mykowiecka	Institute of Computer Science, Polish Academy of Sciences, Poland

Pawel Myszkowski	Wrocław University of Science and Technology, Poland
Fulufhelo Nelwamondo	Council for Scientific and Industrial Research, South Africa
Huu-Tuan Nguyen	Vietnam Maritime University, Vietnam
Loan T. T. Nguyen	Ho Chi Minh City International University, Vietnam
Quang-Vu Nguyen	Vietnam-Korea University of Information and Communication Technology, Vietnam
Thai-Nghe Nguyen	Cantho University, Vietnam
Mariusz Nowostawski	Norwegian University of Science and Technology, Norway
Alberto Núñez	Universidad Complutense de Madrid, Spain
Tarkko Oksala	Aalto University, Finland
Mieczysław Owoc	Wrocław University of Economics and Business, Poland
Marcin Paprzycki	Systems Research Institute, Polish Academy of Sciences, Poland
Panos Patros	University of Waikato, New Zealand
Danilo Pelusi	University of Teramo, Italy
Marek Penhaker	VSB - Technical University of Ostrava, Czech Republic
Maciej Piasecki	Wrocław University of Science and Technology, Poland
Bartłomiej Pierański	Poznań University of Economics and Business, Poland
Dariusz Pierzchala	Military University of Technology, Poland
Marcin Pietranik	Wrocław University of Science and Technology, Poland
Elias Pimenidis	University of the West of England, UK
Jaroslav Pokorný	Charles University in Prague, Czech Republic
Nikolaos Polatidis	University of Brighton, UK
Elvira Popescu	University of Craiova, Romania
Petra Poulova	University of Hradec Kralove, Czech Republic
Om Prakash	University of Allahabad, India
Radu-Emil Precup	Politehnica University of Timisoara, Romania
Małgorzata Przybyła-Kasperek	University of Silesia, Poland
Paulo Quaresma	Universidade de Evora, Portugal
David Ramsey	Wrocław University of Science and Technology, Poland
Mohammad Rashedur Rahman	North South University, Bangladesh
Ewa Ratajczak-Ropel	Gdynia Maritime University, Poland
Alexander Ryjov	Lomonosov Moscow State University, Russia
Keun Ho Ryu	Chungbuk National University, South Korea
Virgilijus Sakalauskas	Vilnius University, Lithuania

Daniel Sanchez	University of Granada, Spain
Rafal Scherer	Częstochowa University of Technology, Poland
S. M. N. Arosha Senanayake	Universiti Brunei Darussalam, Brunei Darussalam
Natalya Shakhovska	Lviv Polytechnic National University, Ukraine
Andrzej Siemiński	Wrocław University of Science and Technology, Poland
Dragan Simic	University of Novi Sad, Serbia
Bharat Singh	Universiti Teknology PETRONAS, Malaysia
Paweł Sitek	Kielce University of Technology, Poland
Krzysztof Ślot	Łódź University of Technology, Poland
Adam Słowik	Koszalin University of Technology, Poland
Vladimir Sobeslav	University of Hradec Kralove, Czech Republic
Zenon A. Sosnowski	Bialystok University of Technology, Poland
Harco Leslie Hendric Spits Warnars	Binus University, Indonesia
Bela Stantic	Griffith University, Australia
Stanimir Stoyanov	Plovdiv University "Paisii Hilendarski", Bulgaria
Ja-Hwung Su	Cheng Shiu University, Taiwan
Libuse Svobodova	University of Hradec Kralove, Czech Republic
Jerzy Świątek	Wrocław University of Science and Technology, Poland
Julian Szymański	Gdansk University of Technology, Poland
Yasufumi Takama	Tokyo Metropolitan University, Japan
Maryam Tayefeh Mahmoudi	ICT Research Institute, Iran
Zbigniew Telec	Wrocław University of Science and Technology, Poland
Dilhan Thilakarathne	Vrije Universiteit Amsterdam, Netherlands
Satoshi Tojo	Japan Advanced Institute of Science and Technology, Japan
Bogdan Trawiński	Wrocław University of Science and Technology, Poland
Maria Trocan	Institut Superieur d'Electronique de Paris, France
Krzysztof Trojanowski	Cardinal Stefan Wyszyński University in Warsaw, Poland
Ualsher Tukeyev	Al-Farabi Kazakh National University, Kazakhstan
Olgierd Unold	Wrocław University of Science and Technology, Poland
Jørgen Villadsen	Technical University of Denmark, Denmark
Eva Volna	University of Ostrava, Czech Republic
Wahyono Wahyono	Universitas Gadjah Mada, Indonesia
Junzo Watada	Waseda University, Japan
Pawel Weichbroth	Gdansk University of Technology, Poland
Izabela Wierzbowska	Gdynia Maritime University, Poland

Krystian Wojtkiewicz	Wrocław University of Science and Technology, Poland
Krzysztof Wróbel	University of Silesia, Poland
Marian Wysocki	Rzeszow University of Technology, Poland
Xin-She Yang	Middlesex University, UK
Tulay Yildirim	Yildiz Technical University, Turkey
Jonghee Youn	Yeungnam University, South Korea
Drago Zagar	University of Osijek, Croatia
Danuta Zakrzewska	Łódź University of Technology, Poland
Constantin-Bala Zamfirescu	Lucian Blaga University of Sibiu, Romania
Katerina Zdravkova	Ss. Cyril and Methodius University in Skopje, Macedonia
Vesna Zeljkovic	Lincoln University, USA
Aleksander Zgrzywa	Wrocław University of Science and Technology, Poland
Jianlei Zhang	Nankai University, China
Maciej Zięba	Wrocław University of Science and Technology, Poland
Adam Ziębiński	Silesian University of Technology, Poland

Program Committees of Special Sessions

ADMTA 2021: Special Session on Advanced Data Mining Techniques and Applications

Tzung-Pei Hong	National University of Kaohsiung, Taiwan
Tran Minh Quang	Ho Chi Minh City University of Technology, Vietnam
Bac Le	VNUHCM-University of Science, Vietnam
Bay Vo	Ho Chi Minh City University of Technology, Vietnam
Chun-Hao Chen	National Taipei University of Technology, Taipei
Mu-En Wu	National Taipei University of Technology, Taipei
Wen-Yang Lin	National University of Kaohsiung, Taiwan
Yeong-Chyi Lee	Cheng Shiu University, Taiwan
Le Hoang Son	VNU-University of Science, Vietnam
Vo Thi Ngoc Chau	Ho Chi Minh City University of Technology, Vietnam
Van Vo	Ho Chi Minh University of Industry, Vietnam
Ja-Hwung Su	National University of Kaohsiung, Taiwan
Ming-Tai Wu	Shandong University of Science and Technology, China
Kawuu W. Lin	National Kaohsiung University of Science and Technology, Taiwan
Ju-Chin Chen	National Kaohsiung University of Science and Technology, Taiwan
Tho Le	Ho Chi Minh City University of Technology, Vietnam
Dang Nguyen	Deakin University, Australia

Hau Le	Thuyloi University, Vietnam
Thien-Hoang Van	Ho Chi Minh City University of Technology, Vietnam
Tho Quan	Ho Chi Minh City University of Technology, Vietnam
Ham Nguyen	University of People's Security, Vietnam
Thiet Pham	Ho Chi Minh University of Industry, Vietnam
Nguyen Thi Thuy Loan	Nguyen Tat Thanh University, Vietnam
C. C. Chen	National Cheng Kung University, Taiwan
Jerry Chun-Wei Lin	Western Norway University of Applied Sciences, Norway
Ko-Wei Huang	National Kaohsiung University of Science and Technology, Taiwan
Ding-Chau Wang	Southern Taiwan University of Science and Technology, Taiwan

CIV 2021: Special Session on Computational Imaging and Vision

Ishwar Sethi	Oakland University, USA
Moongu Jeon	Gwangju Institute of Science and Technology, South Korea
Benlian Xu	Changshu Institute of Technology, China
Weifeng Liu	Hangzhou Danzi University, China
Ashish Khare	University of Allahabad, India
Moonsoo Kang	Chosun University, South Korea
Sang Woong Lee	Gachon University, South Korea
Ekkarat Boonchieng	Chiang Mai University, Thailand
Jeong-Seon Park	Chonnam National University, South Korea
Unsang Park	Sogang University, South Korea
R. Z. Khan	Aligarh Muslim University, India
Suman Mitra	DA-IICT, India
Bakul Gohel	DA-IICT, India
Sathya Narayanan	NTU, Singapore
Jaeyong Kang	Korea National University of Transportation, South Korea
Zahid Ullah	Korea National University of Transportation, South Korea

CoSenseAI 2021: Special Session on Commonsense Knowledge, Reasoning and Programming in Artificial Intelligence

Roland Bouffanais	Singapore University of Technology and Design, Singapore
Ronaldo Menezes	University of Exeter, UK
Apivadee Piyatumrong	NECTEC, Thailand
M. Ilhan Akbas	Embry-Riddle Aeronautical University, USA
Christoph Benzmüller	Freie Universität Berlin, Germany

Bernabe Dorronsoro	University of Cadiz, Spain
Rastko Selmic	Concordia University, Canada
Daniel Stolfi	University of Luxembourg, Luxembourg
Juan Luis Jiménez Laredo	Normandy University, France
Kittichai Lavangnananda	King Mongkut's University of Technology Thonburi, Thailand
Boonyarit Changaival	University of Luxembourg, Luxembourg
Marco Rocchetto	ALES, United Technologies Research Center, Italy
Jundong Chen	Dickinson State University, USA
Emmanuel Kieffer	University of Luxembourg, Luxembourg
Fang-Jing Wu	Technical University Dortmund, Germany
Umer Wasim	University of Luxembourg, Luxembourg

DMPI-4.0 vol. 2 – 2021: Special Session on Data Modelling and Processing for Industry 4.0

Jörg Becker	University of Münster, Germany
Rafał Cupek	Silesian University of Technology, Poland
Helena Dudycz	Wrocław University of Economics and Business, Poland
Marcin Fojcik	Western Norway University of Applied Sciences, Norway
Du Haizhou	Shanghai University of Electric Power, China
Marcin Hernes	Wrocław University of Economics and Business, Poland
Wojciech Hunek	Opole University of Technology, Poland
Marcin Jodłowiec	Wrocław University of Science and Technology, Poland
Marek Krótkiewicz	Wrocław University of Science and Technology, Poland
Florin Leon	Gheorghe Asachi Technical University of Iasi, Romania
Rafał Palak	Wrocław University of Science and Technology, Poland
Jacek Piskorowski	West Pomeranian University of Technology, Szczecin, Poland
Khouloud Salameh	American University of Ras Al Khaimah, United Arab Emirates
Predrag Stanimirović	University of Niš, Serbia
Krystian Wojtkiewicz	Wrocław University of Science and Technology, Poland
Feifei Xu	Southeast University, China

ICxS 2021: Special Session on Intelligent and Contextual Systems

Adriana Albu	Polytechnic University of Timisoara, Romania
Basabi Chakraborty	Iwate Prefectural University, Japan
Chao-Chun Chen	National Cheng Kung University, Taiwan
Dariusz Frejlichowski	West Pomeranian University of Technology, Szczecin, Poland
Erdenebileg Batbaatar	Chungbuk National University, South Korea
Goutam Chakraborty	Iwate Prefectural University, Japan
Ha Manh Tran	Ho Chi Minh City International University, Vietnam
Hong Vu Nguyen	Ton Duc Thang University, Vietnam
Hideyuki Takahashi	Tohoku Gakuin University, Japan
Jerzy Świątek	Wrocław University of Science and Technology, Poland
Józef Korbicz	University of Zielona Gora, Poland
Keun Ho Ryu	Chungbuk National University, South Korea
Khanindra Pathak	Indian Institute of Technology Kharagpur, India
Kilho Shin	Gakashuin University, Japan
Lkhagvadorj Munkhdalai	Chungbuk National University, South Korea
Maciej Huk	Wrocław University of Science and Technology, Poland
Marcin Fojcik	Western Norway University of Applied Sciences, Norway
Masafumi Matsuhara	Iwate Prefectural University, Japan
Meijing Li	Shanghai Maritime University, China
Min-Hsiung Hung	Chinese Culture University, Taiwan
Miroslava Mikusova	University of Žilina, Slovakia
Musa Ibrahim	Chungbuk National University, South Korea
Nguyen Khang Pham	Can Tho University, Vietnam
Nipon Theera-Umpon	Chiang Mai University, Thailand
Plamen Angelov	Lancaster University, UK
Qiangfu Zhao	University of Aizu, Japan
Quan Thanh Tho	Ho Chi Minh City University of Technology, Vietnam
Rafal Palak	Wrocław University of Science and Technology, Poland
Rashmi Dutta Baruah	Indian Institute of Technology Guwahati, India
Sansanee Auephanwiriyakul	Chiang Mai University, Thailand
Sonali Chouhan	Indian Institute of Technology Guwahati, India
Takako Hashimoto	Chiba University of Commerce, Japan
Tetsuji Kuboyama	Gakushuin University, Japan
Thai-Nghe Nguyen	Can Tho University, Vietnam

ISCEC 2021: Special Session on Intelligent Supply Chains and e-Commerce

Arkadiusz Kawa	Łukasiewicz Research Network – The Institute of Logistics and Warehousing, Poland
Bartłomiej Pierański	Poznań University of Economics and Business, Poland
Carlos Andres Romano	Polytechnic University of Valencia, Spain
Costin Badica	University of Craiova, Romania
Davor Dujak	University of Osijek, Croatia
Miklós Krész	InnoRenew, Slovenia
Paweł Pawlewski	Poznań University of Technology, Poland
Adam Koliński	Łukasiewicz Research Network – The Institute of Logistics and Warehousing, Poland
Marcin Anholcer	Poznań University of Economics and Business, Poland

MMAML 2021: Special Session on Multiple Model Approach to Machine Learning

Emili Balaguer-Ballester	Bournemouth University, UK
Urszula Boryczka	University of Silesia, Poland
Abdelhamid Bouchachia	Bournemouth University, UK
Robert Burduk	Wrocław University of Science and Technology, Poland
Oscar Castillo	Tijuana Institute of Technology, Mexico
Rung-Ching Chen	Chaoyang University of Technology, Taiwan
Suphamit Chittayasothorn	King Mongkut's Institute of Technology Ladkrabang, Thailand
José Alfredo F. Costa	Federal University of Rio Grande do Norte (UFRN), Brazil
Ireneusz Czarnowski	Gdynia Maritime University, Poland
Fernando Gomide	State University of Campinas, Brazil
Francisco Herrera	University of Granada, Spain
Tzung-Pei Hong	National University of Kaohsiung, Taiwan
Piotr Jędrzejowicz	Gdynia Maritime University, Poland
Tomasz Kajdanowicz	Wrocław University of Science and Technology, Poland
Yong Seog Kim	Utah State University, USA
Bartosz Krawczyk	Virginia Commonwealth University, USA
Kun Chang Lee	Sungkyunkwan University, South Korea
Edwin Lughofer	Johannes Kepler University Linz, Austria
Hector Quintian	University of Salamanca, Spain
Andrzej Siemiński	Wrocław University of Science and Technology, Poland
Dragan Simic	University of Novi Sad, Serbia

Adam Słowik	Koszalin University of Technology, Poland
Zbigniew Telec	Wrocław University of Science and Technology, Poland
Bogdan Trawiński	Wrocław University of Science and Technology, Poland
Olgierd Unold	Wrocław University of Science and Technology, Poland
Michał Woźniak	Wrocław University of Science and Technology, Poland
Zhongwei Zhang	University of Southern Queensland, Australia
Zhi-Hua Zhou	Nanjing University, China

Contents

Intelligent and Contextual Systems

Natural Language Processing

Network Systems and Applications

Computational Imaging and Vision

Decision Support and Control Systems

Data Modelling and Processing for Industry 4.0

Data Mining and Machine Learning Methods

CNN Based Analysis of the Luria's Alternating Series Test for Parkinson's Disease Diagnostics

Sergei Zarembo[1(✉)], Sven Nõmm[1(✉)], Kadri Medijainen[2], Pille Taba[3], and Aaro Toomela[4]

[1] Department of Software Science, Tallinn University of Technology, Akadeemia tee 15a, 12618 Tallinn, Estonia
{sezare,sven.nomm}@taltech.ee
[2] Institute of Sport Sciences Physiotherapy, University of Tartu, Puusepa 8, 51014 Tartu, Estonia
kadri.medijainen@ut.ee
[3] Department of Neurology and Neurosurgery, University of Tartu, Puusepa 8, 51014 Tartu, Estonia
pille.taba@kliinikum.ee
[4] School of Natural Sciences and Health, Tallinn University, Narva mnt. 25, 10120 Tallinn, Estonia
aaro.toomela@tlu.ee

Abstract. Deep-learning based image classification is applied in this studies to the Luria's alternating series tests to support diagnostics of the Parkinson's disease. Luria's alternating series tests belong to the family of fine-motor drawing tests and been used in neurology and psychiatry for nearly a century. Introduction of the digital tables and later tablet PCs has allowed deviating from the classical paper and pen-based setting, and observe kinematic and pressure parameters describing the test. While such setting has led to a highly accurate machine learning models, the visual component of the tests is left unused. Namely, the shapes of the drawn lines are not used to classify the drawings, which eventually has caused the shift in the assessment paradigm from visual-based to the numeric parameters based. The approach proposed in this paper allows combining two assessment paradigms by augmenting initial drawings by the kinematic and pressure parameters. The paper demonstrates that the resulting network has accuracy similar to those of human practitioner.

1 Introduction

The present paper proposes an approach to support diagnostics of the Parkinson's disease (PD) using convolution neural networks (CNN) to classify the drawing representing the results of digital Luria's alternating series tests (dLAST). Parkinson's disease is the degenerative disorders which most characteristic symptoms rigidity, tremor and non-purposeful motions may severely affect the quality of everyday life of the patient [7,11]. While there is no known cure from the PD,

© Springer Nature Singapore Pte Ltd. 2021
T.-P. Hong et al. (Eds.): ACIIDS 2021, CCIS 1371, pp. 3–13, 2021.
https://doi.org/10.1007/978-981-16-1685-3_1

early diagnoses and proper therapy may relieve the patients from the majority of the symptoms and in turn, improve the quality of the everyday life.

Drawing tests and their digital versions [23] become more popular in the clinical studies targeted to support early diagnosis of the PD. In the area of kinematic [2,13] and pressure parameters based analysis [3] spiral drawing test [1] is one of the most popular. Some times simpler tests like one described in [8] are used. The battery of Luria's alternating series tests (LAST) was proposed by [6] and [12] and later digitised by [15]. Machine learning-based approach to study older LAST tests is described in [22]. Analysis of more complex tests, like a clock drawing test [5] or Poppelreuter's Test [14] require one to involve neural networks (NN) based techniques either to perform complete analysis or to analyse their parts.

The present research differs from the existing results by the procedure used to incorporate kinematic and pressure parameters into the original drawing. First repeating patterns and their elements are extracted from the drawing. For each element or pattern, kinematic (velocity, acceleration, jerk) and pressure parameters are computed. This data is used to colour each segment and change the thickness of its line. Finally, the typical workflow of training and validation of deep neural network models is applied.

The paper is organised as follows. Section 2 explains the symptoms of PD in terms of kinematic and pressure parameters of the fine motor motions, common to the drawing procedure. The same section presents Luria's alternating series tests and their digital version. Formal problem statement is provide by the Sect. 3. Experimental setting is explained in detail in Sect. 4. Transformation technique used to incorporate kinematic and pressure parameters into the original drawing together with the applied data augmentation technique and deep neural network employed are described in Sect. 5. Main results are presented in Sect. 6. Limitations of the proposed approach, together with the interpretation of the achieved results, are discussed in Sect. 7. The final section lists conclusions and possible directions of the future studies.

2 Background

Once one decides to perform a particular action, their brain generates the sequence of impulses to be sent to the spinal cord. Luria referred these sequences as *motion melodies* [12]. One may think about the motion melodies as the programs to be executed. Then motion melody is sent to the spinal cord to execute the motions. These two steps are usually referred to in the literature as motion *motion planning* and *motion execution* functions or phases. Progressing PD may affect any of these phases or both of them. The symptoms of the PD such as tremor, rigidity and bradykinesia [7] are caused by either distorted planning function or problems on the level of motions implementation. If motion planning function is affected, motion melody would not be optimal to reach the target and would require to be corrected during the motion. Disorders on the level of implementation would disrupt the implementation of the motion melody. From the viewpoint of the motion description, these symptoms of PD are reflected by the features describing velocities, accelerations, and pressure applied the stylus tip to the screen of tablet PC [3,15].

2.1 Luria's Alternating Series Tests

LAST tests were proposed in [6] and [12] later their digital version dLAST in [17]. LAST and dLAST require the tested subject to complete, copy and trace the drawing of a repeating pattern. The pattern is designed such that one would have to switch between the motion melodies. Inability to switch between the melodies is referred to as perseveration. Detecting perseveration was the original purpose of the LAST tests. In their digital version battery allows diagnosing PD on the basis of kinematic parameters describing the motion of the stylus tip. Originally the battery consisted of three tests $\Pi\Lambda$, Π and sin wave and three exercises: *continue*, *follow* and *trace* applied to each test. Such a large battery is difficult and time consuming for some elderly subjects to complete. To optimise the testing procedure, within the frameworks of the present contribution employed only $\Pi\Lambda$ and Π tests are considered. Whereas, only $\Pi\Lambda$ test is used in this paper to explain the proposed technique. In Fig. 1 the thin blue line represents the reference patterns shown to the tested subject one by one. In the same Figure, the thick yellow line represents the drawings produced by the subject during the testing. Besides the simplicity of the tests, there is one more advantage of this battery. Namely, in some cases, it allows determining if PD has affected motion planning function. If the patient has no difficulty to complete trace tests but fails on the tests requiring to copy or continue the pattern, it is a clear indicator that motion planning function is affected and motion execution function not. The difference between the *copy* and *continue* tests is in their complexity. Also, $\Pi\Lambda$ differs from the Π pattern by its complexity. Sometimes in the literature theses tests are referred to as *Alternating Sequences Tests*, and slightly different patterns may be studied [4].

3 Problem Statement

The working hypothesis of the present research is that the machine learning (ML) classifier able to use the shape of the drawn lines together with kinematic and pressure parameters would be able to provide high-level predictions to support diagnostics of the PD. This hypothesis leads to the following problem statement. The main goal of this research is to incorporate the kinematic and pressure parameters describing the motions of the stylus tip to the image of the lines drawn to the test. Then train the classifier to distinguish between the PD patients and healthy control (HC) subjects. This primary goal leads the following sub-problems to be tackled.

- Among available kinematic and pressure parameters chose the subset to be incorporated as part of the image.
- Incorporate chosen parameters without altering the shape of the drawn lines.
- Chose proper classifier architecture, train and validate it.

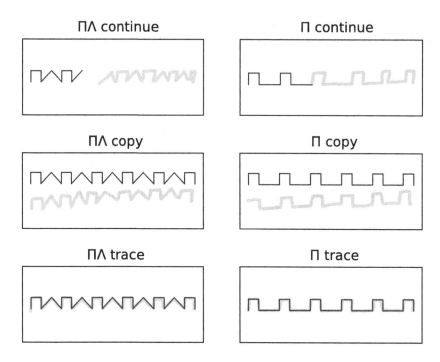

Fig. 1. Reference patterns and patient drawings produced during the testing.

4   Experimental Setting

4.1   Tested Subjects

To answer the problem statement and solve sub-problems identified in the previous section, labelled data-set is required. Following the strict personal data protection laws and with the permission of the ethics committee two groups representing 17 PD patients and similar in age and gender distribution group of 33 HC were chosen among those who volunteered to participate in the trials. Mean age of both groups is 69 years old.

4.2   Data Acquisition

Tablet computer with stylus and special software developed by work-group was used to conduct the tests. The testing software demonstrates the pattern to be completed, copied or traced, and assignment is demonstrated on the screen and duplicated verbally by the practitioner conducting the test. Using stylus pen tested subject continues, copies and traces the patterns. Tablet PC records the position of the stylus pen with respect to its screen surface together with the pressure applied to the screen two hundred times per second. This information is saved in the form of $N \times 4$ matrix, where N is the total number of observation points per test. Four columns of the matrix are the time stamp, x and y coordinates and pressure.

5   Proposed Workflow and Methods

Among all the supervised learning techniques known today, convolution neural networks (CNN) are the most suitable choice for image data classification. This choice immediately poses the requirement to have a large dataset, which in turn requires one to use data augmentation procedure. The proposed workflow is depicted in Fig. 2.

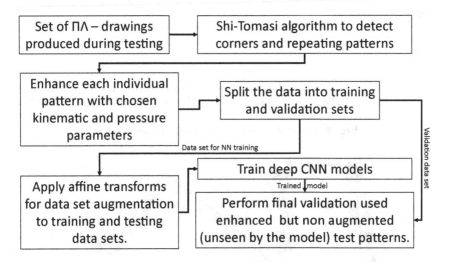

Fig. 2. Research workflow.

5.1   Enhancement and Augmentation

The first step: based on the coordinates and time stamps velocity, acceleration and jerk are computed for each observation point. Together with the pressure, this gives four kinematic parameters to choose from. On the second step, the Shi-Tomasi corner detection algorithm [19] is applied to detect corners of the patterns drawn by the patient. In Fig. 3 thin blue line drawn by the patient and yellow points are the corners detected by Shi-Tomasi algorithm.

Knowledge of the corner coordinates allows extracting straight segments of the drawing. Then combine them into the repeating patterns. Figure $\Pi\Lambda$ depicts one such pattern where line thickness corresponds to the pressure applied by the stylus tip to the tablet screen and colour is generated by the *jet* colour map based on acceleration values. The thin black line represents one repeating

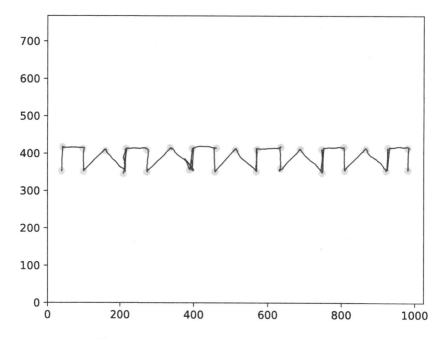

Fig. 3. Corners detect by Shi-Tomasi algorithm.

Fig. 4. Pattern enhanced by the acceleration and pressure

pattern drawn by the tested subject. This method differs from [18] where colour encodes the pressure, and the thickness of the line did not vary at all (Fig. 4).

On the third step, introduce the changes of line width to reflect the pressure. Colour of the line is changed then to reflect acceleration. Then the data is split into training and validation data sets. On the fourth step, the data augmentation procedure [20] is applied to the training set only. Since drawings are not real-life images, there is no need to apply the noise procedure. Also, the colouring of the image was left unchanged. Remaining augmentation transformations belong to the set of affine transforms; stretching and squeezing along the axis and counterclockwise and clockwise rotations. Augmentation parameters then consist of stretching and squeezing parameters and rotation angles. Stretching and squeezing parameters are taken from the interval $(0.85, 1.15)$ whereas rotation angle from $(-5°, 5°)$. For each segment, six values are chosen from each transformation. The number of recognised segments vary between five and four; these lead more around 33000 images to be used for the training and testing. As the last step, each image was resized to 224×224 pixels.

5.2 Workflow of CNN Training Testing and Validation

Augmentation procedure also was used to balance the data-set between (PD) and (HC). Augmented data-set is split for training and validation in the proportion of 70/30. The following deep CNN architectures LeNet-5 [10], AlexNet [9] and Vgg16 [21] were chosen to be evaluated upon their description and suitability for the particular time of the images. These architectures were chosen based on their popularity and description, which is at least in theory, fits the type of images representing drawings of the dLAST.

6 Main Results

LeNet-5 and AlexNet usually converge after four epochs whereas Vgg16 not only took much longer time to converge but also demonstrated low accuracy of just 0.62. Based on its performance for the particular type of task studies in the present paper Vgg16 was excluded from further consideration. More complex AlexNet has demonstrated a better performance. Confusion matrix for the LeNet-5 model is presented by Fig. 5 Confusion matrix for the AlexNet model is presented by Fig. 6. Since each pattern was analysed separately and each test consists of five full patterns (some times subjects draw a lesser or greater number of patterns), it is essential to see how the position of the pattern affects the performance of the classifiers. Figures 5 and 6 depict confusion matrices on the basis of the augmented testing set. Figures 8 and 7 represent numbers of false positives and false negatives (for the separate selection of forty previously unseen by the network segments) for LeNet-5 and AlexNet architectures respectively (Table 1).

Table 1. Goodness of the different deep CNN architectures

	LeNet-5	AlexNet
Accuracy	0.8427	0.9917
Recall	0.9753	0.9976
Precision	0.7709	0.9858
F1 - score	0.8611	0.9917

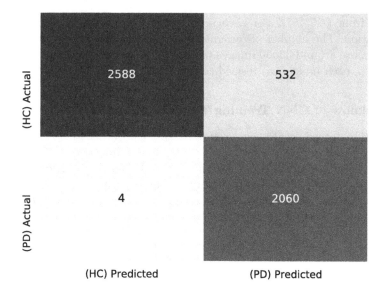

Fig. 5. Confusion matrix for the LeNet-5 model.

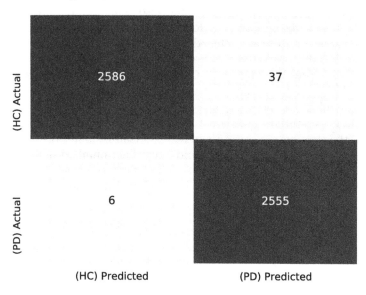

Fig. 6. Confusion matrix for the AlexNet model.

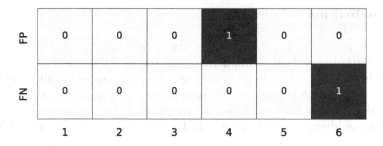

Fig. 7. AlexNet, interval-wise prediction errors.

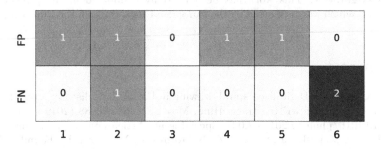

Fig. 8. LeNet-5, interval-wise prediction errors.

7   Discussion

The proposed technique is based on analysing each pattern of the drawing separately, which requires one to summarise classification results for each pattern. Observing classification results for each pattern, one can see that these may be done employing computing the mode of the classes. On the one hand accuracy prediction accuracy for each pattern is in line with [16], which demonstrates that patterns in different positions have different discriminating power. On the other hand, positions of most informative patterns are not the same as in [16], which may be due to the difference between machine learning and deep CNN classifiers.

Performance of only three deep CNN structures was evaluated in this paper, whereas, the architectures were used without any tuning or adjustments. The specific nature of the analysed images may require to adjust or tune one of the existing structures to suit particularities of the dLAST images better.

Among the drawing tests, dLAST tests did not get as much attention as spiral drawing test or sentence writing test. Accuracy of the AlexNet based model exceed the accuracy of the classifiers usually used in the area of statistical machine learning [3, 15] and in pair with the results based on the application of deep learning techniques such as [5].

8 Conclusions

The present paper has proposed a novel way to enhance drawing tests used to diagnose Parkinson's disease. Main results of the paper demonstrate the combined with the image augmentation technique proposed approach allows efficient use of the deep convolution neural networks to support diagnostics of Parkinson's disease. Main results have identified the necessity to pay attention to choosing and tuning architecture of the CNN, which will constitute the subject of future studies.

Acknowledgement. This work has been partially conducted in the project "ICT programme" which was supported by the European Union through the European Social Fund.

References

1. Danna, J., et al.: Digitalized spiral drawing in Parkinson's disease: a tool for evaluating beyond the written trace. Hum. Move. Sci. **65**, 80–88 (2019). https://doi.org/10.1016/j.humov.2018.08.003. Special issue: Articles on graphonomics
2. Drotar, P., Mekyska, J., Smekal, Z., Rektorova, I., Masarova, L., Faundez-Zanuy, M.: Prediction potential of different handwriting tasks for diagnosis of Parkinson's. In: 2013 E-Health and Bioengineering Conference (EHB), pp. 1–4 (2013). https://doi.org/10.1109/EHB.2013.6707378
3. Drotár, P., Mekyska, J., Rektorová, I., Masarová, L., Smékal, Z., Faundez-Zanuy, M.: Evaluation of handwriting kinematics and pressure for differential diagnosis of Parkinson's disease. Artif. Intell. Med. **67**, 39–46 (2016). https://doi.org/10.1016/j.artmed.2016.01.004
4. Fountoulakis, K.N., et al.: Development of a standardized scoring method for the graphic sequence test suitable for use in psychiatric populations. Cogn. Behav. Neurol. **21**(1), 18–27 (2008)
5. Harbi, Z., Hicks, Y., Setchi, R.: Clock drawing test interpretation system. Proc. Comput. Sci. **112**, 1641–1650 (2017). https://doi.org/10.1016/j.procs.2017.08.259. Knowledge-Based and Intelligent Information & Engineering Systems: Proceedings of the 21st International Conference, KES-20176-8 September 2017, Marseille, France
6. Hodges, J.R.: Cognitive Assessment for Clinicians. 2 edn. Oxford Medicine (2007)
7. Kalia, L.V., Lang, A.E.: Parkinson's disease. Lancet **386**(9996), 896–912 (2015)
8. Kotsavasiloglou, C., Kostikis, N., Hristu-Varsakelis, D., Arnaoutoglou, M.: Machine learning-based classification of simple drawing movements in Parkinson's disease. Biomed. Signal Process. Control **31**, 174–180 (2017). https://doi.org/10.1016/j.bspc.2016.08.003
9. Krizhevsky, A., Sutskever, I., Hinton, G.E.: ImageNet classification with deep convolutional neural networks. In: Advances in Neural Information Processing Systems, pp. 1097–1105 (2012)
10. Lecun, Y., Bottou, L., Bengio, Y., Haffner, P.: Gradient-based learning applied to document recognition. Proc. IEEE **86**(11), 2278–2324 (1998)
11. Louis, E.D., Machado, D.G.: Tremor-related quality of life: a comparison of essential tremor vs. Parkinson's disease patients. Parkinsonism Relat. Disord. **21**(7), 729–735 (2015)

12. Luria, A.R.: Higher Cortical Functions in Man. Springer, Heidelberg (1995). https://doi.org/10.1007/978-1-4684-7741-2
13. Marquardt, C., Mai, N.: A computational procedure for movement analysis in handwriting. J. Neurosci. Methods **52**(1), 39–45 (1994). https://doi.org/10.1016/0165-0270(94)90053-1
14. Nõmm, S., Bardõš, K., Mašarov, I., Kozhenkina, J., Toomela, A., Toomsoo, T.: Recognition and analysis of the contours drawn during the Poppelreuter's test. In: 2016 15th IEEE International Conference on Machine Learning and Applications (ICMLA), pp. 170–175, December 2016. https://doi.org/10.1109/ICMLA.2016.0036
15. Nõmm, S., Bardõš, K., Toomela, A., Medijainen, K., Taba, P.: Detailed analysis of the Luria's alternating seriestests for Parkinson's disease diagnostics. In: 2018 17th IEEE International Conference on Machine Learning and Applications (ICMLA), pp. 1347–1352, December 2018. https://doi.org/10.1109/ICMLA.2018.00219
16. Nomm, S., Kossas, T., Toomela, A., Medijainen, K., Taba, P.: Determining necessary length of the alternating series test for Parkinson's disease modelling. In: 2019 International Conference on Cyberworlds (CW), pp. 261–266 (2019)
17. Nõmm, S., Toomela, A., Kozhenkina, J., Toomsoo, T.: Quantitative analysis in the digital Luria's alternating series tests. In: 2016 14th International Conference on Control, Automation, Robotics and Vision (ICARCV), pp. 1–6, November 2016. https://doi.org/10.1109/ICARCV.2016.7838746
18. Rios-Urrego, C., Vásquez-Correa, J., Vargas-Bonilla, J., Nöth, E., Lopera, F., Orozco-Arroyave, J.: Analysis and evaluation of handwriting in patients with Parkinson's disease using kinematic, geometrical, and non-linear features. Comput. Methods Program. Biomed. **173**, 43–52 (2019). https://doi.org/10.1016/j.cmpb.2019.03.005
19. Shi, J., et al.: Good features to track. In: Computer Vision and Pattern Recognition. In: 1994 Proceedings of IEEE Computer Society Conference on CVPR 1994, pp. 593–600. IEEE (1994)
20. Shorten, C., Khoshgoftaar, T.M.: A survey on image data augmentation for deep learning. J. Big Data **6**(1), 60 (2019)
21. Simonyan, K., Zisserman, A.: Very deep convolutional networks for large-scale image recognition. In: Bengio, Y., LeCun, Y. (eds.) 3rd International Conference on Learning Representations, ICLR 2015, San Diego, CA, USA, 7–9 May 2015. Conference Track Proceedings (2015). http://arxiv.org/abs/1409.1556
22. Stepien, P., Kawa, J., Wieczorek, D., Dabrowska, M., Slawek, J., Sitek, E.J.: Computer aided feature extraction in the paper version of Luria's alternating series test in progressive supranuclear palsy. In: Pietka, E., Badura, P., Kawa, J., Wieclawek, W. (eds.) Information Technology in Biomedicine, pp. 561–570. Springer, Heidelberg (2019). https://doi.org/10.1007/978-3-319-91211-0-49
23. Vessio, G.: Dynamic handwriting analysis for neurodegenerative disease assessment: a literary review. Appl. Sci. **9**(21), 4666 (2019)

Examination of Water Temperature Accuracy Improvement Method for Forecast

Yu Agusa[⊠], Keiichi Endo, Hisayasu Kuroda, and Shinya Kobayashi

Graduate School of Science and Engineering, Ehime University, Matsuyama, Japan
g863001b@mails.cc.ehime-u.ac.jp

Abstract. In aquaculture, fishery workers and researchers attach great importance to water temperature. This is because knowing the water temperature makes it possible to take measures against fish diseases and predict the occurrence of red tide. In collaboration with fishery researchers, we constructed a multi-depth sensor network consisting of 16 water temperature observation devices over the Uwa Sea in Ehime Prefecture. In addition, we developed a Web system that can instantly visualize the water temperature measured on this network and provide it to fishery workers in the Uwa Sea. However, the water temperature information provided by this Web system is only current or past information, and fishery workers are requiring the provision of near-future water temperature forecasts for a week or two weeks. Therefore, in this research, as a new function of this Web system, we will implement a function to predict the water temperature in the near future from the past water temperature information and provide the forecast information. This paper examines the steps to be solved for the prediction of seawater temperature, which is the current problem in this system. Moreover, among the steps, this paper reports on the solution method and its evaluation experiment for improving the accuracy of water temperature information.

Keywords: IoT · Big data · Forecast · Outlier removement · Interpolation

1 Introduction

The world population is about 7.7 billion in 2019, and is expected to reach 9.7 billion in 2050 and 10.9 billion in 2100 [1]. The critical problem caused by the increase in the world population is food shortage. Aquaculture is considered an important food supply source because it produces stably.

Ehime Prefecture is one of the leading fisheries prefectures in Japan. (annual production in 2018: about 87.7 billion yen (3rd place in Japan) [2]). The production is particularly supported by the aquaculture in Uwa Sea (annual production in 2018: 65.3 billion yen [3]).

© Springer Nature Singapore Pte Ltd. 2021
T.-P. Hong et al. (Eds.): ACIIDS 2021, CCIS 1371, pp. 14–27, 2021.
https://doi.org/10.1007/978-981-16-1685-3_2

In Ehime University, we developed the system that announced the sea discoloration and plankton concentration information to fishery workers to help aquaculture in Uwa Sea [4]. Moreover, the system that announced the red tide occurrence information sent by the Fisheries Research Center in Ehime to fishery workers was developed [5].

This paper introduces a system that announces water temperature information to fishery workers (Seawater Temperature Announcement System). In addition, this paper examines the steps to be solved for the prediction of seawater temperature, which is the current problem in this system. Moreover, among the steps, this paper reports on the solution method and its evaluation experiment for improving the accuracy of water temperature information.

2 Background

2.1 Importance of Water Temperature

In aquaculture, especially in the Uwa Sea, water temperature in the sea area around the site is very important for the following reasons. First, water temperature zones that cause fish diseases (for example, Red sea bream iridovirus infection [7]) exist. According to interviews with fishery workers, they must reduce the amount of feed to fish when fish disease occurs. If they feeding more than necessary, it will lead to deterioration of fish health and cause death at worst. However, if they can get the current information about the water temperature in their aquaculture ground, they can observe the health condition of their fish and reduce the damage caused by fish diseases. Second, fishery workers can predict the tidal flow in the future based on the water temperature in their aquaculture ground and around. The degree of influence of tidal currents on aquaculture fisheries varies depending on the time, thickness and range of inflow, and it is necessary to adjust the amount of feed according to the tidal current conditions. This issue can be effectively managed by fishery workers based on the accurate and up-to-date information about the current conditions in sea regions in question.

Moreover, fishery researchers also focus to water temperature because they predict the occurrence of red tides using water temperature in an ocean physics perspective.

2.2 Proposed System

2.2.1 Issues

This section describes the situation before developing the proposed system from the viewpoint of fishery workers and researchers in Ehime Prefecture.

Previously, fishery workers had no means of knowing the water temperature at various depths of their aquaculture grounds. Therefore, they were estimating the water temperature from the air temperature. However, in the Uwa Sea, the actual water temperature was significantly different from them estimates because the water temperature changed drastically by a high-temperature tide called "rapid tide" and a low-temperature tide called "bottom-in tide" [8].

Fishery researchers surveys the water temperature every month in the Uwa Sea, at 24 fixed points, against this issue. In addition, five water temperature observation devices set on the sea bottom (Fig. 1 shows the location) to measure the value every hour.

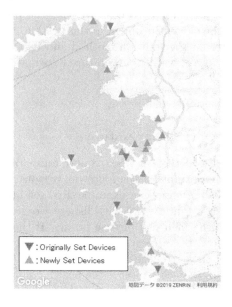

Fig. 1. Location of water temperature observation devices

However, Rapid tides occur at intervals of two weeks so it cannot be captured by a fixed-point survey every month. In addition, the five devices alone do not surround the entire Uwa Sea. Moreover, fisheries workers cannot browse the information measured.

2.2.2 Overview

We proposed a system, as shown in Fig. 2, to address the problem of water temperature information announcement.

First, we constructed a sensor network system enabling measurements on multiple depths (min: 1 m, max: 60 m), covering 17 (as of October 2020) water temperature observation devices (Fig. 1 shows the location of devices). It is possible to surround the entire Uwa Sea and to monitor the situation of the upper and lower layers of the tidal current by constructing such network system. When constructing this network system, we made the observation device that was inexpensive, and sufficiently accurate as information obtained by fisheries workers and fisheries researchers, because setting of devices required a great deal of cost [9].

Table 1. Commercial products vs Our products

	Commercial products	Our products
Cost	High	Low
Measurement interval	30 min or 1 h	30 min
Number of measurement trials	Once	3 times
Resolution	0.001 °C	0.0625(1/16) °C
Data send method	email	Communicate with server

Next, we developed a system called "Seawater Temperature Announcement System" that provided the information collected on using this network system to fishery workers and researchers. The means for fishery workers to browse the measured information was established by implementing this system.

2.3   Water Temperature Observation Device

There are roughly two types of water temperature observation devices: commercially available ones and our originally ones. The Table 1 summarizes the differences between the two.

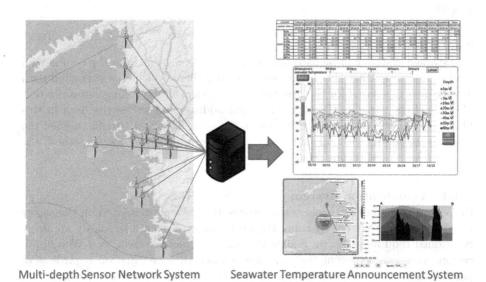

Multi-depth Sensor Network System   Seawater Temperature Announcement System

Fig. 2. Proposed System

In the former, the water temperature is measured once every 30 min or 1 h. In addition, an e-mail containing measurement data is sent to the specified e-mail address from a send-only e-mail address that differs for each measurement point.

In the latter, the water temperature is measured three times in a row every 30 min (It has been verified that the measured values are not always the same) to suppress missing measurements and improve the accuracy of water temperature information. Moreover, unlike the former, the mail does not arrive, but the communication module connects to the server and saves the measurement data on the server. Compared to the former, the resolution is inferior because it is cheaper, but from interviews with fishery workers, it is known that it is sufficient to know the water temperature in units of 0.1 °C, so this resolution is fully satisfied the requirement.

2.4   Seawater Temperature Announcement System

2.4.1   Overview
This system is an information system aimed at promptly announcing the seawater temperature data measured by observation devices to fishery workers and researchers.

This system consists of two subsystems: Storage System and Visualization System. The storage system immediately decrypts the data and store it on the server after measuring the water temperature. The visualization system is designed as a web application that reads the data stored on the server by the storage system and displays it on a web page in the forms of such as tables or graphs.

2.4.2   Storage System
The storage system can be used to decode the e-mails sent from the commercial devices and to extract the data upon receiving them. When the extracted data are the data measured for the first time by a device, this system generates a text file named the same as the sender's e-mail address and writes the data in this file on the server. Otherwise, this system adds the data to a text file named the same as the sender's e-mail address on the server. Using the above method, the data can be stored on the server.

2.4.3   Visualization System
Fishery workers and researchers can browse the data accumulated on the server using the storage system by using the visualization system. To satisfy all aforementioned requirements, we realize the announcement system as a Web application, as it can represent the seawater temperature information regardless of platforms. When this system is launched (accessed), the data stored on the server can be read. Thereafter, this system allows creating such as dynamic tables and graphs based on the registered data and displaying them on the Web page.

This Web page structure contains a public page and special page for fishery researchers, as fishery workers require the simplified information, while fisheries researchers need the detailed information. On each page, users can switch between the displayed information, including the current status and the past status, by pushing button.

2.4.4   Functions

We list the requirements determined based on interviewing fishery workers and fishery researchers as follows:

Requirement 1 The display system should visualize the seawater temperature considering the spatial spread of marine area

Requirement 2 The display system should store the historical water temperature information

Concerning Requirement 1, we implemented five functions defined as follows:

Function (a) Displaying the location of the observation point on the map

Function (b) Displaying the measurement data in a tabular form

Function (c) Displaying the current state of the seawater temperature as a graph

Function (d) Displaying the seawater temperature variation over time as a graph

Function (e) Displaying the seawater temperature variation over time in three dimensions using a distribution chart

To realize Requirement 2, we implemented a function as follows:

Function (f) Saving the measured data in a file in .csv format on the user's terminal

Figure 3 shows an example of display in Function (e). This display format is the result of pursuing a format that is easy for fishery workers to see through interviews with them.

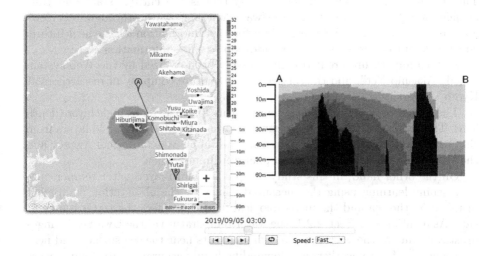

Fig. 3. Example of display in the visualization system

By implementing these functions, the situation has changed from the situation where fishery workers cannot browse the collected water temperature information to the situation where they can browse the latest water temperature information at any time using the Web system. Moreover, for fisheries researchers, visualization of water temperature information has reduced the time and effort required for analysis.

2.4.5   Effectiveness

An information system is an individual and one-time action that can be used by an intended person (owner) to change the activities of individuals or organizations with a certain intention, according to Information Processing Society of Japan, Special Interest Group on Information Systems [10,11]. We investigated trends such as the number of unique accesses to the system and how many times the accessor accessed it in a day to quantitatively evaluate the impact of system implementation in accordance with this guideline.

We confirmed the number of unique accesses to this system in 2018 is 300 to 700 per day. It can be said that the impact on the aquaculture industry in Ehime Prefecture is very large based on this value and the number of aquaculture fisheries management bodies in the Uwa Sea (792 in 2018 [12]). In addition, regarding the access tendency survey of users, we confirmed the users who browse the water temperature information through this system almost every day on time and the users who browse the information more than twice a day on average. Therefore it can be said that the implementation of our system has induced certain changes in individual activities.

2.5   Purpose

The information provided through our system is the current state and time changes of water temperature in the Uwa Sea area. However, through interviews with fishery workers, we feel the need for forecast information for the near future, which is about one to two weeks, in addition to these information.

Water temperature prediction has already been realized by the Web system called "Akashio Net", and the service by the venture company of researchers at Kyoto University [13].

The former provides only water temperature forecasts for a single layer (depth 5 m). The latter is capable of predicting the water temperature of 36 layers from the surface to the seabed, using data from the Japan Meteorological Agency and the US National Oceanic and Atmospheric Administration.

On the other hand, in this study, we predict multi-layer water temperature by machine learning using the measured values by the observation equipment installed on the sea and the meteorological data that affect the water temperature. As mentioned in Sect. 2.2.1, the water temperature of the Uwa Sea changes drastically due to the occurrence of tidal currents near the sea surface and near the seabed, so fishery workers are demanding highly accurate water temperature information at multiple depths. Therefore, this research is expected not only by fishery workers but also by fisheries researchers.

3   Seawater Temperature Prediction

3.1   Issues

The issues we are concerned about this research are as follows:

1. Improving accuracy of water temperature information
   - Implementation and evaluation of outlier detection and removement method
   - Improving accuracy of water temperature interpolation around a measurement point
2. Examination, implementation and evaluation of water temperature prediction method

This paper reports on the examination and evaluation of methods for detecting and removing outliers. In addition, we report on the examination of improvement methods for interpolation.

3.2   Detection and Removement of Outliers

3.2.1   Purpose

The current and past water temperature information used in the forecasts should be accurate to provide highly accurate seawater temperature forecasts. Until now, we tried to improve the accuracy of water temperature information by following methods. First, we developed the observation devices that have multiple sensors at the same depth (increased labor for inspection has become a problem, and it has been found that it can be substituted by the method described later, so it is now abolished). Second, we developed the device that measure continuously at short intervals that can be regarded as multiple measurements at the same time. However, there is a problem that outliers occur due to a bit error during measurement.

In addition, due to he structure of the observation device, the sensor cannot be equipped with a bit error correction function. Therefore, in order to remove the outliers, there is no choice but to judge whether or not the value is outlier from the measured values.

We proposed two methods to detect and remove outliers by comparing with the past information of the water temperature to improve the accuracy of the water temperature information obtained from the observation device.

The strict outlier removal rate (ratio of outlier removements to outliers) is required because the occurrence of outliers greatly affects not only the accuracy of observed values but also the accuracy of interpolation. We aim to reach the removal rate of 99% or more.

3.2.2   Proposed Method 1

The first proposed method is to remove the highest and lowest values as outliers out of the three measurements at the same time. In other words, the median value is regarded as the correct value.

3.2.3   Proposed Method 2

Next, we proposed the following method:

1. This method determines reference value for comparison with each measurement value as follows:
   - If there is data after removing outliers within the past a hour and there is a measured value whose difference from that data is less than 6/16 °C, this method sets the smallest difference from the past data as the reference value.
   - Otherwise, this method sets the median of measurements.
2. This method compares the reference value and the measurements, and removes the measured value whose difference is equal to or greater than the threshold value (2/16 °C).

It is unlikely that a large temperature change will occur in one hour in seawater, so we thought outliers can be detected by comparing with the data measured during that period. Table 2 and Fig. 5 show the frequency distribution of the temperature difference from the time axis viewpoint in the observation device installed in Koike in Uwa Sea (the location is as shown in Fig. 4). The threshold value of 6/16 °C for comparison with past data was determined in consideration of Table 2 and Fig. 5. In addition, the threshold value for comparison with the reference value is set to 2/16 °C, which is stricter than the threshold value for comparison with past data, considering that the measured values are compared with each other at the same time.

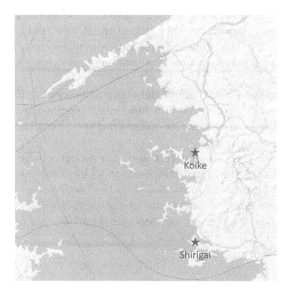

Fig. 4. The locations of Koike and Shirigai in Uwa Sea

Table 2. Temperature difference frequency distribution (table)

Difference	−8/16	−7/16	−6/16	−5/16	−4/16	−3/16	−2/16	−1/16	0	1/16	2/16	3/16	4/16	5/16	6/16	7/16	8/16
1 m	5	6	14	16	25	43	106	1508	9057	1345	109	51	23	8	16	10	5
5 m	3	3	6	19	39	57	170	1201	9452	1101	155	70	46	20	18	6	1
10 m	0	3	9	9	12	32	74	1177	9833	1083	83	36	16	7	7	5	4
20 m	0	1	1	5	12	19	55	1039	10189	964	43	19	13	7	11	3	3
30 m	2	0	7	11	15	14	19	988	10442	839	23	17	11	6	4	7	3
40 m	2	6	4	9	5	11	26	1041	10185	974	24	12	6	9	8	11	7

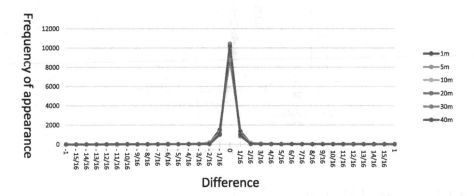

Fig. 5. Temperature difference frequency distribution (graph)

3.3 Improvement of Interpolation Accuracy

3.3.1 Purpose

Our system interpolates the water temperature around each measurement point when running the function shown in Fig. 3. In the current method, we adopted the inverse distance weighted method (IDW) for the plane, and the linear interpolation method for the depth direction. Figure 6 shows a simple interpolation image.

IDW is a method to estimate the data at a certain point by weighting the reciprocal of the square of the distance to the measurement point to the measurement data [14].

Linear interpolation is a method of estimating the estimated value at one point between two measured points as approximately linear expression [15].

However, the interpolation accuracy (maximum error) in this method exceeds the initial target of 0.5 °C. Moreover, fishery workers require more strictical accuracy so that errors in units of 0.1 °C are unacceptable.

In this research, we examine, implement, and evaluate improvement method for the interpolation method.

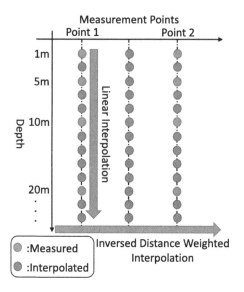

Fig. 6. Current interpolation method

3.3.2 Proposed Method

In this research, we are planning to adopt kriging method as alternative method to IDW, which is an interpolation method on a plane. In Kriging, we create a variogram and a covariance function to find a quantitative statistical dependency (spatial autocorrelation) that depends on the autocorrelation model before calculating the estimate. Thereafter, at the time of estimation, the obtained spatial autocorrelation is weighted together with the distance from the estimation point [16].

Kriging is an advanced method of IDW, and the paper that reports that the accuracy is improved compared to IDW by adopting kriging exists [17]. However, it cannot be said that the successful cases in that report can be applied to our research, so we need to verify.

3.3.3 Analysis of Spatial Structure

In the analysis of the spatial structure, which is the first phase of kriging, we create variogram clouds showing the relationship between the spatial distance of each sample pair and the degree of dissimilarity (the squared difference).

Figure 7 is a variogram cloud formed from water temperature data measured at noon on August 15, 2020.

Most of the variogram clouds formed from past water temperature data, including Fig. 7, are characterized by the dissimilarity diverging for sample pairs over 20000 to 30000 m regardless of the season. Therefore, it is considered appropriate to use a sample for interpolation within 20000 m from the estimated position.

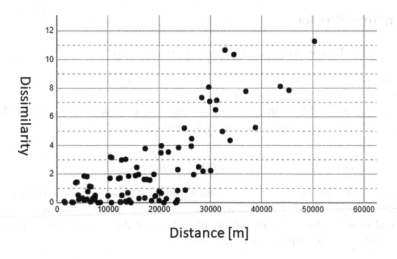

Fig. 7. Variogram clouds

Table 3. Evaluation of proposed method 1

		Outliers	Outliers removed	Removal rate
Total		664	381	57.4%
Breakdown	1 m	269	168	62.5%
	5 m	119	65	54.6%
	10 m	96	50	52.1%
	15 m	92	49	53.3%
	20 m	88	49	55.7%

4 Evaluation of Proposed Method 1

4.1 Evaluation Method

We applied the first proposed method to the data from July 11 to 28, 2019 of the observation device installed in Shirigai (the location is as shown in Fig. 4), and calculated the outlier removal rates.

4.1.1 Result and Consideration

Table 3 shows the overall outlier removal rate and rate at each depth when proposed method 1 is applied. Looking at Table 3, the removal rate is 62.5% at the maximum, which is not sufficient as the outlier removement method.

5   Future Works

5.1   Outlier Detection and Removement Method

As shown in the evaluation results, we proved the proposed method 1 is not a sufficient method for removing outliers.

Proposed method 2 is an improved method of proposed method 1, and it is necessary to evaluate proposed method 2 in the future.

5.2   Improvement of Interpolation Accuracy

We will sample The variogram cloud shown in this paper (generated variogram is called a sample variogram). Thereafter, we will decide the covariance function (spherical model, nugget effect model, etc.) for calculating the weight from the characteristics of the generated sample variogram.

References

1. United Nations: World Population Prospects: The 2019 Revision, p. 1 (2019)
2. Ehime Prefectural Government Office: Ehime Prefecture Fisheries Status in 2018 (Ehime ken gyogyo no chii (Heisei 30)) (2018). https://www.pref.ehime.jp/h37100/toukei/documents/1-1.pdf. Accessed 15 May 2020
3. Ehime Prefectural Government Office: Changes in fishery production by sea area in 1964 to 2018 (Kaiiki betsu gyogyo sansyutsu gaku no suii (Sokatsu : Showa 39 - Heisei 30)). https://www.pref.ehime.jp/h37100/toukei/documents/3-1-2.pdf. Accessed 17 May 2020
4. Ando, K., Okamoto, T., Endo, K., Kuroda, H., Higami, Y., Kobayashi, S.: Development of marine information collection support system for red tide and fish disease estimation. In: Proceedings of the IPSJ 78th National Convention (1), pp. 937–938, March 2016
5. Suehiro, E., Takaichi, R., Fujihashi, T., Endo, K., Kuroda, H., Kobayashi, S.: Development of marine information distribution system for quick countermeasures against red tide damage. In: Proceedings of the IPSJ 80th National Convention (1), pp. 549–550, March 2018
6. Agusa, Yu., Fujihashi, T., Endo, K., Kuroda, H., Kobayashi, S.: Development of seawater temperature announcement system for improving productivity of fishery industry. In: Nguyen, N.T., Gaol, F.L., Hong, T.-P., Trawiński, B. (eds.) ACIIDS 2019. LNCS (LNAI), vol. 11431, pp. 714–725. Springer, Cham (2019). https://doi.org/10.1007/978-3-030-14799-0_61
7. Nakajima, K., Kurita, J.: Red sea bream virus disease. Virus **55**(1), 115–126 (2005)
8. Takeoka, H.: Fisheries support by advanced altitude information (Kodo kaikyo joho ni yoru suisangyo shien). Ehime J. **31**(5), 84–87 (2017)
9. Araki, K., Fujihashi, T., Endo, K., Kuroda, H., Kobayashi, S.: Development of seawater temperature observation device for ocean forecast information service. In: Proceedings of the IPSJ 80th National Convention (1), pp. 553–554, March 2018
10. IPSJ-SIG-IS: Information System Effectiveness Evaluation, Guidelines for Quantitative Evaluation, version 1.1, November 2012

11. IPSJ-SIG-IS: Information System Effectiveness Evaluation, Guidelines for Qualitative Evaluation, version 1.0, September 2013
12. Ehime Prefectural Government Office: Trends in the number of management entitles by type of fishery in 1964 to 2018 (Omo toshite itonanda gyogyo syuruibetsu keieitaisu no suii (Showa 39 - Heisei 30), p. 2 (2019). https://www.pref.ehime.jp/h37100/toukei/documents/2-1-2.pdf. Accessed 17 May 2020
13. Ocean Eyes: OceanEyes Service Overviews. https://oceaneyes.co.jp/en/overview-2. Accessed 27 Oct 2020
14. Meijering, E.: A chronology of interpolation: from ancient astronomy to modern signal and image processing. Proc. IEEE **90**(3), 319–342 (2002)
15. Bailey, T.C., Gatrell, A.C.: Interactive Spatial Data Analysis. Prentice Hall, Upper Saddle River (1996)
16. Wackernagel, H.: Multivariate Geostatistics: An Introduction with Applications. Springer, Heidelberg (1995). https://doi.org/10.1007/978-3-662-03098-1
17. Jedermann, R., Palafox-Albarrán, J., Barreiro, P., Ruiz-Garcia, L., Robla, J.I., Lang, W.: Interpolation of spatial temperature profiles by sensor networks. In: Sensors, pp. 778–781. IEEE (2011)

Assembly Process Modeling Through Long Short-Term Memory

Stefan-Alexandru Precup(✉) [iD], Arpad Gellert(✉) [iD], Alexandru Dorobantiu(✉) [iD],
and Constantin-Bala Zamfirescu(✉) [iD]

Computer Science and Electrical Engineering Department, Lucian Blaga University of Sibiu,
Emil Cioran 4, Sibiu, Romania
{stefan.precup,arpad.gellert,alexandru.dorobantiu,
constantin.zamfirescu}@ulbsibiu.ro

Abstract. This paper studies Long Short-Term Memory as a component of an adaptive assembly assistance system suggesting the next manufacturing step. The final goal is an assistive system able to help the inexperienced workers in their training stage or even experienced workers who prefer such support in their manufacturing activity. In contrast with the earlier analyzed context-based techniques, Long Short-Term Memory can be applied in unknown scenarios. The evaluation was performed on the data collected previously in an experiment with 68 participants assembling as target product a customizable modular tablet. We are interested in identifying the most accurate method of next assembly step prediction. The results show that the prediction based on Long Short-Term Memory is better fitted to new (previously unseen) data.

Keywords: Assembly assistance systems · Training stations · Long Short-Term Memory

1 Introduction

Assembly assistance systems can replace the human trainers in initiating inexperienced workers and increase the efficiency of experienced workers by assisting their manufacturing process. Thus, assistive systems can mitigate some manufacturing knowledge requirements [11]. Such systems must adapt to the worker's needs, characteristics and current behavior [3], but also to the constraints of the task [17]. The adaptability can decrease some human-machine interaction costs [7]. The possibility to dynamically configure the automation levels is also very important [15].

In this work, we analyze the usefulness of a Long Short-Term Memory (LSTM) in modeling assembly processes. Thus, we evaluate the ability of a LSTM to provide a possible next state, considering as input data the current state. In [6], we observed correlations of the assembly score and duration with the following characteristics of the workers: gender, eyeglasses wearer or not, height, and sleep quality in preceding night. Therefore, besides the current assembly state we used all this additional input information in the prediction process. The LSTM is a recurrent neural network and

© Springer Nature Singapore Pte Ltd. 2021
T.-P. Hong et al. (Eds.): ACIIDS 2021, CCIS 1371, pp. 28–39, 2021.
https://doi.org/10.1007/978-981-16-1685-3_3

therefore it could be appropriate in recognizing patterns in the input data. We will start the evaluations with a pre-configured LSTM network and we will perform a design space exploration by systematically varying the main LSTM parameters to determine the optimal configuration for our application.

The rest of this paper is organized as follows. Section 2 reviews the relevant related work applying LSTM. Section 3 describes our proposed LSTM-based next assembly step predictor. Section 4 presents the experimental methodology and discusses the evaluation results, whereas Sect. 5 concludes the paper and suggests further work opportunities.

2   Related Work

There are several companies and research projects that study or implement assistive assembly systems. In [1], Bertram et al. provide a good overview for assistive systems, either from industry (i.e., ActiveAssist developed by Bosch, Der Assistant from Ulixes or Cubu:S from Schnaithmann Maschinenbau AG) or from academia research (i.e., motionEAP or Manual Working Station of SmartFactory). All these systems employ projectors to display information with some added data for context awareness. Dissimilar with these systems, our assistive station [6] is using a touchscreen display to present information.

In our previous work [4], we used two-level context-based predictors to suggest the next assembly state. Furthermore, we used in [5] a Markov predictor and in [6] a Markov predictor enhanced with a padding mechanism to anticipate the next assembly state. The evaluations have shown that the Markov predictor outperformed the two-level context-based predictors. Moreover, the padding mechanism helped the Markov predictor to achieve a higher prediction rate and a better coverage.

Due to its improved performance for time series prediction, in the last decade LSTM has been widely employed to estimate context-dependent human activities with limited number of time steps. Successful examples include prediction of pedestrian movement [9], traffic flow [10], human body movement from sequences of 3D skeleton data [13], emotions recognition from audio-visual data [19], natural language processing [12], etc. For industrial manufacturing with temporal and spatial distribution of time series data, the LSTM predictor becomes a suitable method to predict either pure automatic tasks, such as machine speed to improve the production throughput and minimize energy consumption [2], or even more complex tasks of human-machine collaboration [18] when the mixed-initiative interaction becomes critical factor for effectiveness [16].

However, to be effective, LSTM requires a systematic design space exploration to tune its multiple design parameters. As long as the optimal settings for the parameters significantly vary from problem to problem, no standard approach for their identification exists.

3   Next Assembly Step Prediction Through LSTM

Our goal is to build an adaptive assembly station that can provide tailored assembly experiences towards the user, being able to deliver assembly instructions that better fit the user by taking into consideration various user characteristics (described below).

Assembly and training stations will play a critical role in the factory of the future, as we are seeing a trend towards factory automation and a tighter collaboration between humans and machines. With the increase in highly specialized low demand products, from these assembly stations will benefit both the workers and the factories. Workers can focus more on the quality of their assembly and less on what needs to be assembled, while factories save time and money since those are not required to repurpose their assembly lines nor spend extra money with the training of personnel, as the station can already handle that task.

As in the previous works [4, 5] and [6], the manufactured product is a customizable modular tablet composed of 8 components presented in Fig. 1: the motherboard (to which all the other components will be attached), the screen, two speaker modules (white), two power bank modules (blue) and two flashlight modules (purple). For the correct assembly of the product, there are 7 steps required, with no sequential order predefined. For the prediction of the next assembly step with LSTM, we need to codify the tablet's states. A binary representation has been chosen. A logic value 1 represents a correctly mounted component in its slot on the motherboard, while a logic value 0 represents either an incorrect assembly or that the component has not yet been mounted. The codification and the way it is applied is presented in [4].

Fig. 1. Customizable modular tablet.

Furthermore, with the help of various sensors and self-evaluating questions, we are collecting information relevant in the prediction process. Physical traits like gender, height or if the user is an eye glass wearer, influence the assembly order and correctitude. The sleep quality is another factor in this process. This additional information is used as follows: gender (0 for female, 1 for male), height (value is in centimeters, provided to the LSTM as 0 if it is lower than 174 and 1 if it is higher or equal), wears glasses (0 for No, 1 for Yes), sleep quality (1 to 5, provided to the LSTM as 0 if it is lower than 3 and 1 if it is higher or equal).

Deep learning refers to a special kind of neural networks that are composed of multiple layers. These types of networks are better than classical neural networks at using information from previous events. Recurrent neural networks, shortly RNN, are a type of deep learning architecture. RNN uses the output from the previous prediction steps, together with the current input to produce an output. There are applications in which only recent data is useful for making a prediction while others require a longer history for good accuracy. While RNN can handle a small gap between the information and where it is requested, as this gap widens, it becomes more difficult to train the network.

Fortunately, LSTM networks, a special form of RNN networks, have been introduced [8]. These networks have been developed to deal with the long-term dependency of the information. In its current form, an LSTM unit is composed of a cell, tasked with memorizing information over arbitrary periods of time, and three gates (input gate, output gate and forget gate) tasked with regulating the flow of the information in and out of the cell. Before we will present the components of an LSTM cell, here are the notations used in our formulas:

- $x_t \in R^d$: input vector to the LSTM unit;
- $f_t \in R^h$: forget gate's activation vector;
- $i_t \in R^h$: input gate's activation vector;
- $o_t \in R^h$: output gate's activation vector;
- $h_t \in R^h$: hidden state vector also known as output vector of the LSTM unit;
- $\tilde{C}_t \in R^h$: cell input activation vector;
- $C_t \in R^h$: cell state vector;
- $W \in R^{h \times d}$, $U \in R^{h \times h}$: weight matrices;
- $b \in R^h$: bias vector;
- d: number of input features;
- h: number of hidden units.

The forget gate drops information that the model deems unnecessary in the decision-making process. The previous hidden state and the current input values are passed through a sigmoid function that dictates how much of the previous cell state should be forgotten, with 0 representing drop all and 1 keep all:

$$f_t = \sigma(W_f x_t + U_f h_{t-1} + b_f) \tag{1}$$

The input gate decides what new information should be learned. Like in the forget gate, the hidden state together with the input passes through the sigmoid to determine how much of the new information should be learned (ignore factor):

$$i_t = \sigma(W_i x_t + U_i h_{t-1} + b_i) \tag{2}$$

Then, the hyperbolic tangent function (tanh) squishes the values between -1 and 1:

$$\tilde{C}_t = tanh(W_C x_t + U_C h_{t-1} + b_C) \tag{3}$$

The cell state (long-term memory) persists information from previous timesteps. The new cell state is obtained by multiplying the old state with the forget factor to which the new information from the input gate is added:

$$C_t = f_t \circ C_{t-1} + i_t \circ \tilde{C}_t \tag{4}$$

The output gate decides what the next hidden state should be. Firstly, we pass the current hidden state and the input through a sigmoid function:

$$o_t = \sigma (W_o x_t + U_o h_{t-1} + b_o) \tag{5}$$

Then the newly modified cell state is passed through the tanh function. The information provided by the sigmoid and hyperbolic tangent functions is then multiplied to decide what information should the next hidden state carry:

$$h_t = o_t \circ tanh(C_t) \tag{6}$$

The newly computed hidden state is also the one used for predictions.

Due to the vast assembly possibilities of the tablet and lack of sufficient data, a LSTM network is used to model the assembly process. The network can adapt to new assembly scenarios. For our implementation in Python, we decided to use the TensorFlow library together with Keras. Figure 2 represents the configuration of our network, which consists of 7 layers: the input layer, two LSTM layers, two dropout layers, a flatten layer and a dense layer. Additionally, one dropout layer can be found after each LSTM layer. Most of the network configuration will be presented in the next section, where we will analyze different variations of the parameters.

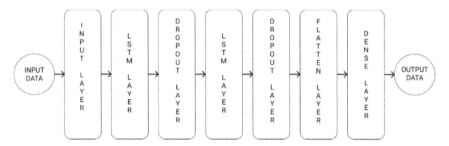

Fig. 2. Network structure.

The input layer, as well as the dense layer have a fixed number of neurons. The number of input layer neurons is set to the amount of features we want to use in our prediction. Since we are using 4 traits of the user, 5 input neurons are required, one for the current state and four for the traits. The number of neurons in the output layer is given by the number of slots we must populate on the motherboard. Since our motherboard has 7 slots, we will have 7 output neurons, each corresponding to one slot. The neuron that has the highest activation value represents the next piece that needs to be assembled.

4   Experimental Results

In this section we determine the configuration that works best for our needs. All the data used in this work has been collected from an experiment, where 68 participants were tasked to assemble the product. The experiment is briefly described in [6] or more in depth in [14]. Out of all participants, 32.4% were females, 48.5% declared that they had slept well in the night before the experiment and 32.3% wore eyeglasses. Female participants needed less time to fully assemble the product, requiring only 69.2 s while male participants needed on average 94 s to assemble it. Although female participants had a faster time of assembly, their assemblies had more errors compared to the male participants. Taller participants had better scores than shorter ones.

The dataset is a csv file with the following structure: <*Current Assembly State, Height, Gender, Wears Eyeglasses, Sleep Quality, Next Assembly State*>. For example, the entry <*1, Tall, Male, Yes, Good, 3*> means that one flashlight module is inserted in the bottom right slot of the motherboard and the user has a height above the average, is male, wears glasses, slept well last night and following assembly state should be 3, meaning that two modules (one flashlight and one speaker) are present on the bottom center and right slots of the motherboard.

The dataset has been split into 2 individual datasets: training and testing. The split percentage is 75/25. In this case, the correct assembly sequences are extracted from the first 75% of the dataset for training while the remaining 25% will be entirely used for testing. In another evaluation approach, denoted 100/100, we extracted the correct assembly sequences from the whole dataset for training and used for testing the dataset entirely. The first evaluation approach indicates which method can adapt to new situations, whereas the second one shows which method can better reproduce a certain model.

We will systematically vary all the relevant parameters of the LSTM in a manual design space exploration, using the 75/25 evaluation approach. We started the evaluation with the LSTM pre-configured with a batch size of 40, dropout rate of 0.28, 64 neurons in the first LSTM layer, 32 neurons in the second layer and 500 epochs. In this stage we did not yet consider a third LSTM layer. Because weight initialization has a major influence over the outputs and, thus, over the prediction accuracy too, for each varied parameter, we did five runs, each with a different seed. The average of those five runs is considered the final result. For comparisons between the implemented methods, we use as evaluation metrics the prediction rate, the prediction accuracy and the coverage. These metrics are calculated as follows: the prediction rate is the number of predictions done divided to the length of the testing dataset, the prediction accuracy is the number of correct predictions divided to the number of predictions done and the coverage is the number of correct predictions divided to the length of the testing dataset.

To get the best results, we tuned the hyperparameters. We now present how the hyperparameters affect the prediction accuracy. We started by varying the learning rate, a parameter representing the step size in each iteration whose role is to find the minimum of the loss surface. We kept the other parameters fixed on the pre-configured values. Figure 3 presents the change in accuracy percentage with a learning rate increasing from 0.01 to 0.05, in steps of 0.01. The accuracy gradually improves up to 0.03, where it peaks, and then a steep decrease in accuracy percentage can be observed. With a large learning rate the network may not converge to the solution, while with a small learning rate it needs a longer training process that might also get stuck in a local optimum. To avoid getting stuck in a local optimum, we also use momentum.

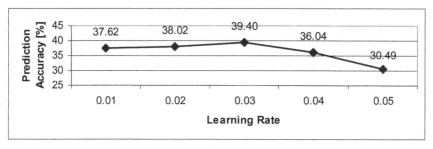

Fig. 3. Learning rate variation.

The second parameter we varied is the batch size representing the number of training examples per batch. After one batch is passed through the network, the weights are adjusted according to the average error from the batch. Depending on the batch size, one or more iterations are required to complete one epoch. We varied the batch size starting from 10 up to 50 with a step of 10, using the optimal learning rate of 0.03 and the preconfigured values for the rest of the parameters.

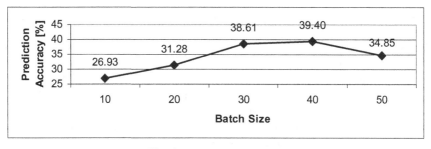

Fig. 4. Batch size variation.

As Fig. 4 depicts, increasing the batch size up to 40, improves the prediction accuracy. If we increase the batch size above 40, a decrease in accuracy can be observed. With a larger batch size, the same number of training epochs is not enough anymore to get the same result (or better). Increasing the number of epochs can lead to overfitting, thus decreasing the prediction accuracy on the test data. The optimal batch size is 40.

We continue our series of tests with the dropout rate. The dropout is a way in which neurons, selected at random, are ignored in the training process, where they do not contribute to the forward propagation nor have their weights updated in the backward propagation. We adjusted the dropout rate in steps of 0.03 starting from 0.16 up to 0.31. In Fig. 5 we can see a small variation in prediction accuracy. However, the highest prediction accuracy is obtained by using a dropout rate of 0.25. Using dropout is another way to prevent overfitting, since it forces pattern learning to be spread between the neurons of the same layer and hinders individual neurons to learn specific training data.

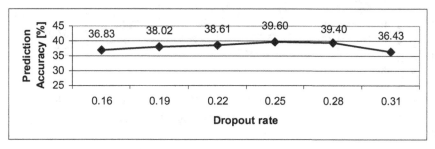

Fig. 5. Dropout rate variation.

After having set the learning rate, batch size and dropout rate, we continue our tests with changing the structure of our network. Firstly, we need to identify the optimal number of neurons in the first LSTM layer. With the number of neurons on the second layer set to 32, the neurons from the first layer have been varied from 40 up to 136 with an increment of 24. In Fig. 6 we can see how the number of neurons affects the prediction accuracy, with a peak at 88 neurons.

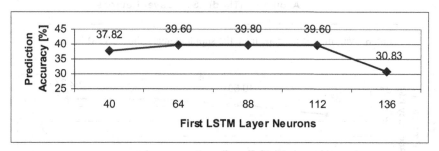

Fig. 6. Varying the number of neurons from the first LSTM layer.

After we have fixed the optimal number of neurons for the first layer, it was time to identify the optimal number of neurons in the second layer. We varied the number of second layer neurons in the range 8 to 40, with an increase of 8 neurons per run. Our tests showed that the optimal number of neurons in the second layer is 24 (see Fig. 7). More neurons on the layers leads to overfitting. Furthermore, we evaluated the possibility of utilizing a third LSTM layer. It can be seen in Fig. 8, that an additional LSTM layer does not contribute to any benefits, the network obtaining a lower prediction accuracy than the one having just 2 layers.

After every parameter has been optimized, it was time to identify the optimal number of epochs to train our network. An epoch represents the propagation of the dataset through the network once. As Fig. 9 depicts, 5000 epochs turned out to achieve the highest prediction accuracy. A high number of epochs will lead to overfitting, while a smaller one does not give the network enough time to learn everything from the provided examples.

Lastly, we wanted to see how a threshold might improve the prediction accuracy. Every time a prediction is made, each activated neuron has a confidence associated

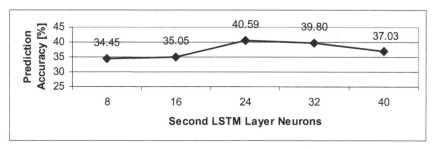

Fig. 7. Varying the number of neurons from the second LSTM layer.

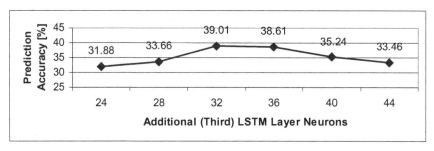

Fig. 8. Varying the neurons from the supplementary third LSTM layer.

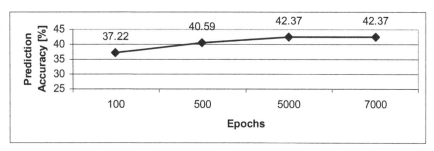

Fig. 9. Varying the number of epochs.

with it. If that confidence is below the threshold, then no prediction is made, otherwise it predicts the next step. We tested the threshold from 0 up to 0.3 in increments of 0.05. In Fig. 10 we can see that a threshold of 0.2 maximizes the prediction accuracy. Applying this threshold means a slightly lower prediction rate and coverage. The highest activation value ranges mostly in the interval [0.15, 0.25]. Thus, a higher threshold leads to a lower prediction rate. Due to highly confident mispredictions, a decrease in accuracy is observed.

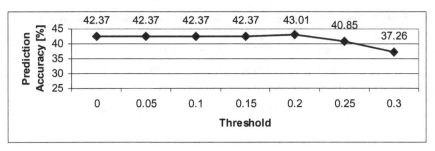

Fig. 10. Varying the value of the confidence threshold.

The best configuration obtained after we have varied all the parameters, is presented in Table 1.

Table 1. Optimal LSTM configuration.

Parameter	Value
Batch size	40
Learning rate	0.03
Dropout rate	0.25
Neurons in the first LSTM layer	88
Neurons in the second LSTM layer	24
Training epochs	5000
Confidence threshold	0.20
Optimizer	Adam
Loss	SparseCategoricalCrossentropy
Activation function	SoftMax

Note that this configuration is suitable for the selected device. For new devices, the methodology for configuring the network and fine-tuning its hyperparameters should be repeated. These hyperparameters can be used as starting values for the new configurations.

Now that the best configuration was presented, we will compare the LSTM to our previous work [6]. For a fair comparison, the Markov predictor presented in [6] has been improved to use complex states containing information about the worker as well (height, gender, sleep quality and if he is an eyeglass wearer). In Table 2, the comparison of the two methods is presented. For each metric we included two columns: 75/25 and 100/100. The 75/25 column means that the network was trained on the first 75% of the dataset and tested on the rest of 25%, while 100/100 means that the network was trained and tested on the same dataset.

As the results show, using complex states in the Markov predictor is not beneficial due to its implementation. If it encounters characteristics of an individual that are not

Table 2. Comparison of the two methods using the 75/25 evaluation approach.

Method	Prediction rate [%]		Accuracy [%]		Coverage [%]	
	75/25	100/100	75/25	100/100	75/25	100/100
Markov (R = 2)	37.62	92.27	50	84.36	18.81	77.84
LSTM	98.02	96.85	43.01	52.08	42.17	50.41

in its knowledge universe, it cannot predict the next assembly step, thus, the Markov model with complex states is not suitable for the task. While on the whole dataset the Markov model does a better job at making predictions, when it comes to new data, the LSTM network seems to be better fitted for the job with prediction accuracy of 43% and a prediction rate of 98%. The coverage is an important metric for us and because of this, the LSTM prediction method is better suited than the Markov model on new data.

5 Conclusions and Further Work

In this paper we analyzed the possibility of using an LSTM to predict the next assembly step in a manufacturing process. We are interested in integrating such a predictor into our assembly assistance system. For our implementation we used the TensorFlow library together with Keras. The LSTM was evaluated and compared with the existing Markov predictor on a dataset obtained in an earlier experiment. We configured the LSTM for our application through a systematic design space exploration. The optimal configuration includes a batch size of 40, a learning rate of 0.03, a dropout rate of 0.25, 88 first layer neurons, 24 second layer neurons, 5000 training epochs and a confidence threshold of 0.20. This optimal LSTM outperforms the earlier developed Markov predictor on new (previously unseen) data, but it is less performant on data known from the training stage. Probably, a combination of these two methods would provide even better results. The development and analysis of such a hybrid predictor will be the subject of further work.

Acknowledgements. This work is supported through the DiFiCIL project (contract no. 69/08.09.2016, ID P_37_771, web: https://dificil.grants.ulbsibiu.ro), co-funded by ERDF through the Competitiveness Operational Programme 2014–2020.

References

1. Bertram, P., Birtel, M., Quint, F., Ruskowski, M.: Intelligent manual working station through assistive systems. In: 16th IFAC Symposium on Information Control Problems in Manufacturing, Bergamo, Italy (2018)
2. Essien, A., Giannetti, C.: A deep learning model for smart manufacturing using convolutional LSTM neural network autoencoders. IEEE Trans. Ind. Inf. **16**(9), 6069–6078 (2020)
3. Funk, M., Dingler, T., Cooper, J., Schmidt, A.: Stop helping me – I'm bored! Why assembly assistance needs to be adaptive. In: 2015 ACM International Joint Conference on Pervasive and Ubiquitous Computing and 2015 ACM International Symposium on Wearable Computers, Osaka, Japan, pp. 1269–1273 (2015)

4. Gellert, A., Zamfirescu, C.-B.: Using two-level context-based predictors for assembly assistance in smart factories. In: Dzitac, I., Dzitac, S., Filip, F.G., Kacprzyk, J., Manolescu, M.-J., Oros, H. (eds.) ICCCC 2020. AISC, vol. 1243, pp. 167–176. Springer, Cham (2021). https://doi.org/10.1007/978-3-030-53651-0_14
5. Gellert, A., Zamfirescu, C.-B.: Assembly support systems with Markov predictors. J. Decis. Syst. (2020). https://doi.org/10.1080/12460125.2020.1788798
6. Gellert, A., Precup, S.-A., Pirvu, B.-C., Zamfirescu, C.-B.: Prediction-based assembly assistance system. In: 25th International Conference on Emerging Technologies and Factory Automation, Vienna, Austria, pp. 1065–1068 (2020)
7. Hancock, P.A., Jagacinski, R.J., Parasuraman, R., Wickens, C.D., Wilson, G.F., Kaber, D.B.: Human-automation interaction research: past, present, and future. Ergon. Des.: Q. Hum. Factors Appl. 21(2), 9–14 (2013)
8. Hochreiter, S., Schmidhuber, J.: Long short-term memory. Neural Comput. 9(8), 1735–1780 (1997)
9. Jiang, X., Lin, W., Liu, J.: A method for pedestrian trajectory prediction based on LSTM. In: The 2nd International Conference on Computational Intelligence and Intelligent Systems, Bangkok, Thailand, pp. 79–84. ACM (2019)
10. Kang, D., Lv, Y., Chen, Y.: Short-term traffic flow prediction with LSTM recurrent neural network. In: IEEE 20th International Conference on Intelligent Transportation Systems (ITSC), Yokohama, Japan, pp. 1–6 (2017)
11. Korn, O., Schmidt, A., Hörz, T.: Augmented manufacturing: a study with impaired persons on assistive systems using in-situ projection. In: 6th International Conference on Pervasive Technologies Related to Assistive Environments, Rhodes, Greece (2013)
12. Nammous, M.K., Saeed, K.: Natural language processing: speaker, language, and gender identification with LSTM. In: Chaki, R., Cortesi, A., Saeed, K., Chaki, N. (eds.) Advanced Computing and Systems for Security. AISC, vol. 883, pp. 143–156. Springer, Singapore (2019). https://doi.org/10.1007/978-981-13-3702-4_9
13. Qiongjie, C., Huaijiang, S., Yupeng, L., Yue, K.: A deep bi-directional attention network for human motion recovery. In Proceedings of the Twenty-Eighth International Joint Conference on Artificial Intelligence, Macao, pp. 701–707 (2019)
14. Precup, S.-A.: Metode de inteligență artificială în modelarea procesului de asamblare asistată a produselor, Editura Universității Lucian Blaga din Sibiu (2020)
15. Romero, D., Noran, O., Stahre, J., Bernus, P., Fast-Berglund, Å.: Towards a human-centred reference architecture for next generation balanced automation systems: human-automation symbiosis. In: Umeda, S., Nakano, M., Mizuyama, H., Hibino, H., Kiritsis, D., von Cieminski, G. (eds.) APMS 2015. IAICT, vol. 460, pp. 556–566. Springer, Cham (2015). https://doi.org/10.1007/978-3-319-22759-7_64
16. Singh, H.V.P., Mahmoud, Q.: LSTM-based approach to monitor operator situation awareness via HMI state prediction. In: IEEE International Conference on Industrial Internet (ICII), Orlando, USA, pp. 328–337 (2019)
17. Stork, S., Schubö, A.: Human cognition in manual assembly: theories and applications. Adv. Eng. Inform. 24(3), 320–328 (2010)
18. Wen, X., Chen, H.: 3D long-term recurrent convolutional networks for human sub-assembly recognition in human-robot collaboration. Assem. Autom. 40(4), 655–662 (2020)
19. Wollmer, M., Kaiser, M., Eyben, F., Schuller, B., Rigoll, G.: LSTM-modeling of continuous emotions in an audiovisual affect recognition framework. Image Vis. Comput. 31(2), 153–163 (2013)

CampusMaps: Voice Assisted Navigation Using Alexa Skills

Revathi Vijayaraghavan$^{(\boxtimes)}$, Ritu Gala$^{(\boxtimes)}$, Vidhi Rambhia$^{(\boxtimes)}$, and Dhiren Patel

VJTI, Mumbai, India
{rvijayaraghavan_b17,rsgala_b17,vsrambhia_b17}@ce.vjti.ac.in,
director@vjti.ac.in

Abstract. Voice assistants have become a common occurrence that are present in our smart-phones, speakers or dedicated devices like the Google Home and Alexa Echo. Over the years, these voice assistants have become more intelligent and the number of tasks that can be performed by them has increased. Of the many voice assistants that exist, Amazon's Alexa is one of the most compatible voice assistants, which can be programmed to suit specific use cases by the use of Amazon's Alexa Skills Kit. Through this paper, we leverage the power of Alexa's voice assistance by designing a navigation system for our college campus, which allows users to request directions in the most intuitive way possible. This is a cost-effective and scalable solution based on Amazon Web Services (AWS) Lambda.

Keywords: Voice assistant · Alexa skill · Navigation system · Routing · Amazon echo

1 Introduction

Voice recognition and assistance has immensely increased in past years, and the use of voice assistants for day-to-day activities and basic tasks have become ubiquitous. Voice assistance has become popular because on an average, humans can type 40 words per minute, whereas they can speak up to 150 words per minute, making it the preferred form of communication. Voice assistants leverage this fact and offer a natural way of interaction, also providing easy access to those who are visually impaired. The most popular voice assistants in today's times are Google's Google Assistant, Amazon's Alexa, Apple's Siri and Microsoft's Cortana. As time has progressed, so have the capabilities of these assistants.

General tasks of voice assistants include setting reminders and alarms, giving weather updates, etc. However, their usage isn't limited to these tasks. They can also be employed for building systems that are tailored for specific purposes. Each of the voice assistants can be leveraged for their own strengths. Among the most popular ones, Amazon's Alexa is one of the most developer friendly and best in terms of compatibility among all voice assistants [2]. Amazon provides developers with the "Amazon Skills Kit" (ASK)

R. Vijayaraghavan, R. Gala and V. Rambhia—Equal Contribution.

T.-P. Hong et al. (Eds.): ACIIDS 2021, CCIS 1371, pp. 40–51, 2021.
https://doi.org/10.1007/978-981-16-1685-3_4

[3] which can be used to build custom skills, which are like apps for Alexa, with self-defined functionalities. The ASK provides self-service APIs that can be used to build skills that range from games and music streaming to multimodal experiences and smart home control.

For this reason, we use Amazon Alexa for augmenting a skill to build a voice-assisted campus navigation system. This skill is based on the multimodal experience provided as part of the Amazon Skills Kit, that gives spoken navigation assistance as well as visual reference for viewing the path. We propose a methodology that has high accuracy while also being comparatively inexpensive. The built skill will be accessible on any device that can run Alexa, but the visual component of the skill is limited to Amazon's Echo devices that come with a built-in display. To thoroughly test our system, it has been deployed on an Amazon Echo Show 8.0. The Echo Show has an 8″ display that can be used to display our multimodal response in the form of images. Any device that can run the Alexa voice assistant will be able to access our skill.

The rest of the paper is organized as follows: Sect. 2 discusses background and basic terminologies. In Sect. 3, the proposed system for Alexa based Navigation is discussed. Section 4 discusses the advantages of the system and how challenges in the system were addressed, and Sect. 5 includes results. Conclusion and future scope are presented in Sect. 6 with references at the end.

2 Background and Terminology

A voice assistant is a software agent that uses natural language techniques to interpret speech input given by users and responds via synthesized voice messages [4]. The first ever voice activated product was the Radio Rex. In the period from 1971 to 1976, DARPA funded R&D in Speech Understanding Research (SUR) program, the outcome of which was CMU's Harpy, which could recognize over 1000 words [1]. The first modern voice assistant was Siri, released by Apple in the year 2011. Following this, Microsoft's Cortana and Amazon's Alexa were released in 2014, and Google's Google Assistant in the year 2016 [5]. In 2017, Google Assistant had achieved a 95% accuracy in recognizing words in the English language.

2.1 Voice Assistants

Alexa. Amazon's Alexa is a voice assistant which runs on smart speakers (Echo devices) or smartphones. In addition to performing basic tasks like setting alarms/reminders, giving weather updates, playing music, etc., other features can be added to Alexa, called skills.

Google Assistant. This is a personal assistant by Google, which is available on smartphones and smart devices like the Google Home. In 2016, Google also launched Google actions, which is a developer platform to build apps for Google Assistant [6].

Cortana. Cortana is a productivity assistant developed by Microsoft. Cortana uses the Bing search engine to answer questions asked by the user. Cortana's key features include reading an overview of mails and enabling voice-dictated replies, creating planners, recommending activities, etc. [7].

Siri. Siri is the voice assistant developed by Apple. Siri was the first voice assistant to be rolled out to the public. A wide range of voice commands are supported by Siri, ranging from basic commands such as searching the internet and setting reminders to more complicated ones like engaging with third party apps on iOS [8].

Comparison. All four voice assistants mentioned above have been popular and made life more convenient. Siri was one of the first voice assistants in the market, yet it was considered inflexible and unnatural compared to other voice assistants. Cortana was removed from the iOS market in 2019 and from the Android market in 2020 [9]. With Google Assistant and Amazon Alexa as the two best choices for building custom applications, we compared the advantages and disadvantages of using either of them [10]. Amazon has the feature of Alexa blueprints, which provides templates for getting started with skill development. Further, Amazon also has easy-to-use APIs for integration with different applications and provides ease of building custom apps in the form of skills. A device with Alexa is claimed to have more compatibility with third-party applications and services compared to its counterparts [2]. The potential of Alexa and Echo devices to take on a range of different roles and functions in multi-user interactions makes it particularly relevant for our use case.

Amazon Echo and Echo Show. Amazon Echo, or Echo, are smart-speakers built by Amazon, that use the Alexa voice assistant [11]. The device can act as a simple voice assistant that does day to day tasks or also be combined with several smart devices like smart bulbs, smart fans and smart switches to name a few, thus facilitating home-automation. The different variants are the first-generation Echo, the Echo Dot, Amazon Tap, Echo Look, Echo Spot, Echo Plus, Echo Connect and Echo Flex. An important Echo variant that has been the primary device from the point of view of this paper is the Echo Show, which comes with an in-built display that aids multimodal responses. All of these devices vary in terms of size and presence of an inbuilt speaker, however the voice recognition software running on each is the same.

2.2 Navigation Systems

The purpose of navigation systems is to give directions to the user for the requested destination from a source location. Before the advent of GPS and navigation systems like Google Maps, navigation would be done using paper-printed maps. The invention of digital navigation was a game-changer providing dynamic routes and assistance. Mainly, directions can be provided by visual, audio and/or haptic means [12]. Outdoor navigation systems can use GPS (Global Positioning System) for positioning. However, radio signals cannot penetrate solid walls rendering GPS ineffective for indoor positioning [13]. In such cases, additional installations like beacons, sensor networks, etc. are required which would increase the system cost by a fair amount.

2.3 Terminology

Some of the terminologies associated with Amazon Alexa that have been used in the paper have been elaborated upon in this section.

Intent. An intent is a representation of an action that responds to a user's spoken request. Intents may have slots (placeholders for arguments) [14]. In our case we have defined 5 Custom Intents which are as follows:

- DirectionIntent
- SourceIntent
- GreetIntent
- WhereAmI Intent
- WashroomIntent

We have also leveraged some of the built-in intents provided by Alexa Skills Kit like YesIntent, NoIntent, HelpIntent, etc.

Utterance. Utterances are phrases which will be most likely used for a particular intent [15]. Utterance sets should be unambiguous and should contain as many relevant phrases as possible. For example, in our use-case, "DirectionIntent" will have an utterance set consisting of phrases like: where is <location>, where can I find <location>, how do I reach <location>, etc.

Custom Slot Types. Custom slot type is a collection of possible values for a slot type. Custom slot types are used to represent items that are not included in Amazon's built-in slot types [14], which in our case, are names of locations in the campus.

Multimodal Response. Multimodal is the addition of other forms of communication, such as visual aids, to the voice experience that Alexa already provides. Multimodal skills [15] can provide more information through visual aids leading to a better user experience. Our skill uses Alexa Presentation Language (APL) to render visual aids with every direction response.

Endpoints and Lambda Function. The endpoint for skill is the service to which Alexa sends requests when users invoke it. A custom skill can be hosted using AWS Lambda or as a web-service. We are hosting our skill's backend service as an AWS Lambda function [14].

3 Proposed System: CampusMaps

We propose a voice assisted solution along with a visual representation of the path to enhance user experience. The visual maps are also integrated with alexa skills, but will be visible only on alexa devices that have an inbuilt display. The overall workflow is shown in Fig. 2. Once the skill has been invoked using the correct invocation *"open campus maps"* the user can ask for directions to a specific location. If the source location (location at which the echo device/location of user) has not been set, the skill will prompt the user to set a source location. Once this has been set, the user can ask for directions and the skill will compute the path and directions at the backend and return a set of directions along with a visual path in the form of an image to the user. We have additionally included features like finding the closest washroom or staircase. The user can also enquire about his own location by asking *"Where am I"* and this will return the preset source location. The built skill uses two languages that are supported by Alexa, namely English (EN) and Hindi.

3.1 Architecture

The user interaction with the system will take place through an Alexa enabled device (Amazon Echo Show). The user can request directions for a location within the campus and this location will be passed as input for the Custom Alexa Skill backend. The skill (backend) then processes this input and generates a path from a pre-set source location to the destination as mentioned by the user along with a map representing the route for the same. The generated directions and image of the map are returned to the alexa device, which then displays the visuals while speaking out directions to reach the given destination. Figure 1 shows user interaction with the custom skill and how it is processed.

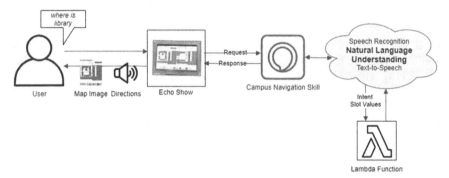

Fig. 1. User interaction with custom skill and how user requests are processed

3.2 Modelling Maps

The present blueprints of the college have been digitized. The campus has been divided into 3 regions and each region has maps representing different floors. An undirected graph data structure has been used to represent the entire college, where each important location on the campus is represented as a node and corridors or pathways connecting these locations are represented as undirected edges. Each node is represented as a JSON object in the following template:

```
{
        node_name: string,
        x_pos: float,
        y_pos: float,
        floor: int,
        map_number: int
}
```

The *node_name* represents the name of the location. The *x_pos* and the *y_pos* are the respective x and y pixel coordinates of that location in the bitmap image. The *floor* represents the floor number of that location. As mentioned previously, the entire blueprint of the college is divided into 3 maps, so the *map_number* attribute represents which of these maps the node belongs to. Undirected edges are the corridors and pathways between the nodes, which have been identified from the blueprints of the college. A unique node index is provided to each node so that pertinent information can easily be accessed. As the input received from the user is the name of the location (*node_name*), a mapping has been created between *node_name* and the *unique_index*. *node_name* by itself cannot be the unique index as there are multiple locations that would have the same node name (e.g. washroom).

3.3 Routing

This forms the crux of the system, and to provide the shortest and most efficient solution, we have used a modified version of Dijkstra Routing Algorithm [15]. In the modified version, we are not only finding the minimum distance, but also the path from the source to destination. For special cases like washroom, our algorithm returns the closest washroom to the user.

Floor Navigation. The algorithm supports navigation on different floors as well. This has been done by adding edges for every staircase present. In the event that there may be multiple staircases to reach a destination, the closest staircase will be chosen.

Directions. Providing the correct directions from a source to the destination is one of the most important parts of the CampusMaps. If the general compass (North, South, East, West) directions are considered, it might be tempting to simply consider that if x coordinate increases then the person is traveling east (assumed to be right) and if the y coordinate increases he goes south (assumed to be straight). However, for someone entering from one end of the college, some places will be on his/her right, but for another person exiting the college from a gate on the opposite end, the same place will be on his/her left. For this reason, it is essential to account for the direction the person is already walking in, and only then compute the next direction. The direction that the person is already walking in will serve as a reference for what would represent the person's left and right.

For the special case where there is no previous direction to reference, i.e. when the user is at the source location, a provision has been provided such that the user is first told to turn in the correct direction, after which the rest of the directions are provided. The directions are provided in a language that sounds natural and conversational to the user.

Distance Calculation. The college plot is more or less a rectangular plot. As a result of this, the total length and total width of the college was found in meters. Existing blueprints (paper-based) of the college were digitized and the same map was now expressed in the form of a digital image. A correlation was created between the length and width of the

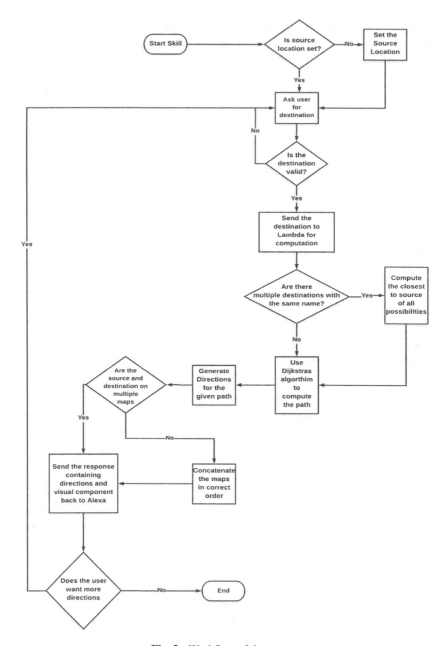

Fig. 2. Workflow of the system

college in pixels, to the length and width of the college in meters respectively. When the path between source and destination is generated, the Euclidean distance is calculated accordingly in pixels. From the above mentioned correlation, the pixel distance is mapped to the distance in meters as an acceptable estimate for the individual.

3.4 Alexa Skill Development

Amazon Developer Console has been used to develop our Alexa skill. The initial configuration consists of setting an appropriate skill name, selecting languages for the skill (English-India and Hindi in our case) and choosing a model to add to the skill (Custom). The next step involves setting up build configurations which includes selecting an invocation name, defining intents along with their respective utterance sets, slots and defining an endpoint which will trigger actions every time the skill is invoked. Finally, the model is trained with the chosen configurations and built.

The endpoint for our skill is a lambda function. The core backend logic for the skill resides in this function. It is responsible for extracting intent, slot values and other metadata from the request object it receives every time the skill is invoked. A response object is constructed which has an appropriate speech output and a reprompt message (if needed). If the device invoking our skill has a screen, a map image is sent using Alexa Presentation Language [14] (APL). The images are pre-constructed to reduce response-time and are hosted on a separate server.

Users can interact with the system by giving voice commands. Amazon's Natural Language Understanding [14] (NLU) module is responsible for identifying utterances and their corresponding intents from the input and sending its findings to the skill's backend (lambda function) in the form of a JSON object. The intents and slot values (if any) are extracted from this response object and an appropriate action is launched. Every intent has a set of associated utterances which should be present in the user's request for the intent to be identified. For instance, *"where is {campusLocation}"* is an utterance

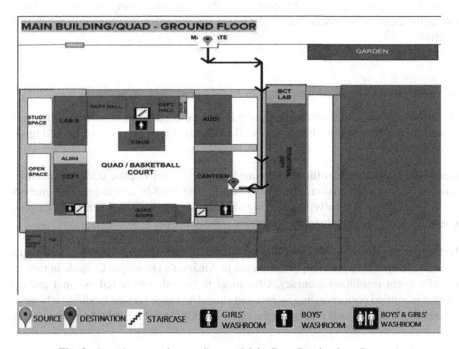

Fig. 3. Sample output image (Source: Main Gate, Destination: Canteen)

which represents *DirectionIntent*. Thus, if a user asks *"where is library"* the intent identified is *DirectionIntent* with value for slot *campusLocation* as library. A function called direction(), which is defined in the lambda function, is called. This function will return text which gives directions to the library, which should be interpreted by the device as speech-output. Figure 3 shows sample output map from Main gate (source location) to Canteen (destination). The directions for the same case are as follows: *"Walk straight. Take the next left. Take the next right. Continue straight. Take the next right. You have arrived at the canteen. You have walked a total of 114 m."*

4 Advantages and Challenges

4.1 Advantages

Cost Effective. The system is built economically, compared to other navigation systems like indoor maps and GPS based systems, that require a huge budget. The only cost incurred in this system is that of the Echo device(s).

Scalable. With the use of Amazon Web Services Lambda and Google Drive to store the generated output of maps for each use case, the system can be scaled easily as the storage requirements are minimum.

Time Required to Build the System. Compared to other systems like indoor maps and GPS based systems, this system is far quicker to build. The map modeling for any campus can be done in a short period of time and since this is the only change required to extend this architecture, this system can be replicated for any other campus or establishment rapidly.

4.2 Challenges

The major challenge that we faced was the incorrect recognition of locations that were spoken by the user in both English as well as Hindi. This was resolved by taking the following steps:

Removing Utterance Conflicts. As our skill had custom intents, utterance conflicts between both custom-custom intents and custom and built-In utterances were encountered. This issue was resolved by making utterances in custom intents more descriptive to remove ambiguity. Redundant utterances were removed.

Testing Intent Resolution Capability. As our skill's interaction model was being built, we used the utterance profiler [16], a feature in Amazon's Developer Console to test our model's intent resolution accuracy. Utterances to be tested were fed as input and the profiler identified corresponding intents and slots. Utterance set was modified whenever the test utterances did not resolve to the right intent followed by re-building the model.

Batch Testing Using NLU Tool. Natural Language Understanding [14] (NLU) evaluation tool in the developer console was used to test batches of sample utterances. The test set (aka an annotation set) consists of utterances mapped to the expected intents and slots. NLU evaluation score with the annotation determines how well the skill's model performs against our expectations [17]. The score was improved by removing ambiguities in the utterance set and adding synonyms for various slot values.

Testing Sample Audio Files Using ASR Tool. The Automatic Speech Recognition [14] (ASR) Evaluation tool allows us to batch test audio files to measure the speech recognition accuracy of our skill. With this tool, we have tested and compared expected transcriptions against generated transcriptions for sample utterances. The issues uncovered during this test have been resolved by spelling out the words in utterance sets considering how they will be actually pronounced.

5   Results

This system underwent both alpha and beta tests. Alpha tests were performed by us in the form of unit tests and integration tests. This system was then rolled out to a variety of students, faculty and administrators to test its efficacy.

5.1   Alpha Tests

Unit Tests. The first step of testing was to test the individual components on our code editor itself. The following factors were considered to gauge correct working with each test case:

- Path is selected corrected
- Directions are given correctly
- Correct nodes are selected on the visual component
- Correct path is shown on the visual component
- All maps are shown in the case of source and destination being on different maps
- The maps are concatenated in the correct order before saving

Manual verification was done for each test case. All the test cases ran successfully.

Alexa Developer Console Test Simulator. Before deploying the skill to the echo devices, we ran the tests on the alexa developer console for testing skill in english and hindi. The output of the tests are shown in Fig. 4 and 5.

Fig. 4. Testing on console (English)   **Fig. 5.** Testing on console (Hindi)

5.2 Beta Tests

To gauge insights on how the system would work in the real world, we performed beta tests by allowing expected users to use the system. Beta testing included 6 students, 2 faculty members and 1 administrative member. The users were allowed to interact with the system and were provided with only a basic instructional guide for the system. A feedback form was then provided to the beta testers for evaluation. The feedback form consisted of the following fields- ease of use, system naturalness, accuracy of results, bug reports and overall satisfaction. The results (rated and averaged out of 10) are shown in Table 1.

Table 1. Beta test results

Field	Rating
Ease of use	8.5
System naturalness	8
Accuracy of results	9.5
Overall satisfaction	9

The beta testers helped identify 3 minor bugs that were fixed before the system was rolled out to the rest of the university.

6 Conclusion and Future Scope

In this work, we have proposed a voice assisted navigation system for campuses and establishments where extensive systems like google maps or indoor maps aren't necessary or feasible. The purpose of the system was to leverage the power of Alexa Skills, which has been done efficiently as supported by the results. This prototype of the system makes the navigation within the campus an easier and more convenient task for new visitors. Currently, the position of the Alexa enabled device is static thus limiting the user to understand the directions in a single-go. As a future scope, this system can be further enhanced by the use of beacons for positioning, which can provide higher accuracy and dynamic navigation.

References

1. Newell, A.: Harpy, production systems, and human cognition. Perception and production of fluent speech, pp. 289–380 (1980)
2. O'Boyle, B.: Google Assistant vs Alexa vs Siri: Battle of the personal assistants. Pocket-Lint, 12 October 2020. https://www.pocket-lint.com/smart-home/buyers-guides/124938-goo gle-assistant-vs-alexa-vs-siri-personal-assistants. Accessed 12 Nov 2020
3. Build Skills with the Alexa Skills Kit I Alexa Skills Kit. https://developer.amazon.com/en-US/ docs/alexa/ask-overviews/build-skills-with-the-alexa-skills-kit.html. Accessed 12 Nov 2020
4. Hoy, M.B.: Alexa, siri, cortana, and more: an introduction to voice assistants. Med. Ref. Serv. Q. **37**(1), 81–88 (2018). https://doi.org/10.1080/02763869.2018.1404391
5. López, G., Quesada, L., Guerrero, L.: Alexa vs. Siri vs. Cortana vs. Google assistant: a comparison of speech-based natural user interfaces. In: Nunes, I.L. (ed.) Advances in Human Factors and Systems Interaction, pp. 241–250. Springer International Publishing, Cham (2018). https://doi.org/10.1007/978-3-319-60366-7_23
6. Google Assistant. https://en.wikipedia.org/wiki/Google_Assistant. Accessed 12 Oct 2020
7. Kapko, M.: Cortana explained: How to use Microsoft's virtual assistant for business. Computerworld, 7 February 2018. https://www.computerworld.com/article/3252218/cortana-exp lained-why-microsofts-virtual-assistant-is-wired-for-business.html. Accessed 12 Nov 2020
8. Siri. https://en.wikipedia.org/wiki/Siri. Accessed 12 Nov 2020
9. Cortana. https://en.wikipedia.org/wiki/Cortana. Accessed 12 Nov 2020
10. Alexa vs. Google Assistant - what's best for workplace apps? https://www.adenin.com/blog/ alexa-vs-google-assistant-whats-best-for-workplace-apps/. Accessed 12 Nov 2020
11. Amazon Echo. https://en.wikipedia.org/wiki/Amazon_Echo. Accessed 12 Nov 2020
12. Fallah, N., et al.: Indoor human navigation systems: a survey. Interact. Comput. **25**(1), 21–33 (2013)
13. Goshen-Meskin, D.R.O.R.A., Bar-Itzhack, I.Y.: Observability analysis of piece-wise constant systems. II. Application to inertial navigation in-flight alignment (military applications). IEEE Trans. Aerosp. Electron. Syst. **28**(4), 1068–1075 (1992). https://doi.org/10.1109/7.165368
14. Alexa Skills Kit Glossary I Alexa Skills Kit. (n.d.). https://developer.amazon.com/en-US/ docs/alexa/ask-overviews/alexa-skills-kit-glossary.html. Accessed 12 Nov 2020
15. Dijkstra, E.W.: A note on two problems in connexion with graphs. Numer. Math. **1**(1), 269–271 (1959)
16. Test Your Utterances as You Build Your Model I Alexa Skills Kit (n.d.). Amazon. https://developer.amazon.com/en-US/docs/alexa/custom-skills/test-utterances-and-imp rove-your-interaction-model.html. Accessed 12 Nov 2020
17. Batch Test Your Natural Language Understanding (NLU) Model I Alexa Skills Kit (n.d.). Amazon. https://developer.amazon.com/en-US/docs/alexa/custom-skills/batch-test-your-nlu-model.html. Accessed 12 Nov 2020

Advanced Data Mining Techniques and Applications

Solving Reduction Problems in Cover Lattice Based Decision Tables

Thanh-Huyen Pham[1,2]([✉]), Thi-Ngan Pham[1,3], Thuan Ho[1,4], Thi-Hong Vuong[1], Tri-Thanh Nguyen[1], and Quang-Thuy Ha[1]

[1] Vietnam National University, Hanoi (VNU), VNU-University of Engineering and Technology (UET), No. 144, Xuan Thuy, Cau Giay, Hanoi, Vietnam
phamthanhhuyen@daihochalong.edu.vn, hothuan@ioit.ac.vn, {hongvtn, ntthanh,thuyhq}@vnu.edu.vn
[2] Halong University, Quang Ninh, Vietnam
[3] The Vietnamese People's Police Academy, Hanoi, Vietnam
[4] Institute of Information Technology (VASC), Hanoi, Vietnam

Abstract. Covering based rough set is an important extension of Pawlak's traditional rough set. Reduction is a typical application of rough sets, including traditional, covering based and other rough set extensions. Although this task has several proposals, it is still an open problem to decision systems (tables) in covering based rough set. This paper focuses on the reduction problem for the condition lattice, and fitting problem for the decision lattice in the decision table based on cover lattice. A corresponding algorithm is proposed for each problem. Two examples to illustrate a covering based decision table and two related problems show the applications of these concepts and problems.

Keywords: Covering rough set · Cover lattice based decision table (system) · Reduction lattice · Fitting lattice

1 Introduction

Covering based rough set theory, initiated by W. Zakowski [10], is an important extension of Pawlak's rough set theory [4]. For describing the approximation space, traditional rough set uses partitions (by equivalent relations), while covering based rough set theory uses covers (by tolerance relations) of the universe [1, 6, 9, 10, 13, 14].

Reduction is a typical application of rough set (including traditional, covering based and other rough set extensions). Though this task has been mentioned, it is still an open problem to decision systems (tables) on covering based rough set. A typical type of covering based decision system is a tuple $(U, \Delta \cup D)$, where the condition part is a family of covers and the decision part is a property [3, 7, 8]. The reduction problem in the rough decision system is presented in relation between the induced cover Δ and the set of equivalence classes on U, according to the value set of decision attribute D.

This paper is an extension of cover-lattice based decision tables [5]. We focus on reduction and related problems as well as application aspects. This paper has following

© Springer Nature Singapore Pte Ltd. 2021
T.-P. Hong et al. (Eds.): ACIIDS 2021, CCIS 1371, pp. 55–64, 2021.
https://doi.org/10.1007/978-981-16-1685-3_5

main contributions: i) the *condition lattice reduction problem* is introduced, and an algorithm that finds all condition lattice reducts is proposed; ii) corresponding to the condition reduction problem, the decision lattice reduction problem (called the *fitting problem*) is presented with an algorithm that finds all fitting decision lattice reducts; iii) we provide sample applications of the proposal to show its potential.

The remainder of this paper is organized as follows. Next section presents the definition of a cover-lattice based decision table, in which decision lattice is used instead of decision attribute. An example illustrating the meaning of a cover-lattice based decision table is presented. Two reduction and fitting problems in the cover-lattice based decision table are presented in Sect. 3. The corresponding algorithms for solving these problems together with an illustrative example in the collaborative filtering-based recommender system are also discussed. Section 4 introduces some related work. The last section presents the conclusion of the paper.

2   Covering Based Decision Tables

2.1   Covering Based Rough Sets and Covering Decision Systems

W. Zakowski defined an approximation space as a pair (U, C), where U denotes an arbitrary nonempty set; C denotes a cover of U [10]. He also defined minimal description, the family of sets bottom approximation, and an equivalence relation $\sim_C$. C. Degang et al. defined the induced cover of a cover, and the induced cover of a family of covers [3]. The task of the attribute reduction of covering decision systems with covering based rough sets was also defined.

Definition 1 (*covering decision system* [3]): A covering decision system is an ordered pair $S = (U, \Delta \cup D)$, where U is the universe; Δ is a family of covers of U; and D is a decision attribute.

Covering decision system is also called covering decision information system [7].

Assume that there exists an information function $D : U \to V_D$ where V_D is the value set of the decision attribute D, $IND(D)$ is the equivalence relation of D, and U/D is the set of equivalence classes by $IND(D)$.

Definition 2 (Δ-*positive region of D* [3]): Let $S = (U, \Delta \cup D)$ be a covering decision system, and $Cov(\Delta)$ is the induced cover of Δ. The Δ-positive region of D is calculated as $POS_\Delta(D) = \bigcup_{X \in U/D} \underline{\Delta}(X)$, where $\underline{\Delta}(X) = \bigcup\{\Delta_x | \Delta_x \subseteq X\}$.

Definition 3 ([3]): Let $\Delta = \{C_i | i = 1, \ldots, m\}$ be a family of covers of U, D is a decision attribute, U/D is a decision partition on U. If for $\forall x \in U$, $\exists D_j \in U/D$ such that $\Delta_x \subseteq D_j$, where $\Delta_x = \bigcap\{C_{ix} | C_{ix} \in Cov(C_i), x \in C_{ix}\}$, then, the decision system (U, Δ, D) is called a *consistent covering decision system*, and denoted as $Cov(\Delta) \preceq U/D$. Otherwise, (U, Δ, D) is called an *inconsistent covering decision system*.

2.2 Cover Lattice Based Decision Table

2.2.1 A Definition of Covering-Lattice Based Decision Table

Let U be the universe. Assume that there exists a partial order relation denoted "$\leq$" in the set all covers of U.

Definition 4. Let A, B be two covers of U. The cover A is smoother than the cover B if and only if for every set AS belonged to the cover A, there exists a set BS belonged to the cover B such that $AS \subseteq BS$.

Definition 5. (*the conformity of the relation* " $\leq$"). The "$\leq$" *is called a conformity relation* if and only if for every pair of two cover sets C_1, C_2 if $C_1 \leq C_2$ then C_2 is smoother than C_1.

In the case of a conformity relation "$\leq$", the cover $\{U\}$ is the minimum cover. In this paper, assume that the relation "$\leq$" is a conformity relation.

Definition 6. (*the lattice of covers*) [5]. L is defined as a lattice of covers of U if and only if L is a set of covers of U $(C_1, C_2, \dots, C_n)$, and for every C_1, C_2 belong to L, there also exist Y_1, Y_2 belonging to L such that $Y_1 \leq C_1, Y_1 \leq C_2$ and $C_1 \leq Y_2, C_2 \leq Y_2$.

Definition 7. (*the top-cover and the bottom-cover of a lattice of covers* [5]). Since the universe U is finite, there exist C_{top}, C_{bottom} such that C_{top}, C_{bottom} belonging to L, and $C \leq C_{top}(C_{bottom} \leq C)$ for every C in L.

Definition 8. [5] A *cover lattice based decision table* (CDT) is a triple $\langle U, CL, DL \rangle$, where CL and DL are two lattices of covers of the universe U; CL and DL are called *condition lattice* and the *decision lattice*, respectively.

Definition 9. (*the induced cover of CL*) [5]. Let $CDT = \langle U, CL, DL \rangle$ be a cover lattice based decision table; $TopCL = \{S_1, S_2, \dots, S_k\}$ be a top-cover of CL. For every $x \in U$, let $CovCL_x = \bigcap\{S_j \in TopCL, x \in S_j\}$, then, the set $Cov(CL) = \{CovCL_x | x \in U\}$ is also a cover of U and it is called *the induced cover of CL*.

$CovDL_x$ and the induced cover $Cov(DL)$ of DL are also defined in the same way.

Definition 10. (*CL-positive region of DL*). Let $CDT = \langle U, CL, DL \rangle$ be a cover lattice based decision table. The *CL*-positive region of DL is calculated as

$$\text{POS}_{CL}(DL) = \bigcup_{x,y \in U} CovCL_x \left(CovCL_y \right) \tag{1}$$

Definition 11. (*the dependence of a cover lattice based decision table*) [5]. Let $CDT = \langle U, CL, DL \rangle$ be a cover lattice based decision table. We say that DL depends on CL at a degree $k(0 \leq k1)$, denoted by $CL \rightarrow_k DL$, and

$$k = \frac{|\text{POS}_{CL}(DL)|}{|U|} \tag{2}$$

We also call the dependence of a cover lattice based decision table by the dependence of its condition lattice CL.

The DL's dependency on CL is an indicator of the performance that supports decisions of cover lattice based decision table.

2.3 An Example: Cover Lattice Based Decision Tables in Transaction Databases

Recall to the lattice of a itemset in the task of discovering the frequent itemsets from a transaction database [2]. For each itemset X in a transaction database D, there exists a lattice of a itemset LX such that X is the top element (node) and the empty set is the bottom element (node) in LX. There is a map from lattice of a itemset to lattice of covers in transaction database D.

We consider the task of discovering strong association rules based on a closed frequent itemset from the transaction database D. Let X be a closed frequent itemset, $s(X)$ be the support of X. Let Y be an itemset, $Y \subset X$, and $Y \neq \varnothing$, then $Y \rightarrow (X - Y)$ is the corresponding association rule, where Y and $(X - Y)$ are two non-empty itemsets.

Example 1. Let $I = \{A, B, G, H, E\}$ be the set all of items in a transaction database. Let D be a transaction database consisting of below transactions:

Tid	Items
10	A, B, G, H, E
20	B, G, H, E
30	A, G, H, F

With a minimum support of 0.66, itemset $X = \{B, G, H, E\}$ is a maximum closed itemset. We consider all association rules of which both the antecedent and the consequent are 2-items, i.e., $\{B, G\} \rightarrow \{H, E\}, \{B, H\} \rightarrow \{G, E\}$, etc.

For example, the association rule $\{B, G\} \rightarrow \{H, E\}$ is corresponded with a cover lattice based decision table $\langle D, C_{\{B,G\}}, C_{\{H,E\}} \rangle$, in which the *condition lattice* and the *decision lattice* are described in Fig. 1.

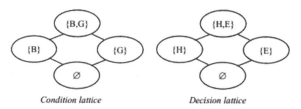

Condition lattice   Decision lattice

Fig. 1. The *condition lattice* and the *decision lattice of* cover lattice based decision table $\langle D, C_{\{B,G\}}, C_{\{H,E\}} \rangle$

Description of the covers in the condition lattice: The empty itemset $\varnothing$ is corresponding to the universe $C_\varnothing = \{D\}$. Two 1-itemsets $\{B\}$ and $\{G\}$ are corresponding to partitions $C_{\{B\}} = \{\{D\}, \{\text{transactions include } B\}\} = \{D, \{10, 20\}\}; C_{\{G\}} = \{\{D\}, \{\text{transactions include } G\}\} = \{D\}$. The 2-itemset $\{B,G\}$ is corresponding to a cover: $C_{\{B, G\}} = \{\{D\}, \{\text{transactions include } B\}, \{\text{transactions include } G\}\} = \{D, \{10, 20\}\}$. The top-cover of the condition lattice $TopCL$ is $C_{\{B,G\}}$ then $CovCL_{10} = \{10, 20\}$, $CovCL_{20} = \{10, 20\}$, and $CovCL_{30} = \{D\}$ then $Cov(CL)$, the induced cover of CL, is $Cov(CL) = \{\{10, 20\}, D\}$.

Similarly, we have the description of covers in the decision lattice: $C_{\{H\}} = \{D\}$, $C_{\{E\}} = \{\{10, 20\}, D\}$, $C_{\{H,E\}} = \{\{10, 20\}, D\}$; *TopDL* is $C_{\{H,E\}}$ then $CovDL_{10} = \{10, 20\}$, $CovDL_{20} = \{10, 20\}$, and $CovDL_{30} = D$ then $Cov(DL) = \{\{10, 20\}, D\}$.

Covers corresponding to 3-itemsets, 4-itemsets, and k-itemsets are set up in the same way.

Note: The dependence degree of a covering decision table and the confidence degree of responding association rule are different. The confidence degree of responding association rules is calculated on a number of transaction, i.e., $s(BGHE)/s(BG) = 2/2 = 1$), while the dependence degree is determined on relations of covers $|POS_{CL}(DL)|/|U|$.

3   The Reduction of the Condition Lattice and the Fitting of the Decision Lattice

Let $CDT = \langle U, CL, DL \rangle$ be a cover lattice based decision table, where CL, DL are the condition and decision lattices, respectively. We consider two tasks of reduction for CL and DL in cover lattice based decision table. Firstly, we introduce the sub-lattice definition and its property.

Definition 12. (*sub-lattice*): Let L be a lattice of covers on U. A lattice of covers SL on U is called by a sub-lattice of L if and only if: (i) the set of covers in SL is a subset of the set of covers in L, and (ii) for every cover C in SL, for every cover X in L satisfies $X \leq C$, then X is also in SL.

The Definition 12 of the sub-lattice is appropriate for association rule mining, where the lattice of an itemset X is a sub-lattice Y such that $X \subset Y$.

Corollary 1. If k is the dependence of cover lattice based decision table $\langle U, CL, DL \rangle$, sk is the dependence of cover lattice based decision table $\langle U, SCL, DL \rangle$, then $k \geq sk$ for every sub-lattice SCL of the condition lattice CL. Similarly, if k is the dependence of cover lattice based decision table $\langle U, CL, DL \rangle$ and sk is the dependence of cover lattice based decision table $\langle U, CL, SDL \rangle$ then $sk \geq k$ for every sub-lattice SDL of the decision lattice DL.

Definition 13 (*child-lattice, parent-lattice*). Let SL be a sub-lattice of L. SL is called a child-lattice of L if only if there is no sub-lattice $SL2$ of L that SL is a sub-lattice of $SL2$. If SL is a child-lattice of L, then L is called by the parent-lattice of SL.

3.1   Reduction in Cover Lattice Based Decision Tables

This section considers the reduction in covering based decision tables. We first introduce a definition of the condition reduction of cover lattice based decision tables, then we describe an algorithm for finding all reducts.

Definition 13 (*condition reduction*). Let $CDT = \langle U, CL, DL \rangle$ be a cover lattice based decision table. A sub-lattice S CL of CL called by a reduced of CL if the dependence of $\langle U, SCL, DL \rangle$ is equal to the dependence of $\langle U, CL, DL \rangle$. A reduct SCL of CL is called a reduction of CL if only if: (i) SCL is a reduced of CL, and (ii) if there exists a reduct $SCL2$ of CL and $SCL2$ is sub-lattice of SCL then $SCL2 = SCL$.

We propose a recursive algorithm Reduct_Finding for finding all the reduction lattices of a cover lattice based decision table. The algorithm has a global constant CL which is the dependence of DL; a global variable $GCRL$ is a set of current reduced lattices. At the beginning of the algorithm, $GCRL$ includes only CL, i.e., $GCRL = \{CL\}$. At the termination of the algorithm, $GCRL$ includes all reduction lattices of CL.

Algorithm Reduct_Finding (*CCL*, *PCCL*)
   // **CCL** is the current condition lattice
   // **PCCL** is the parent of **CCL** (**CL** is the "parent" of itself)
IF **CCL** = **CL**   //starting with **CL**
   THEN
      FOREACH sub-lattice **SCCL** of **CCL**
         Reduct_Finding (**SCCL**, **CCL**)   //running on all sub-lattices of **CL**
      END FOR
   ELSE
      $\rho CCL \leftarrow$ the dependence of $\langle U, CCL, DL \rangle$
      IF $\rho CCL = \rho CL$ //**CCL** is a reduced set
         THEN
            Add **CCL** in **GCRL**
            IF **PCCL** in **GCRL** THEN
               Remove PCCL from GCRL   //**PCCL** is a reduced set but not a reduct
            END IF
            FOREACH sub-lattice **SCCL** of **CCL**   //running all sub-lattices of **CCL**
               Reduct_Finding (**SCCL**, **CCL**)
            END FOR
      END IF
END IF

The operation of Reduct_Finding algorithm is explained as follows. The algorithm is started by calling Reduct_Finding (*CL*, *CL*), in this case, the algorithm works, one by one, on all sub-lattices of the *CL* condition set. With each subsequent call to Reduct_Finding (*CCL*, *PCCL*), the algorithm first calculates the dependency of the current decision table $\langle U, CCL, DL \rangle$. If *CLL* is a reduced condition lattice, then add it into *GCRL*, remove its parent lattice if it is in *GCRL*. Continue searching for all children *SCCL* of *CCL*. At the end of the algorithm, *GCRL* contains all the reducts of the condition lattice *CL*. When no sub-lattice is found, *GCRL* only contains *CL* as at the beginning of the algorithm.

3.2 Decision Fitting in the Cover Lattice Based Decision Table

In cover lattice based decision tables, we consider not only the reduction of the condition lattice but also the fitting of the decision lattice. We first introduce a definition of a decision fitting of a cover lattice based decision table, then we describe an algorithm for finding all decision fittings.

Definition 14 (*decision fitting*): Let $CDT = \langle U, CL, DL \rangle$ be a cover lattice based decision table, let σ be a threshold, which is greater or equal to the dependence of CDT, i.e., $\rho CL < \sigma \leq 1$. The task of decision fitting in CDT is to find all sub-lattices SDL of DL such that: i) the dependence of $\langle U, CL, DL \rangle$ is not less than σ; and ii) the dependence of $\langle U, CL, PSDL \rangle$ is less than σ, where $PSDL$ is the parent-lattice of SDL.

We propose a recursive algorithm called Fitting_Finding for finding all fitting lattices of a cover lattice based decision table. In the algorithm, there is a global variable $GCFL$ containing the set of current fitting lattices.

At the starting of the algorithm, $GCFL$ is empty and at the termination of the algorithm, $GCFL$ includes all fitting lattices.

Algorithm Fitting_Finding (CDL)   //CDL is the current decision lattice
    $\rho CDL \leftarrow$ the dependence of $\langle U, CL, CDL \rangle$
    IF $\rho CDL \geq \sigma$
        THEN
            Add CDL into $GCFL$
        ELSE
            FOREACH sub-lattice $SCDL$ of CDL
                Fitting_Finding ($SCDL$)
            END FOR
    END IF

The operation of the Fitting_Finding algorithm is explained as follows. The algorithm starts by calling Fitting_Finding(DL), which executes, one by one, for all the sub-lattice of the decision lattice DL. For each subsequent call to Fitting_Finding(CDL), the algorithm first calculates the dependency ρCDL of the current decision table $\langle U, CL, CDL \rangle$. If CDL meets the search condition, i.e., $\rho CDL \geq \sigma$, then adds CDL to $GCFL$. Otherwise, further searching is run on all children $SCDL$ of CDL. At the end of the algorithm, $GCFL$ contains all (possibly empty) sub-lattices of the decision lattice DC.

3.3 A Typical Application of Cover Lattice Based Rough Set

Consider a collaborative filtering recommender system with a set U consisting of m users, and a set I consisting of n items. The user-item relationship matrix $P(u, i)$ of size $m \times n$ indicates the ratings of m users on n items. Let s_u, s_{uv}, s_A, respectively, be the set of items evaluated by user u; by both users u and v; and by all users in set A ($A \subseteq U$). Let s_j, s_{jk}, s_B, respectively, be the sets of users who have evaluated item j; both items j and k; all items in set B ($B \subseteq I$).

For an active user u, a cover lattice based decision table $\langle U, CL_u, DL_u \rangle$ can be constructed, where CL_u is the lattice corresponding to the S_u; DL_u is the lattice corresponding to the itemset $I \backslash S_u$. In practice, CL_u (or DL_u) is the lattice corresponding to the real subset of S_u ($I \backslash S_u$).

In this cover lattice based decision table, the reduction problem means to find the subset that have been evaluated by the user instead of the set of all items evaluated by the user. The fitting problem means to find subsets of items (unknown the active user u) to recommend to u with a confidence that is not less than a pre-defined threshold σ. The Reduce-Finding and Fitting_Finding algorithms can be used to get the corresponding results.

Example 2: Let have transaction database with the covers in the condition lattice and covers in the decision lattice as in **Example 1**. We are going to identify reduction in cover lattice based decision tables and decision fitting in the cover lattice based decision table by performing the Reduct_Finding algorimth and Fitting_Finding algorithm above.

Acorrording to the Example 1, the included cover of *CL and DL, Cov(CL)and Cov(DL)are the same, Cov(CL) = Cov(DL) =* {{10, 20}, *D*}

Therefore, *CL-positive region of DL, POS$_{CL}$(DL) is D and the dependence of a cover latice* $\rho CL = \frac{|POS_{CL}^{*}(DL)|}{|U|} = \frac{2}{2} = 1.$

Performing Reduct_Finding algorithm with the initial configurations of CCL = *{BG}*, ***PCCL = {BG}***

// *The cover lattice based decision table is* $\langle U, \{BG\}, \{HE\}\rangle$; *the global varibale is.*

// *GCRL = {CL} = {BG}*

Step	Current condition lattice CCL and its parent PCCL	*TopCCL*	*Cov(CCL)* and *Cov(DL)*	ρCCL	GCRL
Initial step	$(\{B, G\}, \{B, G\})$	*TopCCL =* {{10, 20}, *D*} *TopDL =* {{10, 20}, *D*}	*Cov(CCL) =* {{10, 20}, *D*} *Cov(DL) =* {{10, 20}, *D*} *POS$_{CL}$(DL) =* {{10, 20}, *D*}	1	{B, G}
1	$(\{B\}, \{BG\})$	*TopCCL =* {{10, 20}, *D*} *TopDL =* {{10, 20}, *D*}	*Cov(CCL) =* {{10, 20}, *D*} *Cov(DL) =* {{10, 20}, *D*} *POS$_{CL}$(DL) =* {{10, 20}}	1	{B}
2	$(\{G\}, \{BG\})$	*TopCCL =* {*D*} *TopDL =* {{10, 20}, *D*}	*Cov(CCL) =* {*D*} *Cov(DL) =* {{10, 20}, *D*} *POS$_{CL}$(DL) =* {*D*}	1	{{B}, {G}}

At the termination of the algorithm, we have reduced the lattices of condition lattices *CL* is *GCRL =* {{*B*}, {*G*}}. Because the covers corresponding to *B* are the same as those corresponding to {*B*, *G*}, the reduced lattices of the cover lattice based decision table are $\langle \mathcal{D}, C_{\{B\}}, C_{\{H,E\}}\rangle$ or $\langle \mathcal{D}, C_{\{G\}}, C_{\{H,E\}}\rangle$.

Performing the Fitting_Finding algorithm with the initial configuration of CDL = {HE}

// *Let GCFL = {}; and the dependence threshold of DL,* $\delta = \rho CL = 1$

Step	Current decision lattice CDL	(CL,DL)	TopCCL and TopCDL	Cov(CL) and Cov(CDL)	ρCDL	GCFL
Initial step	{H, E}	({BG},{HE})	$TopCL = \{\{10, 20\}, D\}$ $TopDL = \{\{10, 20\}, D\}$		1	{}
1	{H}	({BG}, {H})	$TopCCL = \{\{10, 20\}, D\}$ $TopCDL = \{D\}$	$Cov(CL) = \{\{10, 20\}, D\}$ $Cov(CDL) = \{D\}$ $POS_{CL}(DL) = \{D\}$	0.5	{}
2	{E}	({BG},{E})	$TopCCL = \{\{10, 20\}, D\}$ $TopCDL = \{\{10, 20\}, D\}$	$Cov(CL) = \{\{10, 20\}, D\}$ $Cov(CDL) = \{\{10, 20\}, D\}$ $POS_{CL}(DL) = \{\{10, 20\}, D\}$	1	{E}

At the termination of algorithm, the fitting lattice of decision lattice DL is $GCFL = \{E\}$. Because the covers corresponding to E are the same as the those corresponding to $\{E, H\}$, the fitting lattices of the cover lattice based decision table are $\langle \mathcal{D}, C_{\{B,G\}}, C_{\{E\}} \rangle$.

4 Related Work

C. Degang et al. investigated the conditional coverage reduction problem Δ of a covering-based decision system $(U, \Delta \cup D)$ [3]. The authors consider reduction problem in two cases, i.e., the system $(U, \Delta \cup D)$ is consistent and inconsistent (according to Definition 3 above). In each case, after determining some of the required properties of the covering based decision system, the discernibility matrix of $(U, \Delta \cup D)$ is defined; procedures to find the collection of all relatively indispensable covers $(Core_D(\Delta))$ and the collection of all relatively small indispensable covers (Red (Δ, D)) are proposed. Experiments on several data sets from UCI Repository of machine learning databases were conducted with different scenarios of varying cover sizes.

Following C. Degang et al., C.Z. Wang et al. improved the definition of discernibility matrix and proposed improved procedures to find $Core_D(\Delta)$ and Red(Δ, D) sets [8]. Via experiments, the authors showed the better effectiveness of the proposed procedures in comparison with the previous ones.

Using the discernibility matrix from C.Z. Wang et al.'s definition, A. Tan et al. proposed a method of solving the reduction problem based on a matrix [7]. Two matrix-based algorithm for calculating the positive region of a covering decision system, a matrix-based algorithm for finding a reduct of a covering decision system were proposed and evaluated by experiments.

This paper uses a special architecture of the cover lattice based on the conformity of the relation of the relation "$\leq$" (cf. Definition 5) to propose uncomplicated algorithms to solve reduction problems. In addition, a decision cover in a cover lattice based decision table provides a different approach from using a decision attribute in a covering based decision system $(U, \Delta \cup D)$ in related studies.

5 Conclusions and Future Work

This paper focuses on developing the research on reduction problems in cover lattice based decision tables. The reduction problem is considered not only in the condition part, but also in the decision part of the decision table. Two illustrative examples have been provided to show the meaning and applicability of the cover lattice based decision table and its reduction problem. Especially, we showed the application of the proposal in content filtering-based recommender systems [11, 12].

In our future studies, experiments should be conducted to verify the performance of the algorithms in practice, and the relevant properties of cover lattice based decision tables need to be carefully investigated.

References

1. Bonikowski, Z., Bryniarski, E., Wybraniec-Skardowska, U.: Extensions and intentions in the rough set theory. Inf. Sci. **107**(1–4), 149–167 (1998)
2. Brin, S., Motwani, R., Ullman, J.D., Tsur, S.: Dynamic itemset counting and implication rules for market basket data. In: SIGMOD Conference, pp. 255–264 (1997)
3. Degang, C., Wang, C., Qinghua, H.: A new approach to attribute reduction of consistent and inconsistent covering decision systems with covering rough sets. Inf. Sci. **177**(17), 3500–3518 (2007)
4. Pawlak, Z.: Rough sets. Int. J. Comput. Inf. Sci. **11**(5), 341–356 (1982)
5. Pham, T.-H., Nguyen, T.C.-V., Vuong, T.-H., Ho, T., Ha, Q.-T., Nguyen, T.-T.: A definition of cover lattice based decision table and its sample applications. In: ICISA 2020 (2020, in press)
6. Skowron, A., Dutta, S.: Rough sets: past, present, and future. Natural Comput. 1–22 (2018)
7. Tan, A., Li, J., Lin, Y., Lin, G.: Matrix-based set approximations and reductions in covering decision information systems. Int. J. Approx. Reason. **59**, 68–80 (2015)
8. Wang, C., He, Q., Chen, D., Qinghua, H.: A novel method for attribute reduction of covering decision systems. Inf. Sci. **254**, 181–196 (2014)
9. Yao, Y., Yao, B.: Covering based rough set approximations. Inf. Sci. **200**, 91–107 (2012)
10. Zakowski, W.: Approximations in the space (U, Π). Demonstratio Math. **XVI**(3), 761–769 (1983)
11. Zhang, Z., Kudo, Y., Murai, T.: Neighbor selection for user-based collabo-rative filtering using covering-based rough sets. Ann. Oper. Res. **256**(2), 359–374 (2017)
12. Zhang, Z., Kudo, Y., Murai, T., Ren, Y.: Improved covering-based collaborative filtering for new users' personalized recommendations. Knowl. Inf. Syst. **62**(8), 3133–3154 (2020). https://doi.org/10.1007/s10115-020-01455-2
13. Zhao, Z.: On some types of covering rough sets from topological points of view. Int. J. Approx. Reason. **68**, 1–4 (2016)
14. Zhu, W.: Basic concepts in covering-based rough sets. In: ICNC, no. 5, pp. 283–286 (2007)

Forecasting System for Solar-Power Generation

Jia-Hao Syu[1]([✉]) , Chi-Fang Chao[2], and Mu-En Wu[2]

[1] Department of Computer Science and Information Engineering,
National Taiwan University, Taipei, Taiwan
f08922011@ntu.edu.tw
[2] Department of Information and Finance Management,
National Taipei University of Technology, Taipei, Taiwan
mnwu@ntut.edu.tw

Abstract. Environmental protection is a highly concerned and thought-provoking issue, and the way of generating electricity has become a major conundrum for all mankind. Renewable or green energy is an ideal solution for environmentally friendly (eco-friendly) power generation. Among all renewable energy, solar-power generation is low cost and small footprint, which make solar-power generation available around our lives. The major uncontrollable factor in the solar-power generation is the amount of solar radiation, which completely dominates the electricity generated by solar panel. In this paper, we design prediction models for solar radiation, using data mining techniques and machine learning algorithms, and derive precision prediction models (PPM) and light prediction models (LPM). Experimental results that the PPM (LPM) with random forest regression can obtain R-squared of 0.841 (0.828) and correlation coefficient of 0.917 (0.910). Compared with highly cited researches, our models outperform them in all measurements, which demonstrates the robustness and effectiveness of the proposed models.

Keywords: Solar-power generation · Data mining · Machine learning

1 Introduction

Environmental protection is a highly concerned and thought provoking issue, and the way of generating electricity has become a major conundrum for all mankind. Currently, most of the electricity comes from thermal power, which produce large amounts of carbon dioxide (greenhouse gases [15]) and other harmful gases [4]. Renewable or green energy is an ideal solution for environmentally friendly (eco-friendly) power generation [6]. Common methods of green-power generation include wind-power and solar-power, [14]. Wind-power utilizes the kinetic energy of wind to drive power generators, which is location dependent, high cost and space needed. Conversely, solar-power generation is low cost and small footprint, which make solar-power generation available around our lives.

At the same time, many governments encourage the investment and construction of solar-power generation, and provides substantial subsidies (such as China,

© Springer Nature Singapore Pte Ltd. 2021
T.-P. Hong et al. (Eds.): ACIIDS 2021, CCIS 1371, pp. 65–72, 2021.
https://doi.org/10.1007/978-981-16-1685-3_6

USA, India, Germany, Australia, and Taiwan etc.) [11]. Literatures show that 40% of global electricity growth comes from renewable energy sources, mainly from solar-power and wind-power [11], which shows that solar-power generation is an emerging topic.

The major uncertainty and uncontrollable factor in the solar-power generation is the amount of solar radiation, which completely dominates the electricity generated by solar panel. Less solar radiation will produce less electricity and profits, which may not be able to cover the depreciation. To sum up, a precise solar-radiation prediction model is required in the solar-power generation. In this paper, we design prediction models for solar radiation, using data mining techniques and machine learning algorithms through the meteorological data, and we derive precision prediction models (PPM) and light prediction models (LPM).

Experimental results show that the prediction algorithms of random forest regression and support vector regression have the best performance. The PPM (LPM) with random forest regression can obtain R-squared of 0.841 (0.828) and correlation coefficient of 0.917 (0.910). Compared with highly cited researches, our models outperform them in all measurements, which demonstrates the robustness and effectiveness of the proposed models (PPM and LPM).

2 Literature Review

In this section, the background of solar-power generation and prediction are first introduced. Then, we survey the literature and mechanism of machine learning algorithms used in this paper.

2.1 Solar-Power Generation and Prediction

Solar-power generation convert the solar radiation into electricity through the photovoltaic effect [7]. By connecting array of photovoltaic cells, the photovoltaic system (solar panel) can generate about 150 to 180 W per square meter [13].

There are considerable researchers focus on solar-power prediction. Jang et al. [8] predicted the solar-power by atmospheric motion vectors from satellite images, which determined the motion vectors of clouds and affected the solar radiation. Zeng and Qiao [16] adopted a least-square support vector machine model, which utilized features of historical atmospheric transmissivity and meteorological variables. The results showed that, compared with the models based on auto-regressive and neural network, the support vector machine has better performance.

2.2 Machine Learning Algorithms

Machine learning algorithms are powerful tools for data analysis and mining. To meet the requirements of the proposed edge system, we survey several efficient and light-computing algorithms used in this paper, including multiple linear regression, support vector machine, and random forest regression.

Multiple Linear Regression. Multiple Linear Regression (MLR) is a linear method to estimate the relationship between a dependent variable and multiple dependent variables, which is also known as the level of correlation [10]. In short, the estimated variable is an intercept added with the inner product of the regression coefficients and the independent variables. MLR is the most commonly used regression model for data analysis with the characteristics of easy-to-use and interpretable. Each independent variable is related to the dependent variable through its own regression coefficient, which makes users to explore relationships intuitively.

Support Vector Regression. Support Vector Machine (SVM) is a supervised learning algorithm, proposed by Cortes and Vapnik in late 1995, which is originally developed for two-type classification problems [5]. SVM aims to find a hyperplane with largest margin to separate data on the nonlinear mapped space (through kernel functions). In addition, SVM is also extended to regression problems, especially nonlinear regression, called Support Vector Regression (SVR) [9].

Random Forest Regression. Random forest is a supervised learning algorithm developed by Leo Breiman in 2001 [2], which can be regarded as the expansion and aggregation of the decision tree. Each decision tree is a weak classifier, and multiple trees constitute a strong classifier. Through bagging and bootstrap techniques, random forests can achieve better performance (more accurate and stable) than decision trees [1].

3   Proposed Prediction Models

We adopt several classic machine learning algorithms (the literatures also demonstrate that classic algorithms are better than neural network-based models). Three machine learning algorithms are adopted as predictive models, including multiple linear regression (MLR), random forest regression (RFR), and support vector regression (SVR). We utilize the sklearn library in python to implement the algorithms, whichs' parameters are almost the default settings.

MLR comes from the function of linear_model.LinearRegression with the default parameters. RFR comes from the function of ensemble.Random ForestRegressor with bootstrap, 500 estimators, and criterion of mean squared error. SVR comes from the function of svm.SVR with scaled gamma, kernel of radial basis function, and maximum iteration of 1,000.

As for predictive features, since these models are designed to predict the hourly or daily solar radiation, several hourly observations are adopted, including observation month and time, sun duration, temperature, relative humidity, wind speed (reference from [8,16]). The description and characteristics of the observations are shown in the Table 1.

In this paper, for the purpose of high precision and low computation, we have designed two branches of models, which are called precise prediction models

Table 1. Description and characteristics of the observations

Observation	Abbreviation	Description	Unit
Observation month	$Month$	The month of observation	Dummy, 1,...,12
Observation time	$Hour$	The time (which hour) of observation	Dummy, 1,...,24
Solar radiation	$SRad$	The radiant energy of sunshine	MJ/m^2
Sun duration	$SDur$	The length of sunshine	hour
Temperature	Tem	Average temperature during the observation	$°C$
Relative humidity	RH	Relative humidity during the observation	%
Wind speed	WS	Average speed of wind during the observation	m/s
Cloud amount	CA	The amount (region) of cloud cover the sky	0,...,10

(PPM) and light prediction models (LPM). The difference between PPM and LPM lies in the number of usage features (5 and 12 features).

LPM only uses the features of historical solar radiation (SRad) for the previous five days (at the same hour as the prediction). Briefly, to predict the SRad at h o'clock on date d, $SRad_{d,h}$, we utilize ($SRad_{d-1,h}$, $SRad_{d-2,h}$, ..., $SRad_{d-5,h}$) as features (5 features).

As for the PPM, the features include solar radiation in previous five days (at the same hour as prediction), $Month$ and $Hour$ of the prediction time, and $SDur$, Tem, RH, WS, CA in the previous day (at the same hour as prediction). To be brief, to predict the SRad at h o'clock on date d, $SRad_{d,h}$, we utilize $SRad_{d-1,h}$, $SRad_{d-2,h}$, ..., $SRad_{d-5,h}$, $Month_{d,h}$, $Hour_{d,h}$, $SDur_{d-1,h}$, $Tem_{d-1,h}$, $RH_{d-1,h}$, $WS_{d-1,h}$, and $CA_{d-1,h}$ as features (12 features).

4   Experimental Results

We first introduce the usage datasets in this paper. Then, evaluate the predictive ability of the proposed models, and compare the performance between PPM and LPM models. Finally, compare the performance with highly cited researches.

4.1   Data Usage

In this paper, we utilize hourly weather data provided by the Central Weather Bureau of Taiwan [3] to construct the solar radiation prediction models, and we adopt 5 large-scale datasets. The selected datasets come from weather stations in different cities, covering the wild range of latitude and longitude (in Taiwan), which are listed in Tables 2. The time interval of the datasets we use is from July 2010 to June 2020, including 10 years.

Table 2. Latitude and longitude of the meteorological stations for training

Station ID	Latitude (N)	Longitude (E)	City
467410	22°99′	120°20′	Tainan City
467650	23°88′	120°91′	Nantou County
467540	22°36′	120°90′	Taitung County
466920	25°04′	121°51′	Taipei City
467770	24°26′	120°52′	Taichung City

4.2 Solar Radiation Prediction

In this section, we demonstrate the prediction results of the models, which are PPM and LPM on datasets with IDs of 467480, 467060, 466900, 467440, and 466940. Two performance measures are utilized, including the R-squared and the correlation coefficient (C.C.) between the predicted results and the ground truth. Among all tables in this section, each row represents the performance of a machine learning model on the five weather stations (five columns). In addition, the values shown in bold are the best performance among all machine learning models (the highest value in the column).

First, Table 3 presents the results of the precise predictive models, PPM. In this experiment, the machine learning models of RFR and SVR obtain the best performance, especially for RFR, which obtain the highest R-squared and C.C among all datasets. The best performance of PPM can reach R-squared of 0.841 and C.C of 0.917.

Table 4 presents the results of the light predictive models, LPM. Surprisingly, the results of LPM are quite close to the results of PPM with minor weaknesses. Compared with PPM, the R-square of LPM is reduced by about 0.026, and the C.C is reduced by about 0.015, however, less than half of features are required for LPM to obtain the excellent results. Among various machine learning algorithms in LPM, RFR and SVR still have the best performance. The best R-squared of RFR is 0.828, and the highest C.C of SVR is 0.911.

Table 3. Performance of PPM with machine learning algorithms

Weather Station ID	467480		467060		466900	
Indicators	R-squared	C.C	R-squared	C.C	R-squared	C.C.
MLR	0.809	0.899	0.685	0.827	0.704	0.839
RFR	**0.827**	**0.910**	**0.726**	**0.852**	**0.748**	**0.865**
SVR	0.824	**0.910**	0.705	0.843	0.735	0.860
Weather Station ID	467440		466940			
Indicators	R-squared	C.C	R-squared	C.C		
MLR	0.820	0.906	0.676	0.823		
RFR	**0.841**	**0.917**	**0.729**	**0.854**		
SVR	0.829	0.916	0.712	0.846		

Table 4. Performance of LPM with machine learning algorithms

Weather Station ID	467480		467060		466900	
Indicators	R-squared	C.C	R-squared	C.C	R-squared	C.C.
MLR	0.806	0.898	**0.679**	**0.824**	0.700	0.837
RFR	**0.810**	0.900	0.678	**0.824**	**0.705**	**0.840**
SVR	0.809	**0.904**	0.659	0.817	0.697	0.838
Weather Station ID	467440		466940			
Indicators	R-squared	C.C	R-squared	C.C		
MLR	0.817	0.904	0.665	**0.816**		
RFR	**0.828**	0.910	**0.666**	**0.816**		
SVR	0.817	**0.911**	0.646	0.807		

In summary, even if the LPM use less than half of PPM features, their prediction performance is quite close. Therefore, the proposed LPM has almost excellent prediction results with few features, and the computation can be significantly reduced by more than half, thereby satisfying the purpose of the design of IoT and hedging system.

Table 5. Comparison with highly cited research

Model	PPM-RFR	LPM-RFR	Zeng [16]	Jang [8]	Long [12]
R^2	0.827	0.810		0.731	
C.C	0.910	0.900			0.895
MAE	0.189	0.197	0.329		0.199

We list in the prediction performance of the highly cited researches in Table 5, including Zeng and Qiao [16], Jang et al. [8], and Long et al. [12], and compare with proposed PPM with RFR (PPM) and LPM with RFR (LPM). Measurements include R-squared (R^2), correlation coefficient (C.C.) and mean absolute error (MAE). Please note that the units of solar radiation are different between the researches (affects MAE), and we converted all of them to MJ/m^2.

In Table 5, our models beat all of the highly cited researches. Compared with Zeng and Qiao [16], using SVM algorithm, our models obtain lower MAE in the unit of MJ/m^2. Compared with Jang et al. [8], using SVM algorithm with real-time satellite imaging, our models can obtain higher R^2 only with simple meteorological observations. Compared with Long et al. [12], using the data-driven approaches, our models obtain a slightly higher C.C. and lower MAE. In summary, the experimental results show the robustness and effectiveness of the prediction models (PPM and LPM) among various datasets and measurements, and outperform researchers with high citations.

5   Conclusions

Environmental protection is a highly concerned and thought-provoking issue, and the way of generating electricity has become a major conundrum for all mankind. Renewable or green energy is an ideal solution for environmentally friendly (eco-friendly) power generation. Among all renewable energy, solar-power generation is low cost and small footprint, which make solar-power generation available around our lives.

The major uncontrollable factor in the solar-power generation is the amount of solar radiation, which completely dominates the electricity generated by solar panel. Less solar radiation will produce less electricity and profits, which may not be able to cover the depreciation. In this paper, we design prediction models for solar radiation, using data mining techniques and machine learning algorithms, and we derive precision prediction models (PPM) and light prediction models (LPM).

Experimental results show that the prediction algorithms of random forest regression and support vector regression have the best performance. The PPM (LPM) with random forest regression can obtain R-squared of 0.841 (0.828) and correlation coefficient of 0.917 (0.910). Compared with highly cited researches, our models outperform them in all measurements, which demonstrates the robustness and effectiveness of the proposed models.

References

1. Babar, B., Luppino, L.T., Boström, T., Anfinsen, S.N.: Random forest regression for improved mapping of solar irradiance at high latitudes. Solar Energy **198**, 81–92 (2020)
2. Breiman, L.: Random forests. Mach. Learn. **45**(1), 5–32 (2001). https://doi.org/10.1023/A:1010933404324
3. The Central Weather Bureau: Central weather bureau observation data inquire system (2020). https://e-service.cwb.gov.tw/HistoryDataQuery/index.jsp
4. Chen, T.L.: Air pollution caused by coal-fired power plant in middle Taiwan. Int. J. Energy Power Eng. **6**(6), 121 (2017)
5. Cortes, C., Vapnik, V.: Support-vector networks. Mach. Learn. **20**(3), 273–297 (1995)
6. Ellabban, O., Abu-Rub, H., Blaabjerg, F.: Renewable energy resources: current status, future prospects and their enabling technology. Renew. Sustain. Energy Rev. **39**, 748–764 (2014)
7. Goetzberger, A., Hoffmann, V.U.: Photovoltaic Solar Energy Generation, vol. 112. Springer, Heidelberg (2005). https://doi.org/10.1007/b137803
8. Jang, H.S., Bae, K.Y., Park, H.S., Sung, D.K.: Solar power prediction based on satellite images and support vector machine. IEEE Trans. Sustain. Energy **7**(3), 1255–1263 (2016)
9. Kaneda, Y., Mineno, H.: Sliding window-based support vector regression for predicting micrometeorological data. Expert Syst. Appl. **59**, 217–225 (2016)
10. Khademi, F., Jamal, S.M., Deshpande, N., Londhe, S.: Predicting strength of recycled aggregate concrete using artificial neural network, adaptive neuro-fuzzy inference system and multiple linear regression. Int. J. Sustain. Built Environ. **5**(2), 355–369 (2016)

11. Lei, Y., et al.: SWOT analysis for the development of photovoltaic solar power in Africa in comparison with China. Environ. Impact Assess. Rev. **77**, 122–127 (2019)
12. Long, H., Zhang, Z., Su, Y.: Analysis of daily solar power prediction with data-driven approaches. Appl. Energy **126**, 29–37 (2014)
13. Murmson, S.: The output watts of solar panels (2018). https://sciencing.com/output-watts-solar-panels-6946.html
14. Patel, M.R.: Wind and Solar Power Systems: Design, Analysis, and Operation. CRC Press, Boca Raton (2005)
15. Shine, K.P., Fuglestvedt, J.S., Hailemariam, K., Stuber, N.: Alternatives to the global warming potential for comparing climate impacts of emissions of greenhouse gases. Clim. Change **68**(3), 281–302 (2005)
16. Zeng, J., Qiao, W.: Short-term solar power prediction using a support vector machine. Renew. Energy **52**, 118–127 (2013)

A Recommender System for Insurance Packages Based on Item-Attribute-Value Prediction

Rathachai Chawuthai$^{(\boxtimes)}$ (iD), Chananya Choosak, and Chutika Weerayuthwattana

Department of Computer Engineering, School of Engineering, King Mongkut's
Institute of Technology Ladkrabang, Bangkok, Thailand
{rathachai.ch,59010272,59010337}@kmitl.ac.th

Abstract. Finding a proper insurance package becomes a challenging issue for new customers due to the variety of insurance packages and many factors from both insurance packages' policies and users' profiles for considering. This paper introduces a recommender model named INSUREX that attempts to analyze historical data of application forms and contact documents. Then, machine learning techniques based on item-attribute-value prediction are adopted to find out the pattern between attributes of insurance packages. Next, our recommender model suggests several relevant packages to users. The measurement of the model results in high performance in terms of HR@K and F1-score. In addition, a web-based proof-of-concept application has been developed by utilizing the INSUREX model in order to recommend insurance packages and riders based on a profile from the user input. The evaluation against users demonstrates that the recommender model helps users get start in choosing right insurance plans.

Keywords: Attribute-value prediction · Data analytics · Insurance package ·
Machine learning · Recommender system

1 Introduction

Insurance is a widely accepted way for the protection of financial loss. It helps people manage their risk against any accidences of their life and properties. In the insurance market especially in a life and health insurance, there are a lot of insurance packages or coverage plans together with riders that insurance companies have provided, for example critical illness, cancer, dismemberment, endowment, medical, etc. [1]. All packages have different coverages, benefits, and premiums depended on the business model of insurance companies. In this case, customers have to compare a lot of packages from agents, websites, or handouts in order to choose ones that have appropriate benefits with the criteria and the budget of the customers.

For this reason, a recommender system for insurance packages becomes a proper solution for customers to find suitable packages and for agents to recommend some packages to customers. As we reviewed, there are pieces of research about recommender models in the insurance domain. Several works used a multi-criteria decision-making technique in their models [2, 3]. In addition, there are papers using machine learning

© Springer Nature Singapore Pte Ltd. 2021
T.-P. Hong et al. (Eds.): ACIIDS 2021, CCIS 1371, pp. 73–84, 2021.
https://doi.org/10.1007/978-981-16-1685-3_7

techniques such as Bayesian networks and neural networks in the recommendation methods [4–6]. The according works used each insurance package as an item in the model learning process, so the recommendation is based on items in the training set. However, in the insurance domain, new insurance packages are released frequently, and most of them have a period of sales. In addition, customers do not buy any insurance often, and they have one or a few packages. This behavior is different from normal product items in any e-commerce systems in terms of lacking user-item interaction and having new item cold-start problem. For this reason, the recommendation process has to emphasize on the prediction of the attributes and give a ranking score to each item based on the similarity between the item's attributes and the predicted attributes [7].

This work aims to create a recommender system for insurance packages named INSUREX. The model is based on item-attribute-value prediction using the datasets from user profiles and insurance policies to train and test the model and measure the performance of the model using HR@K and F1-score. To demonstrate our work, preparation steps are described in Sect. 2, our approach is explained in Sect. 3, results are discussed in Sect. 4, and conclusion and future work are drawn in the last section.

2   Preparation Steps

In the preparation steps, we first reviewed the background knowledge under the insurance domain in order to understand terms and business logics of this domain. After that, we had a closer look into the data for understanding and initially cleaning the datasets. Finally, we studied some methods and techniques to build a recommender model for insurance packages.

2.1   Background Knowledge

Insurance is one form of a financial product that provides life protection when customers face with any accidents having a severe impact on life and finances, including death, permanent disability and loss of income in the elderly. When the customers meet this situation, they will receive compensation in accordance with the conditions of the agreement agreed with the company to alleviate the damage to themselves and their families [1]. There are technical terms in the domain of insurance. In this paper, some used terms are described as follows:

- insurer   an agent (a person or a company) who sells insurance packages and pays compensation following the insurance contract.
- insured   an agent (a person or a company) who is covered by insurance contract. In this paper, it means a person, a customer, or a user who buys and has name in the insurance contract.
- insurance package a plan of insurance including options of protection coverage, insurance condition, benefits, and premium. This paper sometimes uses the term "item".
- rider   is an add-on insurance package. It needs to buy a main insurance package at first, and then buy other riders. It is also one of items in this paper.

- insurance policy	(or policy) a contract between an insurer and an insured who bought an insurance package. The insurance policy is a unique document although two or more insureds bought the same insurance package.
- premium	an amount that the insured has to pay to the insurer for the insurance policy.
- benefit	(or benefit of insurance) an amount that the insurer has to pay to the insured when a condition under the policy is met.
- dividend	some policies allow to pay money back to insureds regularly

Next, some examples of general rider's categories are demonstrated in the following list.

- Cancer Protection	riders that cover expenses from cancer treatments.
- Serious Disease Protection	riders that cover expenses from serious diseases such as strokes, heart attacks, etc.
- Accidence Protection	riders that cover medical expenses, loss of life, dismemberment, loss of sight, or total permanent disability, due to an accident.
- Medical Expense Protection	riders that cover medical expenses

In addition, in order to make a clear understanding about the relationship among according pieces of data, a cardinality relationship among data entities is depicted in Fig. 1. A policy has one insured, one insurer, and one insurance packages; while an insured, an insurer, and a package can be appeared in many policies. For the rider, many riders can be added into one policy, and a rider can be reused in other policies.

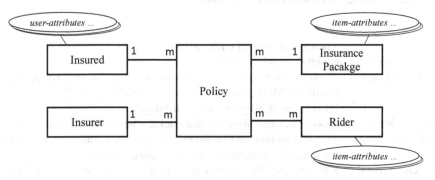

Fig. 1. A cardinality relationship among entities.

2.2 Data Preparation

We collected a dataset having 500 samples of policies from an insurance broker. However, there are 370 samples that are useful for our experiments, because the other ones are long-tail data whose packages are rarely purchased. Thus, in our datasets, there are 7 main packages under two categories: life-insurance and endowment. The frequencies of the 7 main packages are 136, 69, 69, 42, 23, 18, and 13 samples. There are 8 riders that have been additionally purchased with the main packages. The occurrences of the 8 riders are 155, 144, 135, 128, 81, 53, 50, and 50. In addition, there are 308 user profiles of insureds. In this dataset, there are about 1.2 packages by average were purchased by one insured. This statistics confirms the issue of lacking user-item interactions. The schemas of packages and user profiles are described in the following lists. In this work, there are two terms of "attributes" which are "user-attribute" and "item-attribute". The former refers to the attributes of insureds, and the latter refers to the attributes of insurance packages as demonstrated in Fig. 1.

User Profile Attributes (User-Attributes)

- gender	the gender of an insured; male and female.
- age	the age of an insured.
- marital_status	the marital status of an insured; single, marriage, divorced, and widowed.
- smoking	the Boolean value showing that the insured smokes or not.
- job_rate	the dangerous level (1 to 5) of an insured's job. For example; the job rate of safe jobs e.g. CEOs, large business owners, doctors, and nurses are 1; and the dangerous jobs e.g. oilfield workers and firefighters are 5.
- premium_budget	the budget that insured prefers to pay per year

Insurance Package Attributes (item-attributes)

- package_id	a unique identity of each package.
- package_name	a name of each package.
- payment_years	a number of years or a maximum age of an insured for paying a premium. Most values are 5-year10-year, 15-year, 20-year, until-70-year, until-80-year, and until-90-year
- has_dividend	a Boolean value showing that package has dividend or not.
- protection_years	an age of an insured that the package protects. Values in the dataset are until-70-yearuntil-80-year, and until-99-year.
- percent_coverage	a percent that the insured gains at the end of the contract. Values are binned into 0-22-4, and more-than-4.
- times_refund	a number of times that the insures gains the refund. Values are binned into 510, 20, and 30.

2.3 A Review of Techniques

The prediction of the values of insurance package attributes needs machine learning techniques to do. A closer look at the data of insurance packages indicates that most attributes are in form of ordinal data. Thus, some well-known classifiers are focused as follows:

- **Decision Tree (DT)**
  The decision tree technique [8] is a tree of rules which is widely used for classification prediction in machine learning. Each node beginning from the root node splits samples into branches by evaluating attribute's values using its rule. Then, samples are labeled in each leave node.
- **Random Forest (RFo)**
  The random forest technique [9] uses many decision tree models that learn from dissimilar subsets of the whole dataset. Then, each sample is classified by taking the majority vote of each model.
- **K-Nearest Neighbors (KNN)**
  The k-nearest neighbors [8] is a method to classify a sample based on feature similarity between a new sample and training instances. The k is a configured number of most similar training instances. For example, if k is 5, the classification of the new instance is based on the majority class of 5-nearest labels.
- **Support Vector Machine (SVM)**
  The support vector [8] machine is a classification technique that uses a hyperplane with a widest gap to divide instances into classes. The hyperplane has many forms based on mathematic equation as known as kernel.
- **Logistic Regression (LoR)**
  The logistic regression [8] uses a logistic equation to give a label of each instance by evaluating the independent variable or attributes of each sample.
- **Naïve Bayes (NB)**
  The Naïve Bayes [8] is one techniques of probabilistic classifiers using the Bayes' theorem. It is a supervised learning that finds the conditional independence between every pair of features and the class labels.

3   INSUREX Approach

To demonstrate our recommender system for insurance packages, INSUREX, overall flow, item-attribute-value prediction, recommendation model, and evaluation methods are described in the following parts.

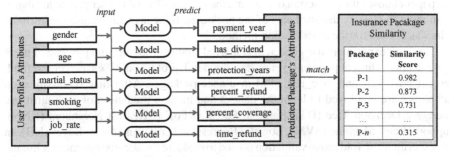

Fig. 2. Overall flow of INSUREX

3.1 Overall Flow

To build a recommender system for insurance package, we introduce the flow of INSUREX in Fig. 2. The flow begins from the left side, which is user profile's attributes. It uses the user attributes as inputs for machine learning models, then each model predicts package's attribute values or item-attribute values. After that, each insurance package is scored using the similarity between its item values and the predicted item values. At last, some packages having high ranking are selected to display to users.

To train the recommender model from the dataset and to measure the performance of the model, it needs an analytics process that describes in the following steps.

1) **Read the Datasets**
   To read user profile attributes and insurance package attributes from the insurance policy dataset.
2) **Do Synthetic Minority Over-Sampling Technique**
   To create more samples under minor classes in order to cope with imbalance classes [10]. This step helps the dataset has balance classes.
3) **Predict Item-Attribute-Values**
   To use machine learning techniques to learn the pattern between user profiles and each insurance package attributes. Then, use input user profiles to predict the value of each insurance package attribute. This method is described in Sect. 3.2.
4) **Recommend Relevant Packages**
   To recommend relevant insurance packages based on the predicted value of thier attribute values. This part is explained in Sect. 3.3.
5) **Evaluate the Recommendation Model**
   To evaluate the recommendation model against F1-score and HR@K which are discussed in Sect. 3.4.
6) **Deploy the Recommendation Model**
   To deploy the built recommendation model into a proof-of-concept application.

3.2 Item-Attribute-Value Prediction

Since, in the insurance business, new packages are often raised; it commonly results in the cold start problem in the beginning when the new packages have a few purchases, and lacking user-item interaction due to the number of packages owned by a person is very low. In this case, using packages as classes for classification is not a proper condition, so this paper chooses the prediction of item-attribute-values by using user profile attributes as an input to classify the attributes of the packages. It means that, we use the attribute values as classes for the classification method.

Thus, our experiment use gender, age, martial_status, smoking, and job_rate of users to predict the values of payment_years, has_dividend, protection_years, percent_refund, percent_coverage, and time_refund of insurance packages. After that, 6 classification techniques are employed to learn the dataset and make a prediction of each attribute. There are Decision Tree (DT), Random Forest (RFo), K-Nearest Neighbors (KNN), Support Vector Machine (SVM), Logistic Regression (LoR), and Naïve Bayes (NB). In this experiment, 4-fold cross-validation is used to evaluate the accuracy of each technique in terms of F1-score.

3.3 Recommendation Model

In the next step, after attributes of packages are predicted, some packages whose feature-vector are closing to the predicted feature-vector are used to recommended. During this task, all attribute values being ordinal labels are transformed into ordinal numbers, for example, until-80-year is changed into 80, 2–4 is changed into 4, etc. After that, all attributes are scaling using standardization as formulated in (1); where x is a particular value of an attribute, $\bar{x}$ is the mean of the attribute, and x_{stddev} is the standard deviation of the attribute.

$$x_{scaled} = \frac{x - \bar{x}}{x_{stddev}} \tag{1}$$

After all attribute are scaled to be feature-vectors, all feature-vectors of insurance products are compared to the predicted feature-vectors using cosine similarity. The equation of this similarity matric is expressed in (2); where $similarity(A, B)$ is a function that return the cosine similarity value between vectors A and B, $A{\cdot}B$ is a dot-product between vectors A and B, and $\|A\|$ is the size of the vector A.

$$similarity(A, B) = \frac{A \cdot B}{A \times B} \tag{2}$$

In order to make a clearly understand, the Fig. 3 depicted the result of the recommendation method. In the figure, it is assumed that each package has 2 features: $f1$ and $f2$. The red-dashed vector is the predicted feature-vector, and the blue-solid vectors are feature-vectors of all insurance packages: *package-1, package-2, package-3,* and *package-4*. As we noticed; the package-2 are closing the feature-vector, so the cosine similarity is closing to 1; while a vector that is far away from the feature-vector like the package-1 has the cosine similarity closing to 0. Thus, the high value of the cosine similarity can be inferred that the insurance packages become highly potential to be purchased based on the analytics of the user preferences.

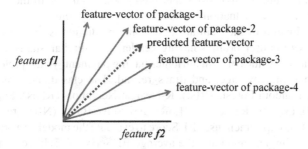

Fig. 3. Example demonstration of the cosine similarity matric

3.4 Evaluation Method

For the evaluation methods, there are 2 main methods used in our experiments. First, the F1-scrore (or F1-measure) is used to evaluate the classification methods for item-attribute-value prediction. The F1-score is a harmonic mean of the precision and recall that is simply expressed in (3).

$$F1 - Score = \frac{2 \times Precision \times Recall}{Precision + Recall} \tag{3}$$

Besides, the evaluation method named HR@K (Hit Rate at Top-K) is employed to evaluate the recommendation model. The equation of HR@K is shown in (4); where K is the number of top items selected, *#hit* is the number of tested items found in the top K items, and n is the number of all tested items.

$$HR@K = \frac{\#hit}{n} \tag{4}$$

The user-defined K is commonly use as a number of recommended items for users to select in an application. In the experiment of this paper, three recommended items are attempted from 7 main packages and 8 riders, and the HR@K is used as HR@3 in the measurement step. Next, the results of our experiment of both classification and recommendation are presented in the following section.

4 Result and Evaluation

After making experiments, the results are demonstrated in the following parts. There are analytical results, a proof-of-concept application, and discussion.

4.1 Analytical Results

For the analytical results, there is two experiments: the evaluation of the item-attribute-value prediction, and the evaluation of the recommender system.

First, the evaluation of the item-attribute-value prediction is the experiment from Sect. 3.2 that uses the user profiles having gender, age, marital_status, smoking, and job_rate to predict the values of item attributes having payment_years, pretection_years, has_dividend, percent_coverage, and times_refund. The classifiers, which are Decision Tree (DT), Random Forests (RFo), K-Nearest Neighbors (KNN), Support Vector Machine (SVM), Logistic Regression (LoR), and Naïve Bayes (NB), are used to learn the training set. This experiment uses F1-Score to evaluate the models. In practice, 4-fold cross-validation is implemented and the average F1-Score of each technique is shown in Table 1. Due to the result, the Random Forests (RFo) technique having the highest score is used to be a classifier technique in the next experiment. For more information of the configuration of the Random Forest in our experiment; there are 100 number of trees in the forest, it uses GINI index as a criterion to measure the quality of a sample split, the minimum number of samples for splitting an internal node is 2, the minimum number of samples for a leaf node is 1, and the minimum weighted fraction is 0.

Second, the evaluation of the recommender system is done by the experiment in Sect. 3.3. It uses the predicted attribute values of insurance packages using the Random Forest, and similarity scores are given to all packages using the cosine similarity index. After that, three top-scored packages are selected, and the HR@3 is used to evaluate the selected packages. The HR@3 results of the main insurance packages and riders are separately demonstrated in Table 2.

Table 1. Experiment result of predicting main insurance packages.

Predicted attributes	F1-score of each classifier					
	DT	RFo	KNN	SVM	LoR	NB
payment_years	.847	**.898**	.787	.781	.745	.377
protection_years	.872	**.875**	.784	.782	.722	.470
has_dividend	.922	**.985**	.886	.871	.862	.810
percent_coverage	.701	**.751**	.667	.650	.669	.397
times_refund	.771	**.780**	.660	.641	.579	.500
Average	*.823*	*.858*	*.757*	*.745*	*.715*	*.511*

Table 2. HR@3 of the recommender models of main insurance packages and riders.

Types of insurance	HR@3
Main insurance packages	.954
Riders	.750
Average	*.852*

4.2 Proof-of-Concept (POC) Application

Our POC, which is a web-based application titled INSUREX, was implemented in order to provide a clearer picture of the outcome of our research. Our recommender model object that was built by Python programming is stored in a Flask web server, and it works with the web application having two simple pages in Fig. 4. The first page, as shown in Fig. 4(a), is an input form for users to fill their profile. The form includes gender, age, marital status, smoking, job, and premium budget. After a button "Recommend" is clicked, the profile is sent to the recommender model, and the model returns several recommended insurance packages having high ranking score. The list of recommended packages is displayed the next page in Fig. 4(b).

4.3 Discussion

This work, INSUREX, pays attention to make a prediction on the attribute values of an item at first, then gives ranking to each item based on the similarity between vectors

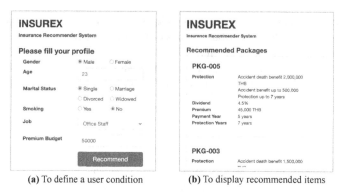

(a) To define a user condition   (b) To display recommended items

Fig. 4. A POC application

of attributes. This method is appropriate for the behavior of selling insurance packages, because any insurers always create new packages and most of them have short product lifetime in the market. Thus, predicting item-attribute values from user-attribute values can help the recommender system be robust for the cold start problem and lacking user-item interaction. The item-attribute-value prediction result shows that the Random Forest classifier provides the highest accuracy based on the collected dataset. After that, all items are ranked based on the cosine similarity between each item features and the predicted features, and the recommendation result has high performance in terms of HR@3. This experiment demonstrates the behavior of users that always choose one main package as summarized from the dataset. Thus, the main contribution of this paper is to use data analytics to predict item-attribute values from use-attribute values, and then to find relevant items to recommend in order to address the issue of new items and lacking user-item interaction.

In addition, a POC application is implemented to demonstrate the feasibility of the INSUREX model. The POC was compared to a related work and discussed with potential insureds and insurance agents. There is an application about insurance recommendation in Thailand named iTAX [11] that suggests insurance plans for tax reduction based on the age and yearly budget of a user. The recommendation model becomes the calculation between premiums and tax benefits in order to answer a specific customer group who concerns about tax. Unlike iTAX, our model provides much more dimensions for recommendation to find relevant products that close to user profiles. Firstly, for our POC, the potential insureds accepted that they took time to compare insurance packages from many companies before making a decision to buy a package. They added that the POC became a starting point for a decision support system for users to filter several packages that relevant to their profile. Moreover, they gave more enhancement points that the recommender system should serve as a guide to users to find out more information of packages. It should not just show a limited number of products; so, when a user views a particle package, the system should suggest other nearby products as an alternative to continue browsing. Secondly, as the viewpoint of the agents, the POC demonstrates that most attributes of both user profiles and insurance packages can be basically used for finding relevant insurance packages, but the recommendation of insurance packages across many companies is not advantage for individual insurer in terms of the business competition. The agent, however, added that the POC application should be enhanced into the full-functioned application for giving high benefit to potential insureds. The application should also support agents to find some relevant packages to discuss with their customers, and it should help the against learn the advantage and disadvantage of relevant packages in order to suggest the insurer to release packages that really meets the customer needs for filling in a market gap.

However, the journey of the INSUREX does not only limited by the item-attribute-value prediction, but some more aspects about users and items are also considered when dealing with the larger size of data under the insurance domain. Personalization, that can be explored from social networking environment including user behaviors, becomes a precise profile rather than an application form [12]. In addition, the content of insurance packages can enhance the accuracy of the recommender system. The semantics-aware content-based recommender systems lead to find out the similarity among insurance contracts using the ontology of key terms and more information from the linked open data cloud [13].

5 Conclusion

This work introduces the recommender system named INSUREX for insurance packages based on item-attribute-value prediction in order to support the issue of the item cold-start problem and lacking user-item interaction which are commonly found in any insurance package markets. Among the clean dataset having 370 transactions, they are 7 main insurance packages and 8 riders to be items that have featured attributes: payment years, protection years, has dividend, percent coverage, and times refund. By average, there are about 1.2 packages purchased by 308 insureds whose profile collected from application forms consists of gender, age, marital status, smoking, job, and premium budget. Our work adopts machine learning techniques to predict each item-attribute value from user-attribute values; and then, uses the cosine similarity to find out relevant packages whose attribute values close to the predicted values. The result of item-attribute prediction is that the Random Forest technique serves the highest performance among other machine learning methods in terms of F1-score as 0.858. It also leads to the high performance of the recommender system in terms of HR@3 as 0.852. In addition, a proof-of-concept application has been implemented to demonstrate the feasibility of the proposed recommendation model. It was used to discuss with users and found that it is suitable to be a decision support tool for potential insureds to find out some relevant insurance packages based on user attributes. Moreover, to consider the personalization and semantics-aware content similarity into a full-functioned application becomes a future challenge. It needs to consider more about rich user experience and user interface to increase user satisfaction; and a business model is also needed to consider in order to enhance and turn this academic work into business.

References

1. Cappiello, A.: Technology and the Insurance Industry: Re-configuring the Competitive Landscape. Springer, Heidelberg (2018). https://doi.org/10.1007/978-3-319-74712-5

2. Hinduja, A., Pandey, M.: An intuitionistic fuzzy AHP based multi criteria recommender system for life insurance products. Int. J. Adv. Stud. Comput. Sci. Eng. **7**(1), 1–8 (2018)
3. Sahoo, S., Ratha, B.K.: Recommending life insurance using fuzzy multi criteria decision-making. Int. J. Pure Appl. Math. **118**(16), 735–759 (2018)
4. Archenaa, J., Anita, E.M.: Health recommender system using big data analytics. J. Manage. Sci. Bus. Intell. **2**(2), 17–24 (2017)
5. Qazi, M., Tollas, K., et al.: Designing and deploying insurance recommender systems using machine learning. In: Wiley Interdisciplinary Reviews: Data Mining and Knowledge Discovery, p. e1363 (2020)
6. Desirena, G., Diaz, A., et al.: Maximizing customer lifetime value using stacked neural networks: an insurance industry application. In: 2019 18th IEEE International Conference on Machine Learning And Applications (ICMLA) (2019)
7. Kang, J., Sim, K.M.: Towards agents and ontology for cloud service discovery. In: 2011 International Conference on Cyber-Enabled Distributed Computing and Knowledge Discovery (2011)
8. Bishop, C.M.: Pattern Recognition and Machine Learning. Springer, Heidelberg (2006)
9. Breiman, L.: Random forests. Mach. Learn. **45**(1), 5–32 (2001)
10. Chawla, N.V., Bowyer, K.W., et al.: SMOTE: synthetic minority over-sampling technique. J. Artif. Intell. Res. **16**, 321–357 (2002)
11. iTAX. https://www.itax.in.th/market. Accessed 16 Aug 2020
12. Zhou, X., Yue, X., et al.: The state-of-the-art in personalized recommender systems for social networking. Artif. Intell. Rev. **37**(2), 119–132 (2012)
13. de Gemmis, M., Lops, P., Musto, C., Narducci, F., Semeraro, G.: Semantics-aware content-based recommender systems. In: Ricci, F., Rokach, L., Shapira, B. (eds.) Recommender Systems Handbook, pp. 119–159. Springer, Boston (2015). https://doi.org/10.1007/978-1-4899-7637-6_4

Application of Recommendation Systems Based on Deep Learning

Li Haiming[✉], Wang Kaili[✉], Sun Yunyun[✉], and Mou Xuefeng[✉]

Shanghai University of Electric Power, Shanghai 201306, China
lhm@shiep.edu.cn

Abstract. With the rapid development of mobile Internet and self-media, it is becoming more and more convenient for people to obtain information, and the problem of information overload has increasingly affected people's sense of use. The emergence of recommendation systems can help solve the problem of information overload, but in the big data environment, the traditional recommendation system technology can no longer meet the needs of recommending more personalized, more real-time, and more accurate information to users. In recent years, deep learning has made breakthroughs in the fields of natural language understanding, speech recognition, and image processing. At the same time, deep learning has been integrated into the recommendation system, and satisfactory results have also been achieved. However, how to integrate massive multi-source heterogeneous data and build more accurate user and item models to improve the performance and user satisfaction of the recommendation system is still the main task of the recommendation system based on deep learning. This article reviews the research progress of recommendation systems based on deep learning in recent years and analyses the differences between recommendation systems based on deep learning and traditional recommendation systems.

Keywords: Recommendation system · Deep learning

1 Introduction

With the rapid development of the mobile Internet and media, it has become more and more convenient for people to obtain information, and the problem of information overload has increasingly affected people's sense of use. The emergence of recommendation systems can well help solve the problem of information overload, but in a big data environment, traditional recommendation system technologies can no longer meet the needs of recommending personalized, real-time, and accurate information to users.

Traditional recommendation methods mainly include content-based recommendation methods, collaborative filtering, and hybrid recommendation methods. The content-based recommendation algorithm extracts the user's interest and preference for the item by analyzing the item content information (such as

T.-P. Hong et al. (Eds.): ACIIDS 2021, CCIS 1371, pp. 85–97, 2021.
https://doi.org/10.1007/978-981-16-1685-3_8

item attributes, description, etc.). The collaborative filtering recommendation method uses interactive information between users and items to make recommendations for users and is currently the most widely used recommendation algorithm. The hybrid recommendation algorithm solves the shortcomings of a single recommendation model by combining the content-based recommendation method and the collaborative filtering recommendation method. Traditional recommendation methods mostly rely on manual extraction of features, and can only learn shallow models of data, and cannot guarantee the accuracy and reliability of recommendations.

Deep learning can characterize massive data related to users and items by learning a deep non-linear network structure. It has a powerful ability to learn the essential characteristics of data sets from samples and can obtain deep-level feature representations of users and items. At the same time, it alleviates the data sparseness and cold start problems in the traditional recommendation system.

Recommendation systems based on deep learning have been extensively studied because they can achieve more personalized and accurate recommendations and meet the needs of more users. This article mainly reviews the application of recommendation systems based on deep learning.

2   Deep Learning

Deep learning has become a boom in big data and artificial intelligence [1]. Deep learning forms a denser high-level semantic abstraction by combining low-level features, thereby automatically discovering distributed features of data, which solves the need for manual design features in traditional machine learning and has made breakthroughs in image recognition, machine translation, speech recognition, and online advertising. In the field of image recognition, in the ImageNet image classification competition in 2016, the accuracy of deep learning exceeded 97%. In the field of machine translation, the Google Neural Machine Translation System (GNMT) based on deep learning has achieved a level that closes to human translation in both English and Spanish and English and French translation [2]. In the field of speech recognition, Baidu, Xunfei, and Sogou all announced that their Chinese speech recognition accuracy based on deep learning exceeded 97%. In the field of online advertising, deep learning is widely used in advertising click-through rate prediction. It has achieved great results in applications such as Google [3,4], Microsoft [5,6], Huawei [7], Alibaba [8], etc. Below we introduce commonly used deep learning models and methods.

2.1   Autoencoder

In 1986, Williams proposed the concept of Autoencoder (AE) and used it for high-dimensional complex data processing [9]. The autoencoder reconstructs the input data through the encoding and decoding process and learns the hidden layer representation of the data. The basic autoencoder can be regarded as a three-layer neural network structure: an input layer, a hidden layer, and an output layer.

The purpose of the autoencoder is to make the input and output as close as possible. If the model is trained only by minimizing the error between input and output, the autoencoder can easily learn an identity function. To solve this problem, researchers have proposed a series of variants of autoencoders, among which the more classic ones include sparse autoencoders and denoising autoencoders [10].

In a recommendation system, autoencoders are mainly used to learn the hidden feature representations of users and items, and then predict users' preferences for items based on this hidden representation. Its application scenarios mainly include scoring prediction, text recommendation, image recommendation, and so on.

2.2 Restricted Boltzmann Machine

Boltzmann machine (BM) is a generative stochastic neural network, proposed by Hinton and Sejnowski in 1986 [11]. BM is composed of some visible units (data samples) and some hidden layer units (hidden layer variables). Both visible variables and hidden layer variables are binary, and their state is 0 or 1. State 0 indicates that the neuron is in an inhibited state and state 1 means that the neuron is active. BM can learn complex rules in data. However, the training process of the BM is very time-consuming. Hinton and Sejnowski [12] further proposed a restricted Boltzmann machine (RBM), by removing all connections between variables in the same layer, which improves learning efficiency.

RBM is the first neural network model used in the recommendation system. The current application is mainly to reconstruct the user's rating data to learn the user's implicit representation, so as to realize the prediction of the unknown rating. The application scenario is mainly user rating prediction.

2.3 Deep Belief Network

Hinton et al. proposed a Deep Belief Network (DBN) in 2006 [13], which is a generative model composed of multiple layers of nonlinear variable connections. The structure of DBN can be seen as a stack of multiple RBM and the hidden layer of the previous RBM in the network is regarded as the visible layer of the next RBM. In this way, in the DBN training process, each RBM can be individually trained using the output of the previous RBM, so compared with traditional neural networks, DBN training is simpler. At the same time, through this training method, DBN can also obtain in-depth feature representation from unlabeled data.

Deep belief networks are currently less used in recommendation systems and are mainly used for music recommendations.

2.4 Convolutional Neural Network

Convolutional neural network (CNN) is a multi-layer perceptron, which is composed of an input layer, convolution layer, pooling layer, fully connected layer and output layer. It is mainly used to process two-dimensional image data. Compared with the traditional multi-layer perceptron, CNN uses pooling operation to reduce the number of neurons in the model and has higher robustness to the translation invariance of the input space.

Convolutional neural networks are widely used in recommendation systems [14]. They are mainly used to extract hidden features of items from images, text, audio, and other content, combined with user implicit representation to generate recommendations for users. Mainly used in image recommendation, music recommendation, text recommendation, etc.

2.5 Recurrent Neural Network

In 1986, Williams et al. [15] proposed the concept of Recurrent Neural Network. Normal fully connected network or convolutional neural network, the layers are fully connected, and the nodes between each layer are disconnected. The nodes between the layers of the RNN neural network are connected. It can calculate the output of the hidden layer at the current moment through the output of the input layer and the state of the hidden layer at the previous moment.

The application scenarios of RNN mainly include scoring prediction, image recommendation, text recommendation, and point-of-interest recommendation in location-based social networks.

3 Application of Deep Learning in Content-Based Recommendation System

The performance of content-based recommendation methods relies heavily on effective data feature extraction. The biggest advantage of deep learning is that it can learn the characteristics of data through a universal end-to-end process, and automatically obtain a high-level representation of the data, without relying on manual design features. Therefore, deep learning is mainly used in content-based recommendations. It is used to extract the implicit representation of the item from the content information of the item, and obtain the implicit representation of the user from the user's portrait information and historical behavior data, and then generate recommendations by calculating the matching degree between the user and the item based on the implicit representation. In the collaborative filtering recommendation algorithm, people only rely on user ratings to model users and items, while ignoring a large amount of review information in the project. Rational use of this review information can improve the quality of recommendations.

3.1   Method Based on Multilayer Perceptron

Taking into account the problem that user characteristics are difficult to obtain in traditional content-based recommendation systems, Elkahky et al. [16] extract user characteristics by analyzing users' historical behavior records, thereby enriching user characteristics. Based on the Deep Structured Semantic Models (DSSM) [17], he proposed a multi-view deep neural network model (Multi-View Deep Neural Network, Multi-View DNN), the model is a very common content-based recommendation method to implement user-item recommendation through semantic matching between the user and item information entities. Its basic idea is to map two types of information entities to a hidden space through a deep learning model, and then calculate the matching degree of the two entities based on the cosine similarity, to further make personalized recommendations.

Deep neural network models are currently widely used in recommendation systems. Covington et al. [18] proposed a deep neural network model for YouTube video recommendation by using user information, contextual information, historical behavior data, and other information. YouTube video recommendation mainly faces three challenges: scalability, freshness, and data noise. The author uses the user's historical behavior data, user characteristics, and context information on YouTube to preliminarily infer the user's personalized preference for videos, and then converts the recommendation problem into a classification problem based on deep neural networks, which simplifies the complexity of the problem to a certain extent Sex. Then we only need to find the N videos closest to the user vector for the recommendation.

3.2   Method Based on Convolutional Neural Network

The deep learning model also plays an important role in the field of music recommendation. In a music recommendation system, a collaborative filtering algorithm often faces the problem of cold start, that is, for some music without user data, the music platform often has no way to recommend, which greatly affects the user's sense of use. To solve this problem, van den Oord et al. [19] studied how to use the deep learning model to solve this problem. First of all, they use the historical listening records and audio signal data of music to project the user and music into a shared hidden space through the combination of weighted matrix factorization and convolutional neural network, to learn the implicit representation of users and songs and make personalized music recommendation. However, for those newly published songs, we can extract the hidden representation of songs from their audio signals by the trained convolutional neural network, to make personalized recommendations for users by calculating the similarity between users and new music, and also help to solve the problem of cold start of items.

3.3   Methods Based on Other Deep Learning Models

Xuejian [20] and others developed a news recommendation model using a dynamic attention depth model to recommend suitable news for target users who have no clear selection criteria. In the process of news push, the editor needs to select news that meets the user's preference from the massive news data to push, which is a complicated process. The author learns the editor's selection criteria for dynamic articles by automatically representing learning and its interaction with metadata, to realize the dynamic selection of appropriate articles from the database.

In general, the content recommendation system based on deep learning can effectively alleviate the problem of user and item feature extraction. It uses deep learning methods to obtain the implicit representation of the user and item and then calculates the matching degree between the user and the item based on the implicit representation. Generate a personalized recommendation list, which brings great convenience to users. But at the same time, there is no good solution for the new user problem of the content-based recommendation method itself, and further research is needed.

4   Application of Deep Learning in Collaborative Filtering Recommendation System

The collaborative filtering algorithm mainly discovers user preferences by mining user historical behavior data, divides users into groups based on different preferences, and recommends products with similar tastes. Traditional collaborative filtering methods have problems such as cold start, data sparseness, and poor scalability. Since deep learning can be adapted to large-scale data processing, it is currently widely used in collaborative filtering recommendation problems. Collaborative filtering based on deep learning mainly uses deep learning to learn the hidden vectors represented by users and items. This article mainly divides the application of deep learning in collaborative filtering into five categories.

4.1   Collaborative Filtering Based on Restricted Boltzmann Machine

Salakhutdinov et al. [21] applied deep learning to a recommendation system for the first time. The proposed collaborative filtering is based on Restricted Boltzmann Machines (RBM), using a two-layer RBM undirected graph model to automatically extract hidden abstract features, and using items during training. The scoring data is used as the input layer, and the hidden vector representation of the hidden layer is calculated through the conditional probability function, and the vector representation of the hidden layer is used in the prediction to obtain the score in reverse. This method has achieved good results on the Netflix data set.

Georgiev et al. [22] extended the RBM model proposed in the literature, constructed two RBM models from the user dimension and the item dimension, and extracted the correlation features between users and the correlation features between items, and the combination of two RBM models is solved as a hybrid model, and the training and prediction process of the model is simplified, and the model can directly process real-valued score data.

Collaborative filtering based on the Restricted Boltzmann Machine has the problems of large parameter scale and long training time, which greatly restricts its practical application.

4.2   Collaborative Filtering Based on Autoencoder

The autoencoder is a neural network with the same input and learning objectives, and its structure is divided into two parts: encoder and decoder. Sedhain et al. [23] proposed autoencoder-based collaborative filtering (AutoRec), which uses an item (or user)-based autoencoder to map part of the observation vector of the item (or user) to a low-dimensional space, and then the output space is reconstructed to score prediction. Compared with the matrix decomposition model, the recommendation algorithm based on the autoencoder only uses the hidden features of the user (or the item) and can learn the non-linear feature representation. Compared with RBM-CF, there are fewer parameters to learn.

Wu et al. [24] proposed a top-N recommendation framework that uses a three-layer (including a hidden layer) denoising autoencoder to extract user and information from the user's feedback information on the item (input layer, excluding auxiliary information). The hidden distributed feature representation (hidden layer) of the item, and then the hidden feature representation is mapped to the input space in the output layer to reconstruct the input vector.

The self-encoder-based collaborative ltering is simple and effective, which can effectively improve the robustness of recommendations. As the stacking depth of the self-encoder deepens, problems such as gradient diffusion will occur.

4.3   Collaborative Filtering Based on Convolutional Neural Network

Convolutional Neural Network is a type of feedforward neural network that includes convolution calculation and has a deep structure, which can classify input information according to its hierarchical structure. To solve the problem of not being able to extract the contextual information of the text, Kim et al. [25] used CNN to extract the contextual semantic information of the document, combined with a probabilistic matrix factorization (PMF) model for scoring prediction, which can handle the sparseness of contextual information well.

In music recommendation, a collaborative filtering algorithm is generally considered to be better than a content-based recommendation algorithm, but collaborative filtering has a cold start problem. To solve the problem of music recommendation cold start (such as recommending new songs and non-popular songs), Oord et al. [19] used convolutional neural networks to extract time-frequency feature representations from the sound signals of music files, and the results were better than traditional bag-of-words models. The article believes that convolutional neural networks can share intermediate features among different features, and the pooling layer can operate on multiple time scales, so it is very suitable for extracting hidden features of audio.

Collaborative filtering based on Convolutional Neural Networks (CNN) can automatically extract features, share convolution kernels, and process high dimensional data without pressure, but there are also problems that the pooling layer will lose a lot of valuable information and ignore the correlation between the part and the whole.

4.4   Collaborative Filtering Based on Recurrent Neural Network

The main idea of collaborative filtering based on Recurrent Neural Network is to use RNN to model the influence of the user's historical sequence behavior on user behavior at the current moment, to realize user's item recommendation and behavior prediction. Wu et al. [26] proposed a deep recurrent neural network model DRNN, which learns user browsing patterns from the sequence of web page visits of a user's one session (from the user entering the website to purchasing the product), and uses the user's historical purchase records to fuse the feedforward neural network (FNN) for real-time product recommendation has achieved obvious results on the koala shopping website.

Wu et al. [27] proposed a Recurrent Recommender Network (RRN) by using recurrent neural networks to model the evolution of user preferences and item features, which can predict the future behavior trajectory of users. Specifically, RRN first uses low-dimensional matrix factorization to learn the static implicit representations of users and items, and at the same time uses the user's historical rating data as input, uses LSTM to learn the dynamic implicit representations of users and items at each moment, and finally achieve a single moment score prediction by aggregating the inner products of two types of implicit representations.

The recommendation system based on Recurrent Neural Network (RNN) can not solve the problem of long-term dependence due to the RNN. To solve this problem, a long short-term memory network (LSTM) is proposed. The collaborative filtering method based on Recurrent Neural Network can effectively model the sequence patterns in user behaviors, and can also incorporate contextual information such as time, as well as various types of auxiliary data to improve the quality of recommendations. The model has high applicability and is used in current recommendations. It has been widely used in the system.

4.5   Collaborative Filtering Based on Other Deep Learning Models

Lee et al. [28] used a combination of recurrent neural networks and convolutional neural networks to learn semantic representations from the context of user conversations, and recommended quotes or common replies to "famous sayings" and "proverbs". The convolutional neural network is used to extract the local semantic representation of each sentence in the conversation context, and the recurrent neural network LSTM is used to model the sentence sequence in the conversation, and it has achieved good results in the Twitter conversation.

Zheng et al. [29] considered the difficulty of optimization in collaborative filtering based on RBM and used NADE to replace RBM for collaborative filtering recommendation. NADE does not need to incorporate any hidden variables, thus avoiding complex hidden variable reasoning processes, thereby reducing model complexity.

In general, applying deep learning to collaborative filtering can obtain more explicit and implicit information about users and items through deep neural networks, avoid the complexity of manually extracting features, and make the recommendation system more intelligent.

5 Application of Deep Learning in Hybrid Recommendation System

The traditional collaborative filtering method only uses the user's display feedback or implicit feedback data and faces the problem of data scarcity. By incorporating auxiliary data such as user profile data, item content data, social annotations, comments, etc., the hybrid recommendation method can effectively alleviate the problem of data scarcity, but the biggest problem this method faces is the representation of auxiliary data. Classic methods such as collaboration (Collaborative Topic Regression CTR) [30] cannot obtain effective auxiliary data representation. Through automatic feature extraction, deep learning can learn effective user and item implicit representations from auxiliary data.

The main idea of the hybrid recommendation algorithm based on deep learning is to combine content-based recommendation methods and collaborative filtering and integrate the feature learning of users or items with the item recommendation process into a unified framework. First, use various deep learning models to learn users or the hidden features of the item are combined with the traditional collaborative filtering method to construct a unified optimization function for parameter training, and then use the trained model to obtain the hidden vector of the user and the final item, and then realize the user's item recommendation.

5.1 Hybrid Recommendation Method Based on Autoencoder

In collaborative filtering, one of the main reasons for inaccurate recommendation results is the sparsity of user rating data and hybrid recommendation algorithm can solve this problem well. Recently, a hybrid recommendation algorithm based on document modeling is to improve the accuracy of rating prediction by using the description documents of items, such as summaries and comments. To use the project description document in the recommended project, we choose to use the document modeling method and stack noise reduction automatic encoder. However, they still have some defects: (1) the bag of words model is used in document modeling, which ignores the relationship between contexts, such as the relationship between sentences and paragraphs. (2) The problem of Gaussian noise is not explicitly considered in a document and project-based latent factor modeling. In order to solve the above problems, Donghyun et al. [31] proposed a convolutional matrix factorization model. Based on the fact that the convolutional neural network (CNN) can effectively capture the context information of documents, a hybrid recommendation algorithm based on document context awareness is proposed. In this method, the convolutional neural network is integrated into the matrix-based collaborative filtering, so that even if the data set is very sparse, the convolution matrix decomposition model can accurately predict the unknown ratings, thus improving the recommendation system performance.

5.2 Hybrid Recommendation Method Based on Other Deep Learning Models

Similar to ConvMF, to grasp the sequence relationship between words in text-aided data, Bansal et al. [32] proposed a text recommendation method based on the cyclic neural network model. While analyzing the rating data, he also used the text information and text label data to make recommendations. Specifically, to better grasp the sequence patterns between sentences in text, the model uses GRU to learn the vector representation of text content and then uses a multi-task learning framework (including text recommendation and label prediction) to construct a joint optimization objective function based on hidden factor model, which optimizes the model parameters under the supervised learning framework.

Li et al. [33] used a deep learning model to study the app's score prediction and abstract tips generation. He used a multi-task learning framework to simultaneously perform score prediction and abstract tips generation. In the scoring prediction task, the author uses the hidden vectors of users and items as input and uses a multi-layer perceptron (MLP) to predict the score. In the task of generating abstract tips, he takes the hidden vectors of users and items as input, uses a GRU to generate text, and finally conducts model training in a multi-task framework.

In general, the deep learning-based hybrid recommendation can effectively alleviate the data scarcity and cold start problems in collaborative filtering by in-corporating auxiliary information. However, most of the current researches adopts different deep learning models for specific auxiliary information, and building a uni ed hybrid recommendation framework for all data is our next goal.

6   Conclusion

With the rapid development of deep learning technology, deep learning can automatically learn complex high-dimensional data. Through deep neural networks, the explicit and implicit features of users and items can be better mined, so that the recommendation system based on deep learning can recommend more personalized and accurate information.

Meanwhile, RNN and CNN methods in deep learning are widely used in the recommendation system. RNN is designed to deal with temporal dependency. RNN can integrate the current browsing history and browsing order in the recommendation system to provide more accurate recommendations. CNN is mainly used to extract potential features from data. When potential factors cannot be obtained from user feedback, CNN can extract hidden features from audio, text, and image data to generate a recommendation.

At present, the research on recommendation systems based on deep learning is still relatively preliminary and the application of deep learning in recommender systems still has some problems, such as poor scalability, lack of interpretability, and so on, which need our further research.

References

1. Silver, D., Huang, A., Maddison, C.J., et al.: Mastering the game of Go with deep neural networks and tree search. Nature **529**(7587), 484–489 (2016)
2. Wu, Y., Schuster, M., Chen, Z., et al.: Google's neural machine translation system: bridging the gap between human and machine translation. arXiv preprint arXiv:1609.08144 (2016)
3. Cheng, H.T., Koc, L., Harmsen, J., et al.: Wide & deep learning for recommender systems. In: Proceedings of the 1st Workshop on Deep Learning for Recommender Systems, pp. 7–10 (2016)
4. Wang, R., Fu, B., Fu, G., et al.: Deep & cross network for ad click predictions. In: Proceedings of the ADKDD 2017, pp. 1–7 (2017)
5. Shan, Y., Hoens, T.R., Jiao, J., et al.: Deep crossing: web-scale modeling without manually crafted combinatorial features. In: Proceedings of the 22nd ACM SIGKDD International Conference on Knowledge Discovery and Data Mining, pp. 255–262 (2016)
6. Zhu, J., Shan, Y., Mao, J.C., et al.: Deep embedding forest: Forest-based serving with deep embedding features. In: Proceedings of the 23rd ACM SIGKDD International Conference on Knowledge Discovery and Data Mining, pp. 1703–1711 (2017)
7. Guo, H., Tang, R., Ye, Y., et al.: DeepFM: a factorization-machine based neural network for CTR prediction. arXiv preprint arXiv:1703.04247 (2017)
8. Zhou, G., Zhu, X., Song, C., et al.: Deep interest network for click-through rate prediction. In: Proceedings of the 24th ACM SIGKDD International Conference on Knowledge Discovery & Data Mining, pp. 1059–1068 (2018)
9. Rumelhart, D.E., Hinton, G.E., Williams, R.J.: Learning representations by back-propagating errors. Nature **323**(6088), 533–536 (1986)
10. Bengio, Y., Lamblin, P., Popovici, D., et al.: Greedy layer-wise training of deep networks. In: Advances in Neural Information Processing Systems, pp. 153–160 (2007)
11. Hinton, G.E., Sejnowski, T.J.: Learning and Relearningin Boltzmann Machines. MIT Press, Massachusetts (1986)
12. Hinton, G.E.: Boltzmann machine. Scholarpedia **2**(5), 1668 (2007)
13. Hinton, G.E., Osindero, S., The, Y.W.: A fast learning algorithm for deep beliefnets. Neural Comput. **18**(7), 1527–1554 (2006)

14. Rawat, Y.S., Kankanhalli, M.S.: ConTagNet: exploiting user context for image tag recommendation. In: Proceedings of the 2016 ACM on Multimedia Conference. Amsterdam, Netherland, pp. 1102–1106 (2016)
15. Williams, D., Hinton, G.: Learning representations by back-propagating errors. Nature **323**(6088), 533–538 (1986)
16. Elkahky, A.M., Song, Y., He, X.: A multi-view deep learning approach for cross domain user modeling in recommendation systems. In: Proceedings of the 24th International Conference on World Wide Web, pp. 278–288 (2015)
17. Huang, P.S., He, X., Gao, J., et al.: Learning deep structured semantic models for web search using clickthrough data. In: Proceedings of the 22nd ACM international conference on Information & Knowledge Management, pp. 2333–2338 (2013)
18. Covington, P., Adams, J., Sargin, E.: Deep neural networks for Youtube recommendations. In: Proceedings of the 10th ACM Conference on Recommender Systems, pp. 191–198 (2016)
19. Van den Oord, A., Dieleman, S., Schrauwen, B.: Deep content-based music recommendation. In: Advances in Neural Information Processing Systems, pp. 2643–2651 (2013)
20. Wang, X., Yu, L., Ren, K., et al.: Dynamic attention deep model for article recommendation by learning human editors' demonstration. In: Proceedings of the 23rd ACM SIGKDD International Conference on Knowledge Discovery and Data Mining, pp. 2051–2059 (2017)
21. Salakhutdinov, R., Mnih, A., Hinton, G.: Restricted Boltzmann machines for collaborative filtering. In: Proceedings of the 24th International Conference on Machine Learning, pp. 791–798 (2007)
22. Georgiev, K., Nakov, P.: A non-IID framework for collaborative filtering with restricted boltzmann machines. In: International Conference on Machine Learning, pp. 1148–1156 (2013)
23. Sedhain, S., Menon, A.K., Sanner, S., et al.: AutoRec: autoencoders meet collaborative filtering. In: Proceedings of the 24th International Conference on World Wide Web, pp. 111–112 (2015)
24. Wu, Y., DuBois, C., Zheng, A.X., et al.: Collaborative denoising auto-encoders for top-n recommender systems. In: Proceedings of the Nineth ACM International Conference on Web Search and Data Mining, pp. 153–162 (2016)
25. Kim, D., Park, C., Oh, J., et al.: Convolutional matrix factorization for document context-aware recommendation. In: Proceedings of the 10th ACM Conference on Recommender Systems, pp. 233–240 (2016)
26. Wu, S., Ren, W., Yu, C., et al.: Personal recommendation using deep recurrent neural networks in NetEase. In: 2016 IEEE 32nd International Conference on Data Engineering (ICDE), pp. 1218–1229. IEEE (2016)
27. Wu, C.Y., Ahmed, A., Beutel, A., et al.: Recurrent recommender networks. In: Proceedings of the Tenth ACM International Conference on Web Search and Data Mining, pp. 495–503 (2017)
28. Lee, H., Ahn, Y., Lee, H., et al.: Quote recommendation in dialogue using deep neural network. In: Proceedings of the 39th International ACM SIGIR Conference on Research and Development in Information Retrieval, pp. 957–960 (2016)
29. Zheng, Y., Tang, B., Ding, W., et al.: A neural autoregressive approach to collaborative filtering. arXiv preprint arXiv:1605.09477 (2016)
30. Wang, C., Blei, D.M.: Collaborative topic modeling for recommending scientific articles. In: Proceedings of the 17th ACM SIGKDD International Conference on Knowledge Discovery and Data Mining, pp. 448–456 (2011)

31. Kim, D., Park, C., Oh, J., et al.: Deep hybrid recommender systems via exploiting document context and statistics of items. Inf. Sci. **417**, 72–87 (2017)
32. Bansal, T., Belanger, D., McCallum, A.: Ask the GRU: multi-task learning for deep text recommendations. In: Proceedings of the 10th ACM Conference on Recommender Systems, pp. 107–114 (2016)
33. Li, P., Wang, Z., Ren, Z., et al.: Neural rating regression with abstractive tips generation for recommendation. In: Proceedings of the 40th International ACM SIGIR Conference on Research and Development in Information Retrieval pp. 345–354 (2017)

Data Reduction with Distance Correlation

K. M. George[✉]

Computer Science Department, Oklahoma State University, Stillwater, USA
kmg@cs.okstate.edu

Abstract. Data reduction is a technique used in big data applications. Volume, velocity, and variety of data bring in time and space complexity problems to computation. While there are several approaches used for data reduction, dimension reduction and redundancy removal are among common approaches. In those approaches, data are treated as points in a large space. This paper considers the scenario of analyzing a topic for which similar multi-dimensional data are available from different sources. The problem can be stated as data reduction by source selection. This paper examines distance correlation (DC) as a technique for determining similar data sources. For demonstration, COVID-19 in the United States of America (US) is considered as the topic of analysis as it is a topic of considerable interest. Data reported by the states of US are considered as data sources. We define and use a variation of concordance for validation analysis.

Keywords: Distance correlation · Data reduction · Concordance

1 Introduction

Distance correlation [6] was defined by Szekely et al. as a new measure of dependence between random variables. Significance of this measure is its simplicity and ease of computation. It is applicable to multidimensional data. This paper proposes distance correlation as a technique for data reduction. Nowadays, it is common to find several data sources (e.g. social media) providing similar data. By identifying data with similar properties, size of data in computations can be significantly reduced. To demonstrate the applicability and usefulness of the technique, it is applied to COVID-19 data reported by the states of US. Selection of these data is influenced by timeliness and free availability. We formulate the problem as the analysis of a topic based on data arriving from many sources as indicators. Data reduction then can be defined as data source (indicator) selection. The goal is to select a subset of sources without losing information available from all indicators for accurate analysis. The source reduction idea is to analyze data from sources pairwise, compute their strength of relationship with respect to the topic. Distance correlation is used as the measure. Also consistency of their relationship should be measured. Cronbach's alpha (α) [3] is a measure used to evaluate internal consistency [1] of different indicators measuring a concept. We propose to use α to measure consistency of relationships built by distance correlation to group sources. Then a representative can be chosen from each group for analysis providing substantial data reduction used

© Springer Nature Singapore Pte Ltd. 2021
T.-P. Hong et al. (Eds.): ACIIDS 2021, CCIS 1371, pp. 98–109, 2021.
https://doi.org/10.1007/978-981-16-1685-3_9

in the analysis stage. Examples of analysis are Hidden Markov Model for forecasting [4] and tweet sentiment based prediction. We also define 'distance concordance', a measure similar to Kendall's τ for validation of pairwise relations obtained by distance correlation.

Contribution of this paper is: development and demonstration of a computationally efficient technique for data reduction when similar data is available from multiple sources (or data combined from different subspaces). Distance correlation (DC) is adopted as similarity measure. As outlined by Szekely et al., the dependence of DC on Mathematics is sound. DC is easily programmable as opposed to many black box techniques. As DC computation depends only on two indicators at a time, computation can be easily parallelized and space needs reduced.

In the next section we specify some related work. We list all required formulae adopted from [6] for computation of distance correlation. Section 3 outlines the method followed in the paper and Sect. 4 analyses the results. Our conclusions and future work description are given is Sect. 5.

2 Related Work

There are numerous research papers on data compression, dimension reduction and data reduction. In this section some papers are referenced. Rehman et al. [5] present a survey of data reduction methods. They group the methods as network theory, big data compression, dimension reduction, redundancy elimination, data mining, and machine learning methods. They describe the methods and provide the underlying model equations and conclude that big data complexity is a key issue and requires mitigation. Weng and Young [8] list several dimension reduction techniques and apply them to survey data analysis. Uhm et al. [7] propose to address the exponential cost of computation due to large data dimension. They combine dimension reduction and dimension augmentation methods for classification, and provide experimental results. Szekely et al. [6] describe the distance correlation method. We describe below the needed formulae for distance dependence statistics for observed random sample as adopted from [6]. Let $(X, Y) = \{(X_k, Y_k) \mid k = 1, \ldots, n\}$ be observed random sample from joint distribution of random vectors X in R^p and Y in R^q. Define:

$$a_{kl} = |X_k - X_l|_p, \quad \bar{a}_{k.} = \frac{1}{n} \sum_{k=1}^{n} a_{kl},$$

$$\bar{a}_{.l} = \frac{1}{n} \sum_{l=1}^{n} a_{kl}, \quad \bar{a}_{..} = \frac{1}{n^2} \sum_{k,l=1}^{n} a_{kl}$$

$$A_{kl} = a_{kl} - \bar{a}_{k.} - \bar{a}_{.l} + \bar{a}_{..}, k, l = 1, \ldots, n$$

Similarly,

$$b_{kl} = |Y_k - Y_l|_p, \quad \bar{b}_{k.} = \frac{1}{n} \sum_{k=1}^{n} b_{kl},$$

$$\bar{b}_{.l} = \frac{1}{n} \sum\nolimits_{l=1}^{n} b_{kl}, \quad \bar{b}_{..} = \frac{1}{n^2} \sum\nolimits_{k,l=1}^{n} b_{kl}$$

$$B_{kl} = b_{kl} - \bar{b}_{k.} - \bar{b}_{.l} + \bar{b}_{..}, k, l = 1, \ldots, n$$

A_{kl} and B_{kl} are normalized pairwise distances of $\{(X_k, Y_k) \mid k = 1, \ldots, n\}$.

Definition 1: The empirical distance covariance $V_n(X, Y)$ is the nonnegative number defined by

$$V^2{}_n(X, Y) = \frac{1}{n^2} \sum\nolimits_{k,l=1}^{n} A_{kl} B_{kl}$$

Similarly,

$$V^2{}_n(X, X) = \frac{1}{n^2} \sum\nolimits_{k,l=1}^{n} A^2{}_{kl}$$

and

$$V^2{}_n(Y, Y) = \frac{1}{n^2} \sum\nolimits_{k,l=1}^{n} B^2{}_{kl}$$

Definition 2: The empirical distance correlation $R_n(X, Y)$ is the positive root of

$$R^2{}_n(X, Y) = \left\{ \begin{array}{ll} \frac{V^2{}_n(X,Y)}{\sqrt{V^2{}_n(X)V^2{}_n(Y)}}, & V^2{}_n(X)V^2{}_n(Y) > 0 \\ 0. & V^2{}_n(X)V^2{}_n(Y) = 0 \end{array} \right\}$$

The statistic $V_n(X) = 0$ if an only if every sample observation is identical. Distance correlation satisfies the condition $0 \leq R \leq 1$ and $R = 0$ if and only if X and Y are independent. Thus a larger R indicates a stronger relation.

Zhou [9] has extended this concept to define an auto distance correlation function (ADCF) to measure the temporal dependence structure of time series. Chaudhuri et al. [2] proposed fast algorithms to compute distance correlation.

Internal consistency of indicators measuring a concept or topic refers to the property that all indicators are good indicators of the concept. Positive correlations between the indicators are a necessary condition for validity [1]. Cronbach's alpha [3] is a measure used to evaluate internal consistency. We adopt the formula: $\alpha = \frac{Nr}{1+(N-1)r}$ where N is the number of indicators and r is the average pairwise correlation among all indicators. A higher alpha indicate higher consistency.

3   Methodology

Our proposed approach to data reduction is outlined in this section. Let us consider the scenario of analyzing a topic/concept based on many sources of data available. An example is sentiment analysis of tweets to predict election outcomes. Sources of data are tweeters, several of whom will be like minded. In this case, we can view data reduction as the action of choosing tweets of representative tweeters which can make computations faster. The data available for analysis can be viewed as multi-dimensional indicators based on which decisions are taken. In the example of tweets, tweet embedding represents a tweet as a vector. An indicator can be viewed as measurement done at every time unit (time series) or an ordered sequence of outcomes. For the sake of convenience we assume the dimensions of all indicators are same (this can be done by adding dummy dimensions if needed). Then we have to identify the subsets if indicators that can produce results of analysis matching the results when all indicators are considered together. If several indicators point to the same results (i.e. similar), then we can reduce data used in the analysis by selecting a representative indicator. The net effect is analogous to dimension reduction. We propose a combination of three measures to achieve this - 1) measure closeness of data sources, 2) measure consistency of groups of data sources, and 3) measure similarity of internal structure of data points of data sources. In this paper we consider distance correlation (DC [6]) as the similarity measure to compare data sources (indicators). Cronbach's alpha is used as measure of consistency. Consistency says that all indicators in a group measure the same concept. The related formulae are provided in the previous section. They are implemented as MatLab functions and used in our research. They are mathematically rigorous and computationally straightforward which is an advantage over many of the machine learning algorithms. Concordance is adapted to measure internal structure of indicators. The proposed methodology to data reduction is outlined as follows:

1. Retrieve indicator values X_i, $X_i \in R^p$
2. For each indicator, build the distance matrix (a_{kl}) where $a_{kl} = |X_k - X_l|_p$, $|.|_p$ is the norm used
3. Compute pairwise DC for indicators (using formulae defined in the previous section)
4. Group indicators based on a threshold value for DC
5. Measure consistency of groups using Cronbach's alpha
6. Validation check using distance concordance measure

To illustrate the method, the topic of analysis considered in this research is the severity of COVID-19 within the USA. This topic is chosen for timeliness and publicly available data. Datasets used for analysis in this paper are downloaded from the website https://covidtracking.com/data/download/ for the period beginning March 15 till September 30, 2020. Daily death count, hospitalization count, negative count and positive count reported by states are the data chosen for analysis as they represent the state of COVID severity in the states. They form the four dimensions of the data variables. The primary objective in this paper is to propose a method and demonstrate DC may be used a valid technique for data reduction as it provides straightforward implementation. So, we selected six states arbitrarily as data sources. Results shown in this paper are based on data from the states of Alabama (AL), Arkansas (AR), Arizona (AZ), California (CA), Colorado (CO) and Connecticut (CT). The graphs in Figs. 1, 2, 3, 4, 5 and 6 show graphs of data used in the analysis per state. Each figure shows the four dimensions for each state. Data cleaning (replacing suspect data) was done before computations of measures. Table 1 shows a snippet of raw data used in this research. Columns 2–5 represent the four dimensions of a four dimensional data variable. We aim to show that data from all these states exhibit similar characteristics as measured by distance correlation and so, one of the states could be used as representative for analyzing severity.

A comparison of death counts from all six states is given in Fig. 7 to illustrate the trajectories of data in different states. The spike down in the figure is possibly an error in data reporting. In computations of DC, we corrected it to 0. Results of these computations of DC are given in the next section as a table. We also compute similarity of internal relation of data from the different sources as a means of validation. The idea of concordance that varies from -1 to $+1$ is used for this purpose. The classical formula of concordance is given in Definition 3. Based on it we define a new formula as Definition 4 to use in our method.

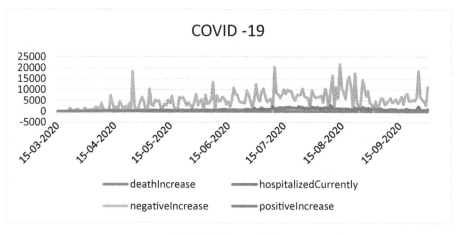

Fig. 1. Alabama data.

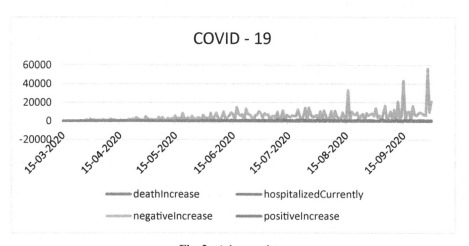

Fig. 2. Arkansas data.

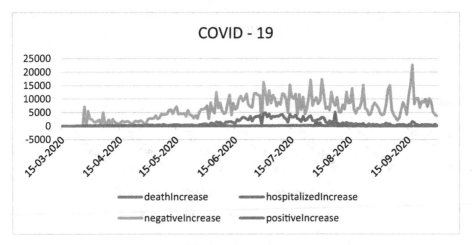

Fig. 3. Arizona data.

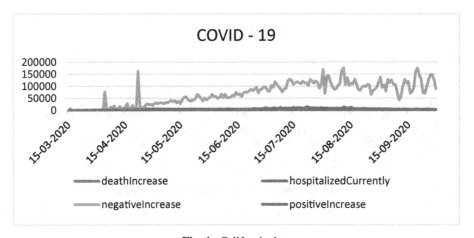

Fig. 4. California data

Definition 3: If $(X, Y) = \{(X_k, Y_k) \mid k = 1, ..., n\}$ sequence of observations, we can define Kendal's τ as

$$\tau(X, Y) = \frac{\sum_{i<j} sign(X_i - X_j) sign(Y_i - Y_j)}{n(n-1)/2}$$

This is a nonparametric measure of association between the variables X and Y. This definition of concordance for the one dimensional case is not useful to multidimensional case. In our case, as X_k and Y_k are multidimensional. So, we adapt this definition as follows:

Definition 4: If $(X, Y) = \{(X_k, Y_k) \mid k = 1, ..., n\}$ sequence of observations where X_k and Y_k are vectors we define distance concordance as.

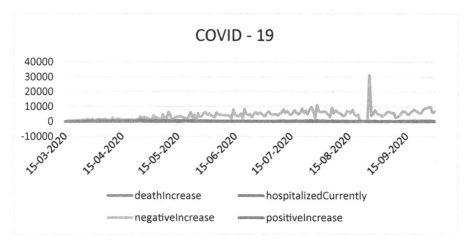

Fig. 5. Colorado data

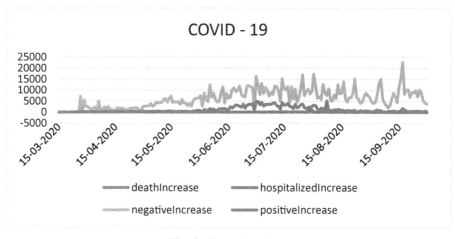

Fig. 6. Connecticut data.

$$d\tau(X, Y) = \frac{\sum sign\left((X_i - X_j)dot(Y_i - Y_j)\right)}{n(n-1)/2},$$ where dot denotes the dot product of two vectors. It is obvious that $-1 \leq \tau \leq 1$. If $\tau = 1$, we have the perfect association.

The definition of Kendall's τ compares the signs of $X_i - X_j$ and $Y_i - Y_j$. This can be interpreted as comparing the directions $(+|-)$ of the values from i to j. With that interpretation as the basis, the definition is extended to vector values. In definition 4, the directions of vectors $X_i - X_j$ and $Y_i - Y_j$ are compared.

Table 1. Sample description of data used for analysis (Alabama data).

Date	Death increase	Hospitalized currently	Negative increase	Positive increase
4/13/2020	6	380	7390	209
4/14/2020	11	380	3793	142
4/15/2020	11	402	723	237
4/16/2020	12	360	2082	232
4/17/2020	11	364	1272	185
4/18/2020	2	359	4565	125
4/19/2020	8	360	2992	182
4/20/2020	13	397	0	188
4/21/2020	10	401	2420	206
4/22/2020	17	440	0	234
4/23/2020	3	406	3568	313
4/24/2020	0	429	0	54
4/25/2020	15	351	18344	305
4/26/2020	4	384	2074	133

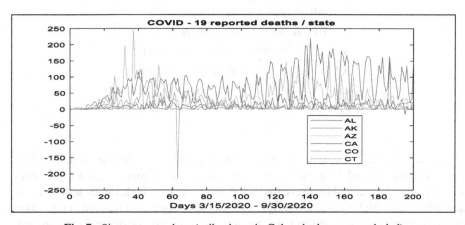

Fig. 7. Six states raw data. (spike down in Colorado data was excluded)

4   Results and Analysis

As stated previously, the primary objective of this research is to propose a method for source reduction. Distance correlation (DC) is a measure to determine similar data and thus add to data reduction methods. DC measures the closeness of two multidimensional data series. Two additional measures are used to determine consistency among similar data sources grouped by DC and internal similarity. Data reduction can be achieved by choosing a representative from similar sources for future analysis. DC for pairs of data sources can be computed in parallel and thus contributing to efficiency. The previous section outlines the six datasets (data sources) we chose for illustrative purpose. Those data are from six US states reporting COVID-19 data. The six states were chosen arbitrarily, but geographically different.

Pairwise distance correlations are computed for the six states' data and are shown in Table 2. In all cases the DC values are 66% or higher. Internal consistency measures the degree of agreement between the indicators with respect to the concept/topic. We computed Cronbach's alpha using the formula $\alpha = \frac{Nr}{1+(N-1)r}$ to validate internal consistency of groups formed by DC measures. In the formula, N is the number of indicators and r is the average correlation. It will be high if all indicators flash the same outcome. As stated earlier, the states' data are viewed as indicators of the concept "severity of COVID in the US". For illustration, we compute alpha values for groups of 2, 3, 4, and 5 with highest correlation values and also all six sources as one group. Results are shown in Table 3. We used the values shown in Table 2 to compute α values. The alpha values are close to 1 indicating internal consistency. It is assumed to validate the case that COVID-19 situation in all six US states are similar in nature. Hence if data reduction is desired, an analyst may consider the data from just one state and derive general conclusions such as predictions.

Table 2. Pairwise distance correlation (DC).

	AR	AZ	CA	CO	CT
AL	0.6600	0.8146	0.8156	0.6921	0.7129
AR		0.7390	0.7435	0.6921	0.7186
AZ			0.8911	0.8109	0.7387
CA				0.8359	0.7910
CO					0.7183

The alpha values indicate consistency considering indicator values as a whole. It does not exhibit the internal differences among indicators. For example, consider the values shown in Fig. 7. One could observe that some indicators have high values at the left end and others have high values at the right end. In other words, the internal composition of the indicators are not uniform. To measure the similarity of internal composition, we use distance concordance measure given in Definition 4. The reason for examining internal differences is to provide an additional level of validity to the result obtained by application of distance correlation.

We computed pairwise distance concordance values. They are shown in Table 4. All distance concordance values are positive which show that there is more agreement than disagreement in internal properties of the data series. The less than 60% values of distance concordance can be attributed to the rises and falls of data at different times. Since all distance concordance values are positive, there is more support than contradiction of the picture presented by the DC and alpha values. Furthermore, concordance could be used as an additional layer of filtering or as a threshold to be met by all members of one group.

Table 3. Cronbach's alpha.

Groups	alpha
AZ, CA	0.9424
AZ, CA, CO	0.9428
AL, AZ, CA, CO	0.9496
AL, AZ, CA, CO, CT	0.9498
AL, AR, AZ, CA, CO, CT	0.9511

Table 4. Pairwise distance concordance.

	AR	AZ	CA	CO	CT
AL	0.2744	0.4171	0.3889	0.3879	0.1525
AR		0.2817	0.3185	0.3034	0.2227
AZ			0.5064	0.4436	0.2273
CA				0.5351	0.3142
CO					0.2521

4.1 Analysis

Previous sections outline the combination of three simple measures to determine groups of similar data sources related to a topic. Expected benefit is that representative sources could be used for future analyses and inferences of related the topic. Results presented in the previous section generally support the method outlined in Sect. 3. Table 3 lists some possible groups of sources based on DC and alpha. However, distance concordance results given in Table 4 indicate that direction of data movement over time could be different in different data sources. That means internal behavior of data sources is not identical. This information could be used as a further threshold in determining groups. Based on all three measures, CA and AZ can be treated as a reliable group.

To validate findings of previous section (which are based on data for the period 3-15-2020 to 9-30-2020), daily positive cases reported by the same six states during the period 11-1-2020 to 12-24-2020 are analyzed. Choice of only one dimension allows the use of classical statistical measures. The data series are plotted individually in Fig. 8. Table 5 shows the pairwise correlations for the six data series. Table 6 shows pairwise concordance measured by Kendall's τ. The CO series shows negative correlation and concordance with the AZ and CA series. Observation of the graphs given in Fig. 8 clearly indicates that CO series values rose during the first part of the time period while others rose in the later part of the time period. This will explain the negative values of correlation and Kendall's τ. Such differences are present in the data sources used in the previous section causing low measures of DC and distance concordance.

A closer examination of the graphs in Fig. 8, the correlations (Table 5) and concordance (Table 6) suggest that AZ and CA group can be validated. If CO data is excluded,

Table 5. Pairwise correlation of daily positive cases Nov 1–Dec 24, 2020.

	AR	AZ	CA	CO	CT
AL	0.7274	0.6274	0.7583	0.0330	0.0217
AR		0.4884	0.5455	0.2350	0.0211
AZ			0.6085	−0.0379	0.0986
CA				−0.3047	0.1101
CO					0.0456

Table 6. Pairwise Kendall's τ values of daily positive cases Nov 1–Dec 24, 2020.

	AR	AZ	CA	CO	CT
AL	0.5648	0.5551	0.6115	0.0580	0.0845
AR		0.4189	0.4027	0.1678	0.1007
AZ			0.5970	−0.0070	0.0316
CA				−0.1125	0.0918
CO					0.0801

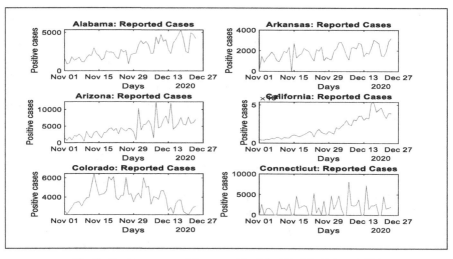

Fig. 8. Positive cases of the sates AL, AR, AZ, CA, CO, and CT.

then the others can be a consistent group. However, if all six sources are considered as a group, another interpretation is possible. CO series increases in the initial period and then decreases. Other series increase during the later period. So, it can be interpreted CO series as prescient warning to other sources in the group .

5 Conclusion

In this research, a combination of measures are proposed and illustrated as a method for data reduction. The significance of the proposed method lies in its simplicity and ease of programming. The method chooses representative data sources/series associated to a topic. Data sources/series are grouped into closely related sets. To achieve data reduction, representatives can be chosen from each group. Data represented by each source/series are assumed to be points in a p-dimensional space or p-dimensional vectors. Distance correlation (DC [6]) is proposed as the measure of similarity of pairs of data sources to form groups of similar sources. Cronbach's alpha is used as a measure of internal consistency (i.e. all sources are indicators of same concept) of members of groups. A newly defined measure of distance concordance is presented as possible threshold to be met by pairs in a similar group. The method is studied using COVID-19 data reported by six states in the USA. Current news stories (cases going up in all states) on COVID-19 in the US states were used as evidence of effectiveness of the method. Comparison with other methods was not attempted as no benchmark data or measures are available in the literature.

As future work, we plan to design programs that apply proposed method to streaming data sources. In this work, vector norm is used as distance measure. Investigation of other measures and their performance is also considered for future work.

References

1. Bollen, K.A.: Multiple indicators: internal consistency or no necessary relationship? Qual. Quant. **18**, 377–385 (1984)
2. Chaudhuri, A., Hu, W.: A fast algorithm for computing distance correlation. Comput. Stat. Data Anal. **135**, 15–24 (2019)
3. Cronbach, L.J.: Coefficient alpha and the internal structure of tests. Psychometrika **16**, 297–334 (1951)
4. Mamon, R.S., Elliot, R.J. (eds.): Hidden Markov Models in Finance. Springer, Heidelberg (2007). https://doi.org/10.1007/0-387-71163-5
5. Habib ur Rehman, M., Liew, C.S., Adbas, A., Jayraman, P.P.: Big data reduction methods: a survey, data science and engineering, January 2017
6. Szekely, G.J., Rizzo, M.L., Bakirov, N.K.: Measuring and testing dependence by correlation of distance. Ann. Stat. **35**(6), 2769–2794 (2007)
7. Uhm, D., Jun, S., Lee, S.: A classification method using data reduction. Int. J. Fuzzy Log. Intell. Syst. **12**(1), 1–5 (2012)
8. Weng, J., Young, D.S.: Some dimension reduction strategies for the analysis of survey data. J. Big Data **4**, 43 (2017). https://doi.org/10.1186/s40537-017-0103-6
9. Zhou, Z.: Measuring nonlinear dependence in time-series, a distance correlation approach. J. Time Ser. **33**(3), 438–457 (2012)
10. https://covidtracking.com/data/download/

Intelligent and Contextual Systems

A Subtype Classification of Hematopoietic Cancer Using Machine Learning Approach

Kwang Ho Park[1] , Van Huy Pham[2] , Khishigsuren Davagdorj[1] ,
Lkhagvadorj Munkhdalai[1] , and Keun Ho Ryu[2,3(✉)]

[1] Database and Bioinformatics Laboratory, School of Electrical and Computer Engineering,
Chungbuk National University, Cheongju 28644, Korea
{khblack,suri,lhagii}@dblab.chungbuk.ac.kr
[2] Faculty of Information Technology, Ton Duc Thang University,
Ho Chi Minh City 700000, Vietnam
{phamvanhuy,khryu}@tdtu.edu.vn
[3] Department of Computer Science, College of Electrical and Computer Engineering,
Chungbuk National University, Cheongju 28644, Korea
khryu@chungbuk.ac.kr

Abstract. Hematopoietic cancer is the malignant transformation in immune system cells. This cancer usually occurs in areas such as bone marrow and lymph nodes, the hematopoietic organ, and is a frightening disease that collapses the immune system with its own mobile characteristics. Hematopoietic cancer is characterized by the cells that are expressed, which are usually difficult to detect in the hematopoiesis process. For this reason, we focused on the five subtypes of hematopoietic cancer and conducted a study on classifying by applying machine learning algorithms both contextual approach and non-contextual approach. First, we applied PCA approach for extracting suited feature for building classification model for subtype classification. And then, we used four machine learning classification algorithms (support vector machine, k-nearest neighbor, random forest, neural network) and synthetic minority oversampling technique for generating a model. As a result, most classifiers performed better when the oversampling technique was applied, and the best result was that oversampling applied random forest produced 95.24% classification performance.

Keywords: Hematopoietic cancer · Gene expression · Subtype classification · Principal component analysis · Synthetic minority oversampling technique

1 Introduction

Recently, in the field of bioinformatics, research has been actively conducted to find meaningful genes or genetic information using machine learning techniques and apply it to preventive medicine based on them [1]. Research through this approach is typical of studies on various diseases that have not been conquered, such as cancer [2–5]. In addition, the impact of these kinds of studies is enormous. For instance, it can be utilized in various area, such as finding new treatments through connection with diseases specific

© Springer Nature Singapore Pte Ltd. 2021
T.-P. Hong et al. (Eds.): ACIIDS 2021, CCIS 1371, pp. 113–121, 2021.
https://doi.org/10.1007/978-981-16-1685-3_10

gene groups by using machine learning or deep learning technique, also being used in individual treatment of patients [2, 5].

In this paper, we applied several machine learning algorithms to make meaningful extraction of hematopoietic cancer to create a model for classifying hematopoietic cancer using genetic information of patients. The hematopoietic cancer is a malignance of cell in immune system. This hematopoietic cancer includes not only blood, but also various elements that make up the blood and cancer that occurs in the bone marrow that produces blood, so it is often called liquid cancer because of these characteristics. According to statistics, these cancers account for 8–10% of all cancer diagnoses, and the mortality rate is also around 10%. As described earlier, this hematopoietic cancer occurs in various blood factors, but the most frequently occurring carcinomas are leukemias, myelomas and lymphomas [6].

Leukemias is one of hematopoietic cancer that results from genetic change in hematopoietic cells in the blood or bone marrow. If an abnormality occurs in the bone marrow, the abnormally generated blood cells are mixed with blood in the body and spread widely into the body. Myelomas is a tumor that occurs in plasma cells which are differentiated from bone marrow, blood, or other tissue. This abnormal plasma cells do not divide into normally, but continue to growing in abnormal life cycle and make a number of antibody. Lymphomas is usually found in distinct stationary masses of lymphocytes, such as lymph node, thymus or spleen. But lymphomas also can be travel through the whole body as like other previous two types of cancer. Previous types of lymphomas is called Hodgkin lymphomas and the other is called non-Hodgkin lymphomas [7].

In this paper, the transcriptome expression data of the representative of hematopoietic cancer: lymphoid leukemias, myeloid leukemias, leukemias nos, mature B-cell leukemias, plasma cell neoplasms patients were collected from TCGA site, and each cancer was clustered using the clustering algorithm principal component analysis (PCA) algorithm, and each cancer was combined with classification algorithm in this clustering result to proceed with the classification of each subtype of hematopoietic cancer. In proceeding with the experiment, there was a total amount deviation of each data, which was used by SMOTE, an oversampling algorithm, to handle the problem of data imbalance also, it is efficient for both contextual and non-contextual system.

Many researchers conducted various studies for this subtype classification. The Liu research team conducted a study using various machine learning algorithms (Support Vector Machine, Naive Bayes, Random Forest, Neural Networks) to classify the subtypes of lung cancer, and the results showed that support vector machine algorithm and random forest algorithms were good at classifying the subtypes of lung cancer [2]. The Muhamed research team used the micro-RNA data to classify subtype of kidney cancer, and used long-short term memory (LSTM) algorithm with two layers as a classification model. Here, to address the class imbalance between subtypes, the oversampling technique was used [3].

Also, many studies were conducted on subtype classification using deep learning. Chen et al. [4], proposed a DeepType framework that classifies subtypes of these two cancers based on multi-layer perceptron techniques using gene expansion data for breast and bladder cancers. Gao et al. [4], also used Deep learning based artificial neural network technique to classify subtypes of colon cancer and breast cancer with 14 subtypes.

The remainder of this paper is organized as follows: Sect. 2 introduces hematopoietic cancer gene expression dataset from TCGA. Also, explains principal component analysis approach for feature selection, synthetic minority oversampling technique for imbalance dataset and four classification algorithms which used in this work. In Sect. 3, the experimental results are provided. Finally, Sect. 4 discusses the experimental results, and addresses our conclusion.

2 Methodology

In this section, we will describe the details of a dataset and the used PCA feature extraction method and machine learning classifiers.

2.1 Dataset

TCGA is a site that has information and data related to many cancers [8]. Currently, there are 47 types of cancer data as of 2020, and these cancers are provided with not only medical data from patients, but also various data such as gene expression data, SNP information, and DNA methylation information. Some raw data are controlled data that have to be approved for using on experiment by the research institute, however, most of the data is freely accessible to users.

We collected gene expression data for various blood cancers through this site. The collected gene expression data consists of a total of five subtypes of hematopoietic cancer: lymphoid leukemias, myeloid leukemias, leukemias nos (not otherwise specified), mature B-cell leukemias, plasma cell neoplasms. The number of each hematopoietic cancer sample is 557 lymphoid leukemias (LL), 818 myeloid leukemia (ML), 104 leukemia nos (NO), 113 mature b-cell lymphomas (MB) and 860 plasma cell neoplasms (PC), consisting of a total 2,457 data. Also this data has 60,453 exons information with the one of gene expression profiling measurement which is called Fragments Per Kilobase per Million (FPKM) mapped measure. The FPKM can be calculated as Eq. 1.

$$FPKM = \frac{Total\ Fragments}{Mapped\ reads\ (Millions) * exon\ length\ (kb)}$$

This FPKM is a normalized estimation of gene expression based on RNA-seq data both considering number of reads and the length of exon which measured by kilo-base unit. That is, a large FPKM means a large amount of expression per unit length, so the FPKM of a certain gene refers to a relative amount of gene expression.

2.2 Principal Component Analysis for Dimension Reduction

Principal component analysis (PCA) is a widely used mathematical algorithm, which reduces a dimensionality while retaining most of the variation in the data [9]. This reduction is achieved by identifying the direction called the principal component (PC_n), and the variation in the data is maximal. This transformed new variables (PC_n) are the linear combination of the original data variable. The first PC (PC_1) accounts for the maximum variability, the remaining PC_{n-1} ($PC_2, PC_3, \ldots PC_n$, n = number of variables) account for the remaining variability in the data. Each PC_n is independent and orthogonal to each other. Only the first few PCs account for most of the total variance in the data expressed as eigenvalues [10]. The value of each PC is calculated by followed equation.

$$PC = a_1 \frac{(x_1 - \bar{x}_1)}{\sigma_1} + \ldots$$

Here, x_1 is the value of the original data point, $\bar{x}_1$ is the mean for the corresponding data, σ_1 is standard deviation for the corresponding variable, and a_1 is the constant of the linear transformation. An important variable is along several directions, and the number of these directions is close to the dimensions of the sample. For data visualization, PCA plots mapped variables and samples through a dimensional space determined by the PC with eigenvalues greater than 1.0 [11]. The plots from the PCA result described the correlation of variables and objectives. The advantage of using these few components is that each sample can be represented in relatively small numbers instead of values for thousands of variables. Then, samples can be plotted to visually assess similarities and differences between samples and determine whether the samples can be grouped.

2.3 Synthetic Minority Oversampling Technique

Oversampling techniques is a method of increasing the number of minority class instances in training set. And one of well-known oversampling technique is named SMOTE in which the minority class be oversampled by generating synthetic samples instead of oversampling with duplicated real variables [12].

This approach can produce satellite data instead of arbitrarily replacing existing samples by generating an oversampled minority class based on the required oversampling ratio, a certain number of samples are selected at random using the KNN approach The composite sample is generated as follows: Calculate the difference between the feature instance under consideration and the nearest feature instance. Then calculate this feature vector difference using a random float value between 0 and 1. As a result, add it to the shape vector corresponding to the new synthetic instance of the minority class [13].

2.4 Classification Algorithms

Classification in data mining and machine learning means that after grasping the characteristics of a new object, one category is allocated by pre-defined class label. This is the approach by which most of objects for classification are generated into a model which explain the class of data or predicting the tendency of the new data. Some of this classification algorithms can be adopting on a part of contextual system. For example, random forest algorithm, using context dependent variables can be used contextual system as a classification model [14]. On the other words, one of classification method, support vector machine, is a representative non-contextual classification algorithm. However, using some kernel function or statistical based estimation theory, this classification algorithm also can be used as contextual classification algorithm on the contextual system [15, 16]. In this research, we performed the experiment on the Intel Xeon E3 1231 v3 processor, 32G memory. And we used python 3.7 for parsing the data and analysis by implementing scikit-learn machine learning library.

Support vector machine (SVM) is non-contextual algorithm, it is based on separates training dataset by binary label with a maximal margin hyper-plane using structure risk minimization technique. As using maximal margin hyper-plane, it can prevent model over-fitting problem. Also, even if SVM is non-contextual in nature [15], SVM can be used both linearly and nonlinearly by mapping the data into the high dimensional space [17, 18]. In this experiment we used linear kernel function (linear-SVM) with maximum iteration = 20,000.

Random forest is the classifier composed a variety of decision tree, and using context –dependent variable [14, 19], the mode class of the results in each classification tree is the output. The inference algorithm is developed by Breiman [20] and the random decision forest by Ho [21]. In this experiment we used the number of tree = 100, impurity criterion = gini-index.

Neural Network (NN) is based one of binary classifier, which is called perceptron, and this NN usually consists of several number of layers, including input layer, hidden layer and output layer. Computation of NN is occurred by several layers' perceptron in hidden layer. Therefore, there are many calculations between each layer. And typically all inputs of neural network are used for processing input vectors [22]. This kinds of things makes NN model to be non-contextual [23, 24]. In this experiment, we set the 1 hidden layer with 100 neurons, activation function = relu, weight optimizer = adam optimizer, learning rate = 0.001, maximum iteration = 20,000.

KNN (K-Nearest Neighbor) is one of well-known algorithm to be simple and easy. KNN algorithm is an algorithm that is often used in classification task. The purpose of KNN is to classify new data based on their features and training data. This KNN algorithm is usally referred contextual model by selecting appropriate k value [25, 26].

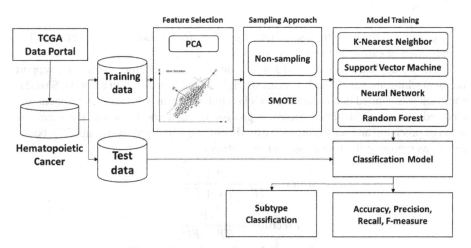

Fig. 1. General workflow of the experiment

Classifying objects based on training data that have the closest distance to the new object is based on the Euclidean distance. In this expriment, we set the k = 5, distance = euclidean distance with minkowski metric.

3 Experiment and Result

Our experimental process was formulated as shown in Fig. 1 above. We collected hematopoietic cancer gene expression data from TCGA website. As mentioned in Sect. 2.1. The original data contains total 2,457 samples. And performed preprocessing step as removing noisy and non-valued instances. This preprocessed data is used for classifying subtypes of hematopoietic cancer. This preprocessed data was divided into 80% of training set and 20% of test set for building the classification model. The simple statistics of preprocessed data is shown in Table 1. As shown Table 1, we can check that the data is distributed unbalanced.

Table 1. Summary of the hematopoietic cancer dataset

Subtype	Number of samples	Train	Test
Lymphoid leukemias (LL)	550	440	110
Myeloid leukemias (ML)	818	654	164
Leukemias NOS (NO)	104	83	21
Mature B-cell luekemias (MB)	113	90	23
Plasma cell neoplasms (PC)	860	688	172
Total	2445	1955	490

We applied PCA feature selection algorithm for extracting 100 features. The visualization of extracted features is shown in Fig. 2.

Then we applied this extracted features on four classification algorithms: support vector machine, neural network, k-nearest neighbor, random forest. In addition, SMOTE approach was applied in consideration of class imbalance, and the same experiment was repeated. And, we calculated accuracy, precision, recall and F1-score (the harmonic mean of precision and recall) on each experiment and compared the results. The below equations represent the four classification matrices.

$$Accuracy = \frac{TP + TN}{TP + TN + FP + FN}$$

$$Precision = \frac{TP}{TP + FP}$$

$$Recall = \frac{TP}{TP + FN}$$

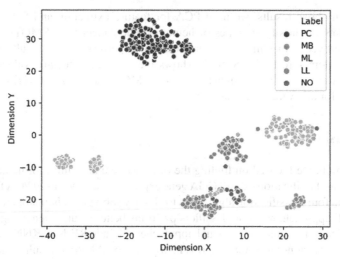

Fig. 2. Visualization of extracted hematopoietic cancers of PCA analysis. LL is indicated lymphoid leukemias, ML is indicated myeloid leukemias, NO is indicated leukemias not other specified, MB is indicated mature B-cell leukemias, and PC is represented plasma cell neoplasms.

$$F1 - measure = \frac{2 \times Precision \times Recall}{Precision + Recall} = \frac{2 \times TP}{2 \times TP + FP + FN}$$

where the TP, TN, FP and FN are the acronym of true-positives, true-negatives, false-positives and false-negatives. A TP and TN is the number of correctly classified whether the subtypes into positive class or negative class. Respectively; A FP represents a number of instances that is incorrectly classified into positive class. Similarly, FN is a number of instances that is incorrectly classified into negative class. And those all classification performance results are shown in Table 2.

Table 2. The classification performance on hematopoietic subtype cancer classification.

Algorithm	Over-sampling method	Accuracy	Precision	Recall	F1-score
RF	Non-SMOTE	94.70%	89.21%	85.33%	86.72%
	SMOTE	**95.24%**	**90.36%**	**88.98%**	**89.60%**
NN	Non-SMOTE	93.61%	86.33%	82.48%	83.67%
	SMOTE	93.61%	85.76%	86.07%	85.86%
KNN	Non-SMOTE	91.71%	82.93%	80.46%	81.36%
	SMOTE	91.71%	85.76%	86.07%	85.86%
SVM	Non-SMOTE	86.14%	87.09%	61.19%	62.02%
	SMOTE	90.63%	79.63%	86.20%	81.64%

As shown above results, we used PCA for feature extraction and four classification algorithms to classify subtypes of hematopoietic cancer with SMOTE approach. Except support vector machine classifier, all classifier performs better result when using SMOTE. Random forest with SMOTE shows the best result for all evaluation measures. In addition, results of neural network, both SMOTE and non-SMOTE shows good performance for all evaluation measures.

4 Conclusion

By this research, we focused on finding the best machine algorithm for hematopoietic cancer subtype classification using TCGA gene expression data. We used PCA for feature extraction and four classification algorithms to classify subtypes of hematopoietic cancer with SMOTE approach. Most of classifiers performs better result when using SMOTE. Also, classification results of contextual approaches, include RF, NN, KNN show better performance than non-contextual approach, which is SVM. As a result, this machine learning based hematopoietic cancer subtype classification can be used on real world medical field for the patients. However, PCA results show that PC and MB were mostly well separated, while the rest of ML, LL, and NO were not exactly separated. To make up for these results a little more, we will proceed with the research to show better performance by adopting a deep learning approach, as in other studies.

Acknowledgement. This research was supported by Basic Science Research Program through the National Research Foundation of Korea (NRF) funded by the Ministry of Science, ICT & Future Planning (No. 2019K2A9A2A06020672 and No. 2020R1A2B5B02001717).

References

1. Xiong, H.Y., et al.: The human splicing code reveals new insights into the genetic determinants of disease. Science **347**, 1254806 (2015)
2. Liu, Y., Wang, X.-D., Qiu, M., Zhao, H.: Machine learning for cancer subtype prediction with FSA method. In: Qiu, M. (ed.) SmartCom 2019. LNCS, vol. 11910, pp. 387–397. Springer, Cham (2019). https://doi.org/10.1007/978-3-030-34139-8_39
3. Muhamed Ali, A., et al.: A machine learning approach for the classification of kidney cancer subtypes using miRNA genome data. Appl. Sci. **8**(12), 2422 (2018)
4. Chen, R., et al.: Deep-learning approach to identifying cancer subtypes using high-dimensional genomic data. Bioinformatics **36**, 1476–1483 (2019)
5. Gao, F., et al.: DeepCC: a novel deep learning-based framework for cancer molecular subtype classification. Oncogenesis **8**(9), 1–2 (2019)
6. Ries, L.A.G., et al.: SEER cancer statistics review 1975–2017. National Cancer Institute (1975)
7. Mak, T.W., Saunders, M.E., Jett, B.D.: Primer to the Immune Response. Academic Cell, Elsevier (2014). (ISBN: 9780123852458)
8. Genomic Data Commons Data Portal. https://portal.gdc.cancer.gov. Accessed 14 Aug 2020
9. Jolliffe, I.T.: Principal Component Analysis. Springer Series in Statistics. Springer, New York (1986). https://doi.org/10.1007/978-1-4757-1904-8

10. Kent, M.: Vegetation Description and Data Analysis: A Practical Approach. Wiley, Hoboken (2011)

11. Chawla, N.V., et al.: SMOTE: synthetic minority over-sampling technique. J. Artif. Intell. Res. **16**, 321–357 (2002)

12. Davagdorj, K., et al.: A machine-learning approach for predicting success in smoking cessation intervention. In: 2019 IEEE 10th International Conference on Awareness Science and Technology (iCAST). IEEE (2019)

13. Sutera, A., et al.: Context-dependent feature analysis with random forests. arXiv preprint arXiv: arXiv:1605.03848 (2016)

14. Bovolo, F., Bruzzone, L.: A context-sensitive technique based on support vector machines for image classification. In: Pal, S.K., Bandyopadhyay, S., Biswas, S. (eds.) PReMI 2005. LNCS, vol. 3776, pp. 260–265. Springer, Heidelberg (2005). https://doi.org/10.1007/11590316_36

15. Negri, R.G., Da Silva, E.A., Casaca, W.: Inducing contextual classifications with kernel functions into support vector machines. IEEE Geosci. Remote Sens. Lett. **15**(6), 962–966 (2018)

16. Li, D.-C., Liu, C.-W.: A class possibility based kernel to increase classification accuracy for small data sets using support vector machines. Expert Syst. Appl. **37**(4), 3104–3110 (2010)

17. Hearst, M.A.: Support vector machine. University of California, Berkeley (1998)

18. Ghimire, B., Rogan, J., Miller, J.: Contextual land-cover classification: incorporating spatial dependence in land-cover classification models using random forests and the Getis statistic. Remote Sens. Lett. **1**(1), 45–54 (2010)

19. Ho, T.K.: The random subspace method for constructing decision forests. IEEE Trans. Pattern Anal. Mach. Intell. **20**(8), 832–844 (1998)

20. Abraham, A.: Artificial neural networks. In: Handbook of Measuring System Design, pp. 901–908 (2005)

21. Huk, M.: Non-uniform initialization of inputs groupings in contextual neural networks. In: Nguyen, N., Gaol, F., Hong, T.P., Trawiński, B. (eds.) ACIIDS 2019. LNCS, vol. 11432, pp. 420–428. Springer, Cham (2019). https://doi.org/10.1007/978-3-030-14802-7_36

22. Huk, M., Mizera-Pietraszko, J.: Context-related data processing in artificial neural networks for higher reliability of telerehabilitation systems. In: 2015 17th International Conference on E-health Networking, Application & Services (HealthCom). IEEE (2015)

23. Chehreghani, M.H., Chehreghani, M.H.: Efficient context-aware K-nearest neighbor search. In: Pasi, G., Piwowarski, B., Azzopardi, L., Hanbury, A. (eds.) ECIR 2018. LNCS, vol. 10772, pp. 466–478. Springer, Cham (2018). https://doi.org/10.1007/978-3-319-76941-7_35

24. Denoeux, T., Kanjanatarakul, O., Sriboonchitta, S.: A new evidential k-nearest neighbor rule based on contextual discounting with partially supervised learning. Int. J. Approx. Reason. **113**, 287–302 (2019)

25. Agrawal, R.: K-nearest neighbor for uncertain data. Int. J. Comput. Appl. **105**(11), 13–16 (2014)

26. Peterson, L.E.: K-nearest neighbor. Scholarpedia **4**(2), 1883 (2009)

Design of Text and Voice Machine Translation Tool for Presentations

Thi-My-Thanh Nguyen, Xuan-Dung Phan, Ngoc-Bich Le,
and Xuan-Quy Dao$^{(\boxtimes)}$

Eastern International University, Binh Duong, Vietnam
{thanh.nguyenthimy,dung.phan,bich.le,quy.dao}@eiu.edu.vn

Abstract. In this paper, a machine translation tool for presentations was presented. This virtual translation tool is a novel approach for generating text or voice in other languages. The proposed system is expected to assists audiences in understanding foreign language content in the live presentations. In this study, the conventional translator was taken over by neural machine translation and human-machine interaction was improved significantly by using text to speech and speech recognition. Experimental results in Vietnamese-English pair showed the effectiveness of the proposed system design and deployment approach.

Keywords: Text to Speech · Automatic speech recognition · Machine translation · Human-machine interaction · Seq2Seq

1 Introduction

In international conferences, a translator is needed when the speakers present in other languages to audiences. In this way, the translator converts content from the source language to a target language. The source language is the language being translated from, while the target language, also called the receptor language is the language being translated into [11]. The accuracy of this approach highly depends on the qualifications of the translator in terms of major, knowledge, experience. Moreover, in this approach, a time delay is unavoidable when the translators convert the content. Another approach is to utilize a machine translation (MT) system which automatically translates from source language to target language. The accuracy of this machine-based method depends on the MT system and the considered language pairs. Fortunately, in recent years, with the rapid development in deep learning techniques, the neural machine translation (NMT) has been significantly increasing its accuracy. In a recent report, Google translate has increased its accuracy by 31% [3]. Furthermore, the source language can be easily converted to multiple target languages at the same time. Therefore, in this study, the utilization of NMT in a text and voice MT which supports live translation in the conferences was proposed.

The recent research results of automatic speech recognition (ASR) systems using deep learning neural network enables the deployment of real applications

© Springer Nature Singapore Pte Ltd. 2021
T.-P. Hong et al. (Eds.): ACIIDS 2021, CCIS 1371, pp. 122–133, 2021.
https://doi.org/10.1007/978-981-16-1685-3_11

with great accuracy. Specifically, the best AI systems in the world such as Siri (Apple), Google Assistant (Google), Cortana (Microsoft), Alexa (Amazon), and Watson (IBM) are able to communicate with people via speech based on ASR systems. The accuracy of the ASR systems such as Siri, Google Assistant, Cortana, Alexa, and Watson has been proved by users. For English recognition, the accuracy of Google ASR system, IBM ASR system is up to 95%. Besides, for Vietnamese recognition, the accuracy of Vais ASR system and Viettel ASR system is up to 95%. To communicate like humans, MT needs Text to Speech (TTS) technologies. The recent research results of TTS systems demonstrate that the voice of TTS technologies has become more and more natural. For English recognition, in the recent results discussed in [19], the quality indicator MOS has reached 4.53 (natural voice - 4.58) [19]. Whereas in Vietnamese [21], MOS has reached 3.94 (natural voice - 4.44). These results inspired us to deploy a real application based on TTS systems.

Based on MT, TTS and ASR systems, a text and voice Vietnamese-English MT tool was proposed for live interpreting in conferences/seminaries. This tool allows us to convert the source language to the target languages in either text or voice. The rest of the paper is organized as follows. Section 2 introduces background and related work. Section 3 describes text and voice MT tool for live interpreting in conferences/seminaries. Section 4 presents experimental results. And Sect. 5 addresses some conclusion and perspective future work.

2   Background and Related Work

2.1   Sequence to Sequence Model

There are a number of sequence to sequence models such as: Connectionist Temporal Classification (CTC), Recurrent Neural Network Transducer (RNN-T), RNN Transducer with Attention, and Attention-based Model. The result of comparison of these models in [16] showed that the attention-based model with two decoder layers is the single best sequence to sequence model. Therefore, we describe the attention-based encoder-decoder neural network which were developed and improved in previous works [4–7,10,17,20,22,24] as follows.

First of all, we describe a RNN which is natural generation of feedforward neural networks to sequences. A standard RNN transforms a sequence of input $(x_1, ..., x_L)$ to a sequence of outputs $(y_1, ..., y_N)$ by iterating the following equation [4,20]:

$$h_t = \text{sigm}(W^{hx}x_t + W^{hh}h_{t-1}) \tag{1}$$

$$y_t = W^{yh}h_t \tag{2}$$

And the Long Short Term Memory (LSTM) is succeed to map the input sequence to a fixed-sized vector using one RNN, and then to map the vector to the target sequence with another RNN which is a simple strategy for general learning. The LSTM computes the conditional probability by the following equation:

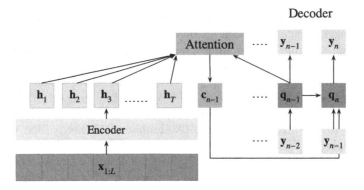

Fig. 1. Attention-based Model (Figure comes from [24])

$$p(y_1, ..., y_N | x_1, ..., x_L) = \prod_{n=1}^{N} p(y_n | v, y_1, ..., y_{n-1}) \tag{3}$$

where v is the fixed-dimensional representation of input sequence $(x_1, ..., x_L)$.

Figure 1 shows the Attention-based Encoder Decoder Model. The encoder RNN transforms the input signal $\mathbf{x} = (x_1, x_2..., x_L)$ into the high level representation $\mathbf{h} = (h_1, h_2, ..., h_T)$ where $T \leq L$. The Attention module summarizes $\mathbf{h}_{1:T}$ into the context vector $\mathbf{c}_n$ in order to predict the output y_n. The decoder RNN predicts the output y_n with the context vector $\mathbf{c}_n$ and the previous output $y_{1:n-1}$. Hence, the Attention-based Encoder-Decoder Model can be formulated as follows [24]:

$$\mathbf{h}_{1:T} = \text{Encoder}(\mathbf{x}_{1:L}) \tag{4}$$

$$\alpha_{n,t} = \text{Attention}(\mathbf{q}_n, \mathbf{h}_t) \tag{5}$$

$$\mathbf{c}_n = \sum_t \alpha_{n,t} \mathbf{h}_t \tag{6}$$

$$\mathbf{q}_n = \text{Decoder}(\mathbf{q}_n, [y_{n-1}; \mathbf{c}_{n-1}]) \tag{7}$$

$$y_n = \arg\max_{\mathbf{y}} (\mathbf{W}^f \mathbf{q}_n + \mathbf{b}^f) \tag{8}$$

And the attention mechanism is used with the attention function given by the equation as follows [4, 17, 24]:

$$\omega_{n,t} = \mathbf{v}^T \tanh(\mathbf{W}^q \mathbf{q}_n + \mathbf{W}^h \mathbf{h}_t + \mathbf{b}^f) \tag{9}$$

$$\alpha_{n,t} = \text{softmax}(\omega_{n,t}) \tag{10}$$

where the matrices $\mathbf{W}^f$, $\mathbf{W}^q$, $\mathbf{W}^h$, and the vectors $\mathbf{v}$, $\mathbf{b}$ are parameters of the model. Next, we describe the Sequence to Sequence model to applying for ASR, MT and TTS.

2.2   Speech Recognition

The Attention-based Encoder-Decoder Model is applied to ASR where the input is the speech utterance $\mathbf{x}$ and the output is character $\mathbf{y}$. The state of art result in [7] minimizes the expected word error rate function, $\mathcal{L}_{\text{MWER}}$, given by the following equation

$$\mathcal{L}_{\text{MWER}} = \mathbb{E}_{P(\mathbf{y}|\mathbf{x})}[\mathcal{W}(\mathbf{y},\mathbf{y}^*)] + \lambda\mathcal{L}_{\text{CE}} \tag{11}$$

where $\mathcal{W}(\mathbf{y},\mathbf{y}^*)$ denotes the number of word errors in the hypothesis, $\mathbf{y}$, compared to the ground-truth label sequence, $\mathbf{y}^* = (y_0^*, y_1^*, ..., y_{N+1}^*)$. The cross-entropy (CE) loss function, $\mathcal{L}_{\text{CE}}$, is defined in [17] which maximizes the log-likelihood of the training data

$$\mathcal{L}_{\text{CE}} = \sum_{\mathbf{x},\mathbf{y}^*} \sum_{u=1}^{N+1} -\log P(y_u^*|y_{u-1}^*, ..., y_0^* = <sos>, \mathbf{x}) \tag{12}$$

The transcript $\mathbf{y}^*$ is given by the minimum expected word error rate (MWER)

$$\mathbf{y}^* = \arg\max_{\mathbf{y}} \log P(\mathbf{y}|\mathbf{x}) + \lambda\log P_{\text{LM}}(\mathbf{y}|\mathbf{x}) + \gamma\mathbf{len}(\mathbf{y}) \tag{13}$$

where $\log P_{\text{LM}}$ is provided by the language model, $\mathbf{len}(\mathbf{y})$ is the number of words in $\mathbf{y}$, and λ and γ are tuned on a development set. The result in [7] demonstrated that ASR system achieved a WER of 4.1% compared to 5% for the conventional system.

2.3   Machine Translate

Similarly to ASR, MT model uses the Attention-based Encoder Decoder Model where $(\mathbf{y}, \mathbf{z})$ is a source and target sentence pair. The object function of Language Model (LM) may be express as the sum of log probabilities of the ground-truth outputs given the corresponding inputs [25]

$$\mathcal{O}_{\text{ML}} = \sum_{i=1}^{M} \log p(\mathbf{z}^{*(i)}|\mathbf{y}) \tag{14}$$

And the object function of model refinement using the expected reward objective which can be expressed as

$$\mathcal{O}_{\text{RL}} = \sum_{i=1}^{N} \sum_{\mathbf{z}\in\mathcal{Z}} p(\mathbf{z}|\mathbf{y}^{*(i)})r(\mathbf{z}, \mathbf{z}^{*(i)}) \tag{15}$$

where $r(\mathbf{z}, \mathbf{z}^{*(i)})$ is the per-sentence score. In order to further stabilize training, the result in [25] demonstrated that we optimize a linear combination of ML (Eq. 14) and RL (Eq. 15) objectives as follows:

$$\mathcal{O}_{\text{Mixed}} = \alpha * \mathcal{O}_{\text{ML}} + \mathcal{O}_{\text{RL}} \tag{16}$$

where this model is first training using the maximum likelihood objective (Eq. 14) until convergence. This model is refined by using a mixed maximum likelihood and expected reward objective (Eq. 16), until BLEU score on a development set is no longer improving. The second step is optional.

2.4   Text to Speech

In this section, we introduce Tacotron 2, state-of-the-art end-to-end speech synthesis model, which synthesizes speech directly from character. Tacontron 2 also uses Attention-based Encoder-Decoder Model described above. Tacontron 2 TTS architecture contains an encoder, an attention-based decoder and postprocessing net, and wavenet vocoder. The encoder aims to convert a character sequence into a hidden feature presentation while the decoder aims to consume this hidden feature presentation to predict a spectrogram. Between the encoder and the decoder is a location-sensitive attention from [8].

WaveNet vocoder which converts mel spectrogram to speech is a new generative model operating directly on the raw audio waveform [15]. In Tacotron 2, WaveNet vocoder is modified by inverting the mel spectrogram feature representation into time-domain waveform sample. PixelCNN++ [18] is used to produce audio samples. The joint probability of a waveform $\mathbf{s} = \{s_1, ..., s_T\}$ is factorized as a product of conditional probabilities as follows:

$$p(\mathbf{s}) = \prod_{t=1} p(s_t|s_1, ..., s_T) \tag{17}$$

where each audio sample s_t is conditioned on the samples at all previous timesteps.

3   System Description

We design a text and voice MT tool illustrated in Fig. 2. First of all, we use ASR system to convert speech to text which is shown in subtitle or translated to target language. Secondly, the target language is shown in subtitle or transformed to speech by TTS system.

- Step 1: ASR system convert instructor's speech to text;
- Step 2: Show text in subtitle;
- Step 3: Text is translated to another language;
- Step 4: Show translated text;
- Step 5: Transform text to speech.

3.1   Speech Recognition

Figure 3 shows the ASR progress where there are a number of ASR techniques such as Hidden Markov Models/Gaussian Mixture Models (HMM/GMM), Hidden Markov Models/Artificial Neural Network (HMM/ANN), Connectionist

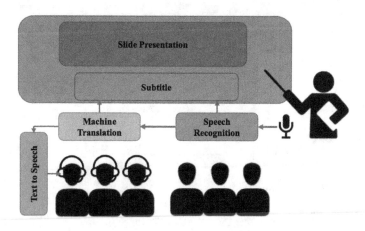

Fig. 2. Text and voice assistant architecture.

Temporal Classification (CTC) and Sequence to Sequence (Seq2seq). The results in this figure showed that recent ASR systems tend toward the human performance. There are several ASR systems including IBM, Microsoft, Apple, Google, CMU Sphinx. However, which of them is the best English ASR system is an important question. According to [13], in 2017, Google English ASR API reached the best result of English ASR system compared with Microsoft Api and CMU Sphinx. In the recent result discussed in [9], Google English ASR Api obtained again the best results of English ASR systems compared with IBM English ASR Api and Wit in 2020. A breakthrough came in 2017 when Google used deep learning neural network algorithms for the ASR system. Google ASR system tends toward human as shown in Fig. 3. Indeed, Google uses the most advanced deep learning neural network algorithms for ASR system. In addition, Google ASR Api supports real-time speech recognition with audio input streamed from a microphone or sent from a prerecorded audio file. Several ASR systems support Vietnamese such as Vais, Viettel, FPT, and Google. Specifically, first of them, Vais is a startup company in natural language processing in Vietnam. Although founded in 2017, Vais won twice of Vietnamese ASR system in Vietnamese Language Signal Processing 2018 and Vietnamese Language Signal Processing 2019. Second, Viettel is the largest telecommunications service provider in Vietnam. In the international workshop on Vietnamese Language and Speech Processing 2019, Viettel ASR performance was ranked behind Zalo ASR and Vais ASR system. Last but not least, FPT is the largest information technology service company in Vietnam. FPT ASR Api has been released since 2017. In our previous study [14], Vietnamese speech recognition platforms including Vais, Vtcc (Viettel), Fpt, and Google have been evaluated. The results demonstrated that Vais outperforms compared with Viettel, Google and Fpt.

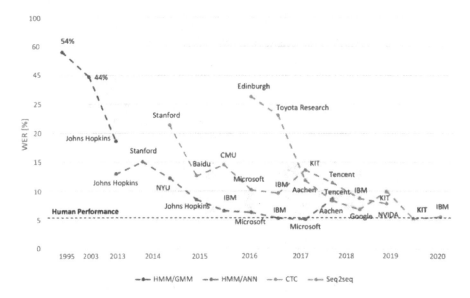

Fig. 3. ASR progress (Figure comes from [23]).

3.2 Machine Translation

There are several online services that support MT including Google Translate, Bing Microsoft Translator, Amazon Translate, IBM Watson Language Translator. Google Translate is one of the most used MT systems because of its conveniences. Specifically, there are 140 billion words translated, and 1 billion users daily; supports 109 languages. Google Translate was launched in 2006 as a statistical MT from 2006 to 2017 and a NMT from 2017 to now. A big breakthrough came in 2016 when Google used NMT technology (deep neural network) [25] that allowed machines to understand the context of an entire sentence as well as helped improving accuracy as shown in Fig. 4 (Figure extracted in [1]). Figure 4 shows a comparison of perfect translation, human, and GNMT in terms of translation quality. This result demonstrates that GNMT tends toward the human. According to Google, the accuracy of the translation depends on the translated content and language. Google translate has been improving its translation accuracy [3] especially the Vietnamese-English language pair.

An example of English-Vietnamese translation is shown: English → Vietnamese: Vietnamese text to speech technology → Công nghệ chuyển văn bản thành giọng nói tiếng Việt; Vietnamese → English:Vais có kết quả tốt nhất → Vais has the best results. The translated result is good enough that we do not need to revise manually.

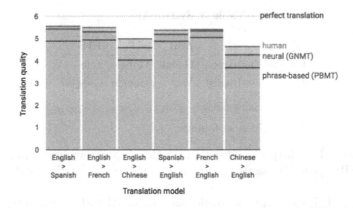

Fig. 4. Google neural machine translation (Figure comes from [1]).

3.3 Text to Speech

Table 1 shows the MOS (Mean Opinion Scale) of Google TTS system which reaches 4.53 compared to 4.58 of human [19]. This result shows that Google TTS technology generates speech with human-like intonation. Built based on DeepMind's speech synthesis expertise, Google TTS API delivers voices that are near human quality. The MOS of Vietnamese TTS systems is shown in Table 2 (this result was extracted from [21]). As described, almost companies achieve good results. Zalo archives the best result with 3.94 of MOS (4.44 of human).

Table 1. English TTS

	Parametric	Tacotron	Concatenative	WaveNet	Tacotron2	Human Speech
MOS	3.67	4.0	4.17	4.34	4.53	4.58

Table 2. Vietnamese TTS

	VTCC	VNGGRD	SUN	Zalo	Human Speech
MOS	2.9	3.85	3.89	3.94	4.44

4 Experiments and Results

MT Tool for presentation is built based on APIs from Google, Vais, Zalo described as follows. Table 3 shows an experimental setup in Vietnamese-English pair.

Table 3. System Api

System	ASR	MT	TTS
English-Vietnamese	Google	Google	Zalo
Vietnamese-English	Vais	Google	Google

The system APIs are given as follows:

– Google translate: https://translation.googleapis.com/language/translate/v2
– Vais ASR: https://vaisapis.vais.vn/analytic/v1/digitalization/audio-upsert-execute
– Google ASR: https://speech.googleapis.com/v1p1beta1/speech:recognize
– Zalo TTS: https://api.zalo.ai/v1/tts/synthesize
– Google TTS: https://texttospeech.googleapis.com/v1beta1/text:synthesize

English-Vietnamese MT tool was developed on Python as in the following.

4.1 Vietnamese-English Translation

– Step 1: Vais ASR API converts speech to text;
– Step 2: Google translate API converts Vietnamese to English text from the step 1;
– Step 3: English text is shown as subtitle or English speech is played using Google English TTS API.

Figure 5 shows the result of Vietnamese-English Translation in Presentations where the system input is Vietnamese speech of the speaker and the system output is English text or/and voice.

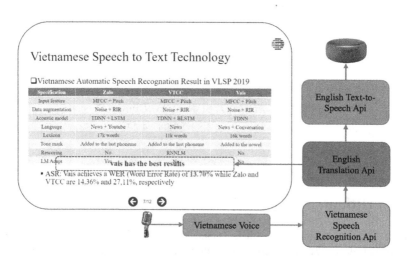

Fig. 5. Vietnamese-English translation in presentations.

4.2   English-Vietnamese Translation

– Step 1: Google English ASR API converts speech to text
– Step 2: Google translate API converts English to Vietnamese text from the step 1.
– Step 3: Vietnamese text is shown or Vietnamese speech is played using Zalo Vietnamese TTS API

Figure 6 shows the result of English-Vietnamese Translation in Presentations where the system input is English speech of the speaker and the system output is Vietnamese text or/and voice.

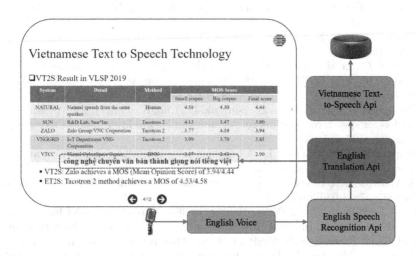

Fig. 6. English-Vietnamese translation in presentations.

5   Conclusion

In this study, a machine translation tool for live interpreting was introduced and described, a novel system for generating text or voice which helps audiences translate the speaker's speech to text or voice in other languages during the presentation at conferences. Furthermore, the design and deployment of English-Vietnamese machine translation using APIs from Google, Vais, and Zalo were conducted owing to their best performance in text to speech, speech recognition, and machine translation in English and Vietnamese. Experiments were carried out to validated our proposal. Experimental results in Vietnamese-English pair showed the effectiveness of the proposed system design and deployment approach.

In the future work, we will focus on applying real time voice cloning which introduced in [2,12] into Vietnamese-English machine translation tool to levitate user experience by using source voice to target voice in Text to Speech is a top priority of face-to-face translation.

References

1. A neural network for machine translation, at production scale. https://ai.googleblog.com/2016/09/a-neural-network-for-machine.html
2. Real time voice cloning. https://github.com/CorentinJ/Real-Time-Voice-Cloning
3. Aiken, M., Wong, Z.: An updated evaluation of Google translate accuracy. Stud. Linguist. Lit. **3**(3), 253–260 (2019)
4. Bahdanau, D., Cho, K., Bengio, Y.: Neural machine translation by jointly learning to align and translate. arXiv preprint arXiv:1409.0473 (2014)
5. Chan, W., Jaitly, N., Le, Q., Vinyals, O.: Listen, attend and spell: a neural network for large vocabulary conversational speech recognition. In: 2016 IEEE International Conference on Acoustics, Speech and Signal Processing (ICASSP), pp. 4960–4964. IEEE (2016)
6. Chan, W., Jaitly, N., Le, Q.V., Vinyals, O.: Listen, attend and spell. arXiv preprint arXiv:1508.01211 (2015)
7. Chiu, C.C., et al.: State-of-the-art speech recognition with sequence-to-sequence models. In: 2018 IEEE International Conference on Acoustics, Speech and Signal Processing (ICASSP), pp. 4774–4778. IEEE (2018)
8. Chorowski, J.K., Bahdanau, D., Serdyuk, D., Cho, K., Bengio, Y.: Attention-based models for speech recognition. In: Advances in Neural Information Processing Systems, pp. 577–585 (2015)
9. Filippidou, F., Moussiades, L.: A benchmarking of IBM, Google and wit automatic speech recognition systems. In: Maglogiannis, I., Iliadis, L., Pimenidis, E. (eds.) AIAI 2020. IAICT, vol. 583, pp. 73–82. Springer, Cham (2020). https://doi.org/10.1007/978-3-030-49161-1_7
10. Guo, J., Sainath, T.N., Weiss, R.J.: A spelling correction model for end-to-end speech recognition. In: 2019 IEEE International Conference on Acoustics, Speech and Signal Processing (ICASSP), ICASSP 2019, pp. 5651–5655. IEEE (2019)
11. Hatim, B., Munday, J.: Translation: An Advanced Resource Book. Psychology Press (2004)
12. Jia, Y., et al.: Transfer learning from speaker verification to multispeaker text-to-speech synthesis. In: Advances in Neural Information Processing Systems, pp. 4480–4490 (2018)
13. Këpuska, V., Bohouta, G.: Comparing speech recognition systems (Microsoft API, Google API and CMU Sphinx). Int. J. Eng. Res. Appl. **7**(03), 20–24 (2017)
14. Nguyen, T., Diep, H., Le, B., Dao, Q.: Comparing Vietnamese speech recognitions. In: 2021 5th International Conference on Machine Learning and Soft Computing (ICMLSC). ACM (2021, accepted)
15. van den Oord, A., et al.: WaveNet: a generative model for raw audio. arXiv preprint arXiv:1609.03499 (2016)
16. Prabhavalkar, R., Rao, K., Sainath, T.N., Li, B., Johnson, L., Jaitly, N.: A comparison of sequence-to-sequence models for speech recognition. In: Interspeech, pp. 939–943 (2017)
17. Prabhavalkar, R., et al.: Minimum word error rate training for attention-based sequence-to-sequence models. In: 2018 IEEE International Conference on Acoustics, Speech and Signal Processing (ICASSP), pp. 4839–4843. IEEE (2018)
18. Salimans, T., Karpathy, A., Chen, X., Kingma, D.P.: PixelCNN++: improving the pixelcnn with discretized logistic mixture likelihood and other modifications. arXiv preprint arXiv:1701.05517 (2017)

19. Shen, J., et al.: Natural TTS synthesis by conditioning WaveNet on Mel spectrogram predictions. In: 2018 IEEE International Conference on Acoustics, Speech and Signal Processing (ICASSP), pp. 4779–4783. IEEE (2018)
20. Sutskever, I., Vinyals, O., Le, Q.V.: Sequence to sequence learning with neural networks. In: Advances in Neural Information Processing Systems, pp. 3104–3112 (2014)
21. Trang, N.T.T., Tung, N.X.: Text-to-speech shared task in VLSP campaign 2019: evaluating Vietnamese speech synthesis on common datasets. In: Vietnamese Language Signal Processing. VLSP (2019)
22. Vaswani, A., et al.: Attention is all you need. In: Advances in Neural Information Processing Systems, pp. 5998–6008 (2017)
23. Waibel, A.: Organic machine learning (2021)
24. Wang, Y., Fan, X., Chen, I.F., Liu, Y., Chen, T., Hoffmeister, B.: End-to-end anchored speech recognition. In: 2019 IEEE International Conference on Acoustics, Speech and Signal Processing (ICASSP), ICASSP 2019, pp. 7090–7094. IEEE (2019)
25. Wu, Y., et al.: Google's neural machine translation system: bridging the gap between human and machine translation. arXiv preprint arXiv:1609.08144 (2016)

Hybrid Approach for the Semantic Analysis of Texts in the Kazakh Language

Diana Rakhimova$^{(\boxtimes)}$![ORCID], Asem Turarbek$^{(\boxtimes)}$![ORCID], and Leila Kopbosyn ![ORCID]

Al-Farabi Kazakh National University, Almaty, Kazakhstan
di.diva@mail.ru, turarbek_asem@mail.ru, leila_s@list.ru

Abstract. In this paper authors propose a hybrid approach for semantic analysis of text resources and documents in the Kazakh language. An overview and difficulties of analysis for the Kazakh language are presented. The developed approach consists of two main parts. The first definition of keywords (phrases) from the text, and the second, based on the data obtained, will build an annotated summarization of the text. To implement the first part of the approach, the TF-IDF algorithm was applied to extract keywords and phrases from texts. The cosine similarity of the sentence data in the Kazakh language was calculated to determine the similarity. With the help of certain similarities semantic links in the text are determined. On the basis of the data obtained, the second part is performed - the abstraction of texts. The number of annotations directly depends on the size of the document. The linguistic corpus of the Kazakh language was collected for carrying out experiments and calculations. A study of various approaches and a hybrid approach for the semantic analysis of the Kazakh language was carried out. The practical part was implemented in Python. The article presents the results of experimental calculations.

Keywords: Kazakh language · Semantic analysis · Keywords · Summarization

1 Introduction

The Kazakh language belongs to the Turkic group of languages and the agglutinative class of languages, it has a complex morphological structure and a rich semantic vocabulary. Unfortunately, at the moment, the Kazakh language is a low-resource language, which hinders the development and conduct of scientific research. For the Kazakh language, the problem of semantic analysis and identification of data or facts is relevant. There are no universal approaches and methods that allow for high-quality semantic analysis, to identify data and facts from texts, etc.

Computer semantic analysis is closely related to the problem of text understanding by a machine. There are many interpretations of the concept "meaning of the text" and the task of understanding it. For example, according to D. A. Pospelov [1], the system understands the text entered into it if, from the point of view of a person (or a group of experts), it correctly answers questions related to the information contained in the text.

© Springer Nature Singapore Pte Ltd. 2021
T.-P. Hong et al. (Eds.): ACIIDS 2021, CCIS 1371, pp. 134–145, 2021.
https://doi.org/10.1007/978-981-16-1685-3_12

2 Related Works

There are various scientific approaches and methods for solving the problem of semantic analysis for a particular language. Some of them will be presented below. Of course, no software can replace the analysis that a human can think of. However, the programs that are currently being developed can reduce the time spent on studying large databases. In this regard, the work of the following programs for solving problems of semantic text analysis is considered. Software offered by various manufacturers, such as Semantic LLC, Tomita-parser (Yandex), Semantic Analyst JHON, SummarizeBot API, TextAnalyst 2.0, Galaktika-ZOOM, NLP ISA Natasha»Etc. is used in different subject areas and for different languages [2–9].

For example, "Semantic LLC" is a program for editing unstructured text. The semi-conductor line is graphically oriented, each node is a semantic element, and the walls represent the elements of the elements. Each attribute of a node is of great importance, the set of attributes depends on the type of element.

Tomita Parser (Yandex) is a program that allows you to extract facts from structured text. Separation of facts is based on context-independent grammar rules. And the program requires a dictionary of keywords. The parser will write its own grammar.

SummarizeBot API - The web service offers a RESTful API to handle all text and image processing tasks. It uses over 100 languages including Russian, English, Chinese, Japanese, and uses machine learning technology. The current version uses the following parameters: 1) automatically link to text; 2) Selection of keywords and conceptual documents; 3) Analysis of a sample of documents and selection of material objects and attributes; 4) Automatically detect the language of the document; 5) Obtaining unpublished data: the main text of articles, forums, forums, etc.; 6) Image processing: identification and recognition of objects in images.

"TextAnalyst 2.0" - a program developed by the research and production innovation center MicroSystems as a tool for text analysis. Text links allow you to create a semantic web of comments, expressed in processed text. The request has the ability to semantic search for fragments of text taking into account the semantic links hidden in the text. Allows you to parse text by constructing a hierarchical tree/heading topics containing text.

The scientific works [10–14] describe the basic ideas of using semantic analysis in information retrieval systems. Various options for finding text statistics are presented, which include counting the number of occurrences of words in documents and the frequency of word contiguity, and new model architectures for computing continuous vector representations of words from very large datasets. The quality of vector representations of words obtained by various models was studied using a set of syntactic and semantic language problems. In [15], the application of language models of a neural network to the problem of calculating semantic similarity for the Russian language is shown. The tools and bodies used and the results achieved are described.

The above presented software products are designed for many resource languages such as English, Spanish. Russian, etc. Unfortunately, for the Turkic languages (Kazakh, Kyrgyz, Turkish, Uzbek, etc.) there is currently no software implementation in the open access. The disadvantage of the developed systems is that they cannot be applied to the

Turkic languages, since they are agglutinative with complex morphological and lexical forms, and semantics dependent sentence structure.

The analysis of a huge amount of data can be simplified if we have keywords or keyphrases that can provide us with the basic characteristics, concept, etc. of a document. The relevant keywords and keyphrases can serve as a summary of the document and help us easily organize documents and extract them based on their contents [16]. It is necessary to distinguish two main approaches to solving the problem of automating the selection of keywords and keyphrases: the assignment of keywords and keyphrases and their extraction [17, 18]. The main difference is that the first approach allows to select only those keywords and keyphrases that are contained in some provided dictionary, and the second approach involves the selection of key information directly from the text.

Keywords can be assigned manually or automatically, but the first approach is very time-consuming and expensive. Thus, there is a need for an automated process that extracts keywords from documents. There are ready-made software solutions to this problem for common languages (English, Russian, Spanish, etc.), and for the Kazakh language there are only a few and they are not in open access.

Below are some approaches and works for carrying out summarization for different languages:

The most common is the superficial approach, which takes into account title words and cue-words (ie, "important", "best" etc.) To extract response results [19].

The paper [20] presents automatic free text processing using material extraction using agent verification. For data processing, the Kmeans algorithm was used as a basis.

There is a common summarization approach based on the structural removal of parts from the text corpus. For example, the WordNet system [21].

The paper [22] presents the Cohesive Approaches, which define and consider the cohesive relationships between concepts within the text. These include synonyms, antonyms, lexical data of the language, etc.

It should be noted that at the moment one of the most popular methods of summation is graphical approaches. Two methods can be attributed to this type: LexRank [23] and TextRank [24].

In [25], the graph approach of summarization a text document is also presented. The difference between this approach is that it simultaneously takes into account local coherence, importance, and redundancy.

The next type of approach is based on machine learning. With this approach, the resulting document results can be transformed into a controlled or semi-controlled learning task. This method requires big data to conduct training.

In the article [26], a new Seq2Seq model is presented for abstract and extractive generalization. A comparative analysis of existing approaches is carried out and it is shown that RNNs and other Seq2Seq models represent a good practical result. The main difference of this approach is at the first-time step during encoding the sequence of adding contextual information using the agent.

3   A Semantic Analysis Based an Algorithm for Extracting Annotation and Keywords

During digital technologies, given the constant growth of the volume of digital data, an important role is played by improving the quality of information retrieval using new semantic approaches and methods.

To work with big data, various algorithms and methods are being developed for the machine solution of this problem, since the amount of data does not allow for manual analysis. Any natural-language is complex, unique, and multifaceted in its own way, therefore, extracting data from documents and text resources is a large and time-consuming work that requires preliminary processing.

This part will present a hybrid approach to the semantic analysis of text resources and documents in the Kazakh language. The developed approach consists of two main parts. The first definition of keywords (phrases) from the text, and the second, based on the data obtained, will build an annotated generalization of the text.

The developed hybrid approach of semantic analysis of the text in the Kazakh language consists of two main stages:

- identify keywords and phrases in the text;
- making semantic annotation of the text based on keywords.

For the first stage, it is necessary to prepare the text. To do this, lemmatization and marking by morphological properties are performed on the texts. The main task of the keyword detection algorithms is the task of finding suitable candidates, identifying attributes and ranking [29].

To rank and determine the frequency, the TF-IDF (Term Frequency - Inverse Document Frequency) indicator was used [28]. With TF-IDF, you can determine the weights for each word relative to the entire document. The words with the highest scores and are the main keywords of the text.

TF-IDF was calculated using the formula below

$$TF * IDF = TF(t, D) * IDF(t) = \frac{n_{t,D}}{\sum_k n_{k,D}} * \log\left(\frac{|TS|}{|\{d : t \in d\}|}\right) \tag{1}$$

where $n_{t,D}$ is the number of occurrences of the word t in the target collection D, $\sum_k n_{k,D}$ is the sum of the occurrences of all words in the target collection D, $|TS|$ is the number of documents in all used collections, $|\{d : t \in d\}|$ is the number of all documents that include the word t at least once.

According to this formula, the weight of the word is calculated. The higher the weight of a word, the higher its relative frequency of use in the collection of text. Based on this algorithm for determining keywords and properties and linguistic resources of the Kazakh language, a modified algorithm for extracting keywords and phrases was developed [13].

To find the similarity of the elements (sentences) of the text and the evaluation, the cosine similarity was applied. To calculate the cosine similarity between sentences, you need to perform the following steps: first, you need to identify all the individual words.

Then the identification of the frequency of occurrence of these words in sentences is formed and is defined as a vector. That is, the sentence itself will be represented as a set of vectors. Next, the cosine similarity function is applied to these vectors, and the cosine of the angle between the vectors is subtracted [14, 15].

x and y are sentence vectors. Their scalar product and the cosine of the angle θ between them are related by the following relation

$$\langle x, y \rangle = ||x|| ||y|| \cos(\theta) \tag{2}$$

Accordingly, the cosine distance is defined as

$$\rho_{cos}(x, y) = \arccos\left(\frac{\langle x, y \rangle}{||x|| ||y||}\right) = \arccos\left(\frac{\sum_{i=1}^{d} x_i y_i}{\left(\sum_{i=1}^{d} x_i^2\right)^{\frac{1}{2}} \left(\sum_{i=1}^{d} y_i^2\right)^{\frac{1}{2}}}\right) \tag{3}$$

Based on the data obtained from formula 3, a matrix of the similarity values of the sentences is constructed. Next, all the offers are ranked according to the similarity matrix. The sentences with the highest weight, which are defined by keywords or phrases, will form the annotation of the document.

This proposed approach takes into account the grammatical properties and rules of the Kazakh language. The next section presents the practical results of the developed hybrid approach to semantic analysis.

4 Application of Approaches and Experimental Results

At the first stage, 2 tasks are solved: preliminary word processing; and the division of the text into separate words and keyphrases.

The first task is language-dependent, therefore, the Kazakh language morphological feature is taken into account here. To solve this problem, a system of complete endings of the Kazakh language is used (through the morphological analyzer of the Kazakh language developed on the platform Apertium [30], we perform markup of the document), the algorithm for stemming and lemmatization for the Kazakh language [31] (implemented in the Python3 programming language). Then, a simple approach was used - the tokenization procedure, which helps to divide the whole text into separate words.

The developed algorithms and approach for hybrid semantic analysis are implemented using the Python programming language and NLTK libraries. To test the program, we have prepared a marked corpus, which consists of more than 120 text documents of various sizes and topics. First, keywords and phrases with the Tf-idf metric were defined for each text. Table 1 below shows an example of the keywords found for texts in the Kazakh language (Figs. 1 and 2).

Table 1. Experimental data of the obtained keywords from texts in the Kazakh language.

Keywords and keyphrases	Tf-idf metric
Document: arabazathistory.txt, Number of words in the text: 1876	
Ливан (Lebanon)	0.03753761448295349
Көтеріліс (revolution)	0.014962316253101847
Француз (French)	0.014881951295324384
Франция (France)	0.011384757884540301
француз үкімет (the French government)	0.013168923967413456
1920 жыл (1920 year)	0.008728017814309411
келісім шарт (agreement)	0.00827156782972209
Document: okushi.txt, Number of words in the text: 3450	
Сабақ (lesson)	0.010324737893214916
Физика (physics)	0.006381991500464335
Ауылшаруашылық (agriculture)	0.003477428443091718
Мұғалім (teacher)	0.003398202016653529
сынып физика (class physics)	0.0037965546730691483
…	…

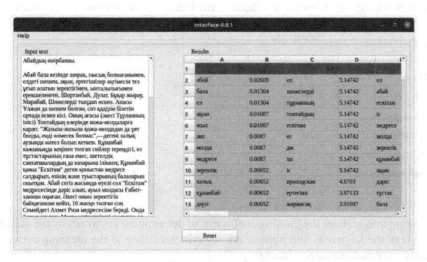

Fig. 1. An example of the operation of the algorithm for determining keywords and phrases (the measure TF and IDF are shown separately).

Table 2 presents the practical results of the developed algorithm for determining keywords and phrases in Kazakh texts.

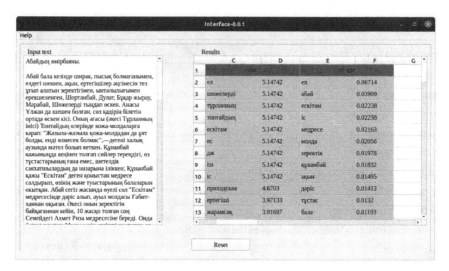

Fig. 2. An example of the operation of the algorithm for determining keywords and phrases (the measure TF-IDF is shown).

Table 2. Experimental results of the developed algorithm for determining keywords for the Kazakh language

Document's name	Document volume (number of sentences)	Borderline coefficient keywords	Number of keywords	Accuracy finding
Sport.txt	87	3–8	8	84,31%
books.txt	79	3–8	8	84%
almaty.txt	96	3–8	8	79,5%
Psychology.txt	298	9–12	10	93,4%
2018biznesmen.txt	320	12–15	12	63,43%
computersciense.txt	415	15–17	12	95,03%
geoinformatika.txt	885	15–17	13	98,3%

Taking into account the limiting coefficient of determining keywords by the volume of the text, the keywords and phrases are selected according to the meaning correctly and has a not bad indicator of accuracy.

To test the operation of the developed algorithm for extracting keywords in the Kazakh language, practical experiments were conducted. In practice, two approaches were compared: the first simple summarization, the second summarization with keywords and phrases. In the experiment, more than 120 documents in the Kazakh language with various topics and volumes were processed. The time spent on identifying the text annotation directly depended on the volume of the input text. The resulting annotations are shown in Table 3.

Table 3. Examples of the work of summarization approaches for texts in the Kazakh language.

Document: computer.txt		Translate
Summa-rization based on keywords	Компьютер (ағылшынша: computer—«есептегіш»), ЭЕМ (электрондық есептеуіш машина)—есептеулерді жүргізуге, және ақпаратты алдын ала белгіленген алгоритм бойынша қабылдау, қайта өңдеу, сақтау және нәтиже шығару үшін арналған машина. Компьютер шеше алмайтын есептерді ағылшын математигі Аланом Тьюринг сипаттаған болатын. Бұл ерекшелікті алғаш рет 1965 жылы «Intel» компаниясының басшыларының бірі Гордон Е Мур сипаттаған болатын. Көптеген ғалымдар компьютерді адамға ыңғайлы ондық санау жүйесінде жасап шығаруға тырысты	Computer (English: computer - "counter"), computer (electronic computer) - a machine designed to perform calculations, and to receive, process, store and output information according to a predetermined algorithm. Problems that a computer cannot solve were described by the English mathematician Alan Turing. This feature was first described in 1965 by Gordon E. Moore, one of the leaders of Intel. Many scientists have tried to build a computer in a human-friendly decimal number system
Simple summa-rization	Компьютер (ағылшынша: computer—«есептегіш»), ЭЕМ (электрондық есептеуіш машина)—есептеулерді жүргізуге, және ақпаратты алдын ала белгіленген алгоритм бойынша қабылдау, қайта өңдеу, сақтау және нәтиже шығару үшін арналған машина. Компьютер тек қана бағдарламада көрсетілген сызықтар мен түстерді енгізу-шығару құрылғыларының көмегімен механикалық түрде көрсетеді. 1946 жылы бұл сөздікте цифрлық компьютер, аналогтық есептеуіш машинасы және электронды компьютер түсініктерінің мағынасы ажыратылып көрсетілді. Бұл ерекшелікті алғаш рет 1965 жылы «Intel» компаниясының басшыларының бірі Гордон Е Мур сипаттаған болатын. Компьютерлер көлемінің кішіреюю процессі де осындай жылдамдықпен жүріп келеді. Алғашқы электрондық есептеуіш машиналар көптеген тонна салмағы бар. Егер цифрлық компьютерлер дискретті сандық және таңбалық айнымалылармен жұмыс жасайтын болса, аналогтық компьютерлер келіп түсетін мәліметтер ағынын үзіліссіз өңдеуге арналған	Computer (English: computer - "counter"), computer (electronic computer) - a machine designed to perform calculations, and to receive, process, store and output information according to a predetermined algorithm. The computer displays the lines and colors shown in the program only mechanically with the help of I/O devices. In 1946, the dictionary differentiated between the concepts of digital computer, analog computer and electronic computer. This feature was first described in 1965 by one of the leaders of Intel, Gordon E. Moore. The process of reducing the size of computers is going at the same speed. The first electronic computers weighed many tons. If digital computers work with discrete numeric and symbolic variables, analog computers are designed for continuous processing of incoming data streams
Document: moon.txt		Translate
Summa-rization based on keywords	Біздің планетамызда жоқ заттардың орнын алмастыру қажет. Сол себепті адамдар Айға көз жүгіртеді. Ай топырағынан оттегі ал технологиясы жердегі зертханаларда пайдаланылған. Айдағы энергетиканы дамытудың басты бағыты. Адамдардың Айды игеруі – бұл жүзеге асыратын іс екенін көрсетті	We need to replace things that do not exist on our planet. That is why people look at the moon. Oxygen from lunar cancer and technology have been used in terrestrial laboratories. The main direction of lunar energy development. The fact that people have mastered the Moon has shown that it is a work in progress

(*continued*)

Table 3. (*continued*)

Document: computer.txt		Translate
Simple summa-rization	Геостационарлыорбита дегеніміз - бұл Жерден шамамен 35800 км биіктіктегі шеңберлер экваторлы орбита. Айдағы энергетиканы дамытудың басты бағыты, бұл күн энергиясын электр энергиясына өзгерту. Луноход -1» аппараты рентген телескопымен жабдықталған еді, ол арқылы галактика аралық рентген сәулелерінің ұзындықтары өлшенді. Адамдардың Айды игеруі – бұл жүзеге асыратын іс екенін көрсетті. Жердің экологиясын тазалау. Айдан әкелінген тас-топырақты зерттеу барысында, онда жер бетінде сирек кездесетін металдардың, пироксеннің, ильмениттің т.б. Жерді аса зиянды қалдықтардан тазарту проблемасын шешу жолында, осы жұмыста көрсетілген бағыт, көңіл аударатындай ерекше болып отыр	A geostationary orbit is an equatorial orbit with circles at an altitude of about 35,800 km above the Earth. The main direction of lunar energy development is the conversion of solar energy into electricity. Lunokhod-1 was equipped with an X-ray telescope, through which the lengths of intergalactic X-rays were measured. The fact that people have mastered the Moon has shown that it is a work in progress. Cleaning the earth's ecology. During the study of rocks and soils brought from the moon, they found rare metals, pyroxene, ilmenite, etc. In addressing the problem of land degradation, the direction outlined in this paper is particularly noteworthy

The Table 3 shows examples of text processing using two summarization methods. From the results obtained, it can be seen that the received annotations convey the semantic concept of the text. In experiments on texts with a small volume, there were cases when the results of the two approaches were very approximate.

Figure 3 below shows the interface of the software solution for defining text annotations. The upper yellow window shows the original text in Kazakh. The total number of words and sentences are also indicated. Further down in the yellow window, you will see the specific keywords and phrases that will be used in the text. The left blue window shows the result of the simple summarization, and the right blue window shows the result of the summarization based on keywords.

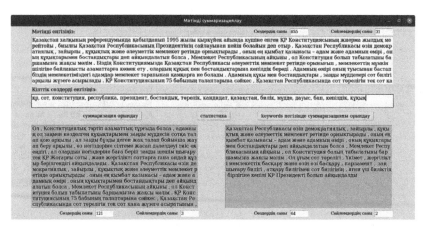

Fig. 3. An example of the program for determining summarization (two approaches) for the Kazakh language

Figure 4 shows the percentage of the results of the two summarization approaches. The horizontal values show the number of words in the document. And vertically, the percentage of the accuracy of determining the annotations of these texts. The analysis and accuracy of the results were carried out manually by three experts (a specialist linguist of the Kazakh language). Then the average value of the experts' assessments was calculated.

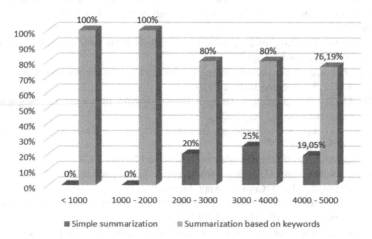

Fig. 4. The percentage of the results of the two summarization approaches.

The best result for defining the annotation of full-text documents is given by the keyword-based summarization approach. This is because keywords are used to cover sentences that have some meaning to the text, rather than simple introductory sentences. The above-developed algorithms and the method of the module are interconnected and provide an integrated approach for processing and analysis of big data in the Kazakh language.

5   Conclusion and Future Work

According to the results of scientific research work, the following results were obtained:

Methods and modern approaches to semantic analysis and abstraction of texts are investigated. Taking into account the peculiarity of the grammar of the Kazakh language, a hybrid semantic analysis of full-text documents was developed. This approach is based on the definition of keywords\phrases and the construction of the text annotation. The practical results of the text analysis show that this approach reveals the contextual meaning of the text. This approach can also be applied to other low-resource Turkic languages. Because it does not require large data for processing.

In the future, it is planned to use this approach in the implementation of machine translation and post-editing systems for Kazakh language.

Acknowledgments. This research is funded by the Science Committee of the Ministry of Education and Science of the Republic of Kazakhstan (Grant No. AP08052421 Project title: «Research and development of the post-editing system o of the Kazakh language in machine translation»).

References

1. Pospelov, D.A.: Ten hotspots in research on artificial intelligence intelligent systems (MSU). (Resource language – Russian), vol. 1, no. 1–4, pp. 47–56 (1996)
2. Semantic: https://semantick.ru/. Accessed 14 July 2020
3. Tomita parser: https://api.yandex.ru/tomita/. Accessed 14 July 2020
4. In the foothills of semantics: https://dworq.com/: 05/29/2020. 5. AI Data Analysis Technologies for Business. https://www.summarizebot.com/summarization_business.html. Accessed 27 May 2020
5. TextAnalyst ver. 2.0: Program for personal text analysis. https://offext.ru/library/data/datakeeping/51.aspx. Accessed 19 Apr 2020
6. Galaktika-Zoom: analytical system for respectable clients. https://www.itweek.ru/themes/detail.php?ID=52215. Accessed 16 June 2020
7. Best Out-Of-The-Box Sentiment Analysis Tools. https://monkeylearn.com/blog/sentiment-analysis-tools/. Accessed 25 July 2020
8. Automatic text analysis technologies (resource language – Russian). https://nlp.isa.ru/. Accessed 26 Apr 2020
9. GitHub Natasha. https://github.com/natasha. Accessed 26 Apr 2020
10. Sonawane, S.S., Kulkarni, P.A.: Graph based representation and analysis of text document: a survey of techniques. Int. J. Comput. Appl. **96**(19), 1–8 (2014)
11. Cicekli, I., Korkmaz, T.: Generation of simple Turkish sentences with systemic-functional grammar. https://doi.org/10.3115/1603899.1603928
12. Manning, Ch.D., Raghavan, P., Schütze, H.: Introduction to Information Retrieval. University Press, Cambridge, p. 210 (2008)
13. Efficient estimation of word representations in vector space. https://arxiv.org/pdf/1301.3781.pdf. Accessed 10 July 2020
14. Word2vec parameter learning explained. https://arxiv.org/pdf/1411.2738.pdf. Accessed 10 July 2018
15. Texts in, meaning out: neural language models in semantic similarity tasks for Russian. https://arxiv.org/ftp/arxiv/papers/1504/1504.08183.pdf. Accessed 20 Apr 2020
16. Sheremeteva, S.O., Osminin, P.G.: Methods and models for automatic keyword extraction (resource language – Russian). Bull. S. Ural State Univ. №. 1, T. 12, pp. 76–81 (2015)
17. Effective approaches for extraction of keywords. https://www.ijcsi.org/papers/7-6-144-148.pdf. Accessed 25 July 2019
18. Keyword extraction a review of methods and approaches. https://langnet.uniri.hr/papers/beliga/Beliga_KeywordExtraction_a_review_of_methods_and_approaches.pdf. Accessed 05 July 2019
19. Nastase, V.: Topic-driven multi-document summarization with encyclopedic knowledge and spreading activation. In: Proceedings of the Conference on Empirical Methods in Natural Language Processing, pp. 763–772 (2008)
20. García-Hernández, R., Montiel, R., Ledeneva, Y., Rendón, E., Gelbukh, A., Cruz, R.: Text summarization by sentence extraction using unsupervised learning. In: Gelbukh, A., Morales, E.F. (eds.) MICAI 2008. LNCS (LNAI), vol. 5317, pp. 133–143. Springer, Heidelberg (2008). https://doi.org/10.1007/978-3-540-88636-5_12
21. Miller, G.A.: Wordnet: A lexical database for English. Commun. ACM **38**(11), 39–41 (1995)
22. Barzilay, R., Elhadad, M.: Using lexical chains for text summarization. In: Advances in Automatic Text Summarization, pp. 111–121 (1999)
23. Erkan, G., Radev, D.R.: LexRank: graph-based lexical centrality as salience in text summarization. J. Artif. Intell. Res. **22**, 457–479 (2004)

24. Mihalcea, R., Tarau, P.: Textrank: Bringing order into texts. Association for Computational Linguistics (2004)

25. Parveen, D., Strube, M.: Integrating importance, non-redundancy and coherence in graph-based extractive summarization. In: Proceedings of the Twenty-Fourth International Joint Conference on Artificial Intelligence (IJCAI 2015), pp. 1298–1304 (2015)

26. Khatri, C., Singh, G., Parikh, N.: Abstractive and extractive text summarization using document context vector and recurrent neural networks (2018). https://arxiv.org/abs/1807.08000

27. Zeng, B., Xu, R., Yang, h, Gan, Z., Zhou, W.: Comprehensive document summarization with refined self-matching mechanism. Appl. Sci. **10**, 1864 (2020). https://doi.org/10.3390/app100 51864

28. TF-IDF. https://en.wikipedia.org/wiki/Tf%E2%80%93idf. Accessed 15 July 2020

29. Hanumanthappa, M., Narayana, S.M., Jyothi, N.M.: Automatic keyword extraction from dravidian language. Int. J. Innov. Sci. Eng. Technol. **1**(8), 87–92 (2014)

30. Rakhimova, D., Turganbayeva, A.: Auto-abstracting of texts in the Kazakh Language. In: Proceedings of the 6th International Conference on Engineering & MIS, pp. 1–5 (2020). https://doi.org/10.1145/3410352.3410832

31. Diana, R., Assem, S.: Problems of semantics of words of the Kazakh language in the information retrieval. In: Nguyen, N.T., Chbeir, R., Exposito, E., Aniorté, P., Trawiński, B. (eds.) ICCCI 2019. LNCS (LNAI), vol. 11684, pp. 70–81. Springer, Cham (2019). https://doi.org/10.1007/978-3-030-28374-2_7

Natural Language Processing

Short Text Clustering Using Generalized Dirichlet Multinomial Mixture Model

Samar Hannachi$^{(\boxtimes)}$, Fatma Najar$^{(\boxtimes)}$, and Nizar Bouguila$^{(\boxtimes)}$

Concordia Institute for Information and Systems Engineering (CIISE),
Concordia University, Montreal, QC, Canada
s_annach@live.concordia.ca, {fatma.najar,nizar.bouguila}@concordia.ca

Abstract. The Artificial Intelligence field is under the spotlight as of its wide use and efficiency in solving real world problems. As of this decade, a notable rise in the amounts of data collected, which were made available to the public, is witnessed. This allowed the emergence of many research problems among which working with short texts and their different challenges. In this paper, we propose the collapsed Gibbs Sampling algorithm for the generalized Dirichlet Multinomial Mixture model for short text clustering (GSDMM). The proposed approach has been evaluated on the Google News dataset. Our approach proved to be more efficient than the related-works and succeeded into overcoming the common challenges that come with short texts.

Keywords: Short text clustering · Generalized Dirichlet Multinomial Mixture · Gibbs sampling

1 Introduction

This last decade witnessed the re-emergence of the Artificial Intelligence field after its withdrawal for several years due to its expensive computational costs and lack of concrete results. As of now, its easy availability to the public and the high-performing computers opened the door to many research works. Indeed, in the last years, the big corporates started making use of the amounts of data they collected through the decades, which was digitized by the use of computers and the rise of the Internet. Depending on the source, the data can be in the form of texts, images, or videos addressing different issues depending on the context of application. The specific sub-field that deals with the data in the form of texts is called Natural Language Processing (NLP) as it grasps the interaction between humans and computers using the human natural language. Some common tools are usually used to address this kind of problems such as the generative statistical models. They are commonly used for documents classification such as the latent Dirichlet allocation (LDA) [1]. The LDA generative process allows the extraction of similarities between different documents and assigning them to unobserved groups using the Dirichlet distribution [2]. This same process is used to estimate the parameters of the multinomial distributions which describe the distribution of the topics and the words of the vocabulary [1]. As for this era and

© Springer Nature Singapore Pte Ltd. 2021
T.-P. Hong et al. (Eds.): ACIIDS 2021, CCIS 1371, pp. 149–161, 2021.
https://doi.org/10.1007/978-981-16-1685-3_13

for more than a decade now, we use short texts to express ourselves on social media using tweets or Reddit and Facebook posts. This informs on the amounts of data available now and which can be used to improve the results from previous research. Nevertheless, this type of data does present its own challenges. Due to their nature, they have many constraints such as data sparsity which hinders an accurate modeling of the language [3]. They also present a large-volume of characteristics which increases the calculations which enhances the complexity of the problems to be addressed. For short texts challenges, authors in [4] focused on using different data representation methods such as Bag of Words (BOW) and Term Frequency-Inverse Document Frequency (TF-IDF). Others in [5] proposed a term weighting scheme and the author in [6] proposed classification and feature weighting using MAP and stochastic complexity. In [7], the authors proposed a topic memory network. Others introduced external knowledge [8] while deep learning methods are applied as short text classifiers in [9,10]. Another very well-known technique to deal with short texts is the use of statistical generative models. Those models generally rely on distributions like the Dirichlet and multinomial distributions to extract the latent topics to which short texts can be assigned. In that sens, similar texts will be grouped under the same topics and different ones will be assigned different topics. Authors in [11] investigated the problem of discrete data by applying finite mixture models. Others in [12] worked on spam and image categorization using support vector machines for training. Bouguila and Ziou used in [13] the principle of MML (Minimum Message length). Another work in [14] used a generative model to classify spatial color image databases while in [15] use was made of a variational Bayes model in the generative process. In [16], authors used the leave-one-out likelihood when estimating the parameters of the statistical model. The work in [17] modeled the trust of web service using Bayesian networks and a mixture model. The work in [18] proposed a new distribution replacing the commonly used Dirichlet distribution by the Scaled Dirichlet distribution. The goal of this paper is to improve the approach used in a previous work that proposed the Dirichlet multinomial distribution to cluster short texts [19]. Our contribution consists on introducing the generalized Dirichlet distribution instead of the simple Dirichlet into our mixture model. This allows a more flexible description of the data when estimating the parameter of the multinomial distribution that gives the distribution of each document over the collection of topics.

2   Background

2.1   Latent Dirichlet Allocation

LDA is a statistical generative model used in text modeling that generates documents according to a fixed number of latent topics [1]. Each document is represented as a distribution over the topics and each topic is represented as a multinomial distribution over the words in the vocabulary. We can generate a document by sampling a mixture of topics from which we sample words. The generation of a document starts by randomly choosing one of the distributions over topics and assigning it to a document. Then, to each word in that document, a topic is assigned randomly from the previously chosen distribution. To assign the new topic to the word, the topics present in a document are monitored and the number of times that same word was assigned a certain topic across all the other documents is counted. This process is repeated for all the words in the different documents [1]. The distribution that assigns to each word of a document a topic is generated by a multinomial distribution with parameter θ. This parameter is estimated using Dirichlet prior with parameter α. The distribution of the topics over the words of parameter ϕ that counts the number of times topics are assigned to words across all the documents is estimated by another Dirichlet distribution with a parameter β. So, estimating the parameters θ and ϕ comes down to estimating which are the words that compose a certain topic and which are the topics that can be the most representative of a document. Those parameters θ and ϕ being intractable, can be approximated using the Gibbs sampling algorithm. It is a Markov chain Monte Carlo algorithm effectively used to estimate the posterior distribution in probabilistic models [20].

2.2 Gibbs Sampling for Dirichlet Multinomial Model

When working with mixture models, it is common use to rely on the Markov chain Monte Carlo algorithm called the Gibbs sampler [21]. The model that we will follow on our work is known as GSDMM (Gibbs Sampling Dirichlet Multinomial Mixture) which was designed as a model for short text clustering [19]. This model can be seen as a rectified LDA given that it assumes that each document can be assigned only one topic. A very known analogy to this model is the "Movie Group Approach". The documents are assimilated to students having each a list of favorite movies representing the words. At first, the students are randomly assigned to K tables. The instruction while shuffling from one table to another is to always take into consideration two factors. The first one is to always choose a table with the highest number of students. The second one is that the film interests of the people in the same table must coincide. This is repeated until the number of clusters becomes unchanged. This same process can be found under the name of "Chinese Restaurant Process" [22].

3   Proposed Model

3.1   Generalized Dirichlet Multinomial Mixture Model

Dirichlet distribution is usually used as a conjugate prior of the multinomial distribution. But, many emerging studies have shown that using the generalized Dirichlet distribution instead of the simple one when dealing with count data clustering gives solution to many limitations [23]. It allows a more general covariance taking into account positive and negative correlations in the counts of categories. It also offers a better flexibility into the description of the composition of the data.

In a formal way, we have a set of documents $D = (d_1, \ldots, d_D)$, a set of words which represent the words in the vocabulary $(x_1, \ldots, x_K)$ and for a document i we have $\boldsymbol{X_i} = (x_{i1}, \ldots, x_{iV})$, where x_{iw} is the number of times the word x_w appears in the document d_i. As for the generalized Dirichlet, it is chosen as a prior to the multinomial distribution giving the density of the MGDD as:

$$\mathbb{P}(\boldsymbol{X_i}|\alpha, \delta) = \mathbb{P}(\boldsymbol{X_i}, c|\alpha, \delta)$$

$$= \frac{\Gamma((\sum_{w=1}^{V} x_{iw}) + 1)}{\prod_{w=1}^{V} \Gamma(x_{iw} + 1)} \prod_{w=1}^{V-1} \frac{\Gamma(\alpha_w + \delta_w)}{\Gamma(\alpha_w)\Gamma(\delta_w)} \prod_{w=1}^{V-1} \frac{\Gamma(\alpha_w')\Gamma(\delta_w')}{\Gamma(\alpha_w' + \delta_w')} \quad (1)$$

where $\Gamma(.)$ is the Gamma function, α_w and δ_w are the parameters of the generalized Dirichlet and $\alpha_w' = \alpha_w + x_{iw}$ and $\delta_w' = \delta_w + x_{i(w+1)} + \cdots + x_{iV}$.

3.2 Collapsed Gibbs Sampling for Generalized Dirichlet Multinomial Mixture Model

This section will present the main contribution of this paper which is the introduction of the generalized Dirichlet distribution when estimating the parameters of the multinomial distribution using the Gibbs sampling algorithm. Indeed, the generalized Dirichlet distribution as presented in the previous section results from the multinomial over the latent parameter of the multinomial distribution giving the probability of selecting a cluster k_i characterized by a generalized Dirichlet distribution as:

$$\mathbb{P}(k_i|\alpha, \delta) = \int \mathbb{P}(k_i|c)\mathbb{P}(c|\alpha, \delta)dc \quad (2)$$

where $\mathbb{P}(k_i|c)$ is a multinomial distribution and $\mathbb{P}(c|\alpha, \delta)$ is a generalized Dirichlet distribution.

Algorithm 1 shows the functioning of the collapsed GSGDMM where as a first step the variables m_k: the number of documents in cluster k, n_k: the number of words in cluster k and n_k^w: the number of occurrence of word w in cluster k are initialized to zero. Then, each document will be assigned a cluster randomly while the variables previously mentioned will be incremented respectively by 1, N_d: the number of words in document d and N_d^w: the number of occurrences of word w in document d. Then, a number of iterations will be chosen to iterate the operation of recording the actual cluster of a document, decrement the parameters by the same amounts mentioned, generate new cluster to each document following the conditional probability using the generalized Dirichlet multinomial distribution, and then increment the variables again.

As shown in Algorithm 1, the sampling of a document d follows two major steps:

1. Selection of an initial cluster to be assigned to a document using the multinomial distribution.
2. Sampling of the cluster of a document d from the conditional distribution $\mathbb{P}(k|k_{\neg d}, \boldsymbol{d})$.

Algorithm 1. GSGDMM

1: Initialization m_k, n_k and n_k^w
2: **for** each of the documents **do**
      Generate $k_d \sim$ Multinomial(1/K)
      Increment m_k, n_k and n_k^w
3: **end for**
4: **for** each of the iterations **do**
      Record k_d
      Decrement m_k, n_k and n_k^w
      Generate $k_d \sim P(k|k_{\neg d}, \boldsymbol{d})$
      Increment m_k, n_k and n_k^w
5: **end for**

$\mathbb{P}(k|k_{\neg d}, \boldsymbol{d})$ is derived from the mentioned generalized Dirichlet multinomial mixture model which confirms two assumptions about the movie group process analogy. The first assumption is that tables having a lot of students will get more students and the second one is that students in the same table will share the same interests as the number of iterations grows. In that sense, only a portion of the K clusters, which will gather the students having same interests, will remain full.

As shown in the Algorithm 1, the hidden cluster of a document d is estimated using the conditional probability given the parameters of the Dirichlet. Indeed, the first one which is a generalized Dirichlet having the parameters α and δ will approximate the parameter θ of the multinomial that will give the distribution of the documents. The second one which is a simple Dirichlet with parameter β will give the distribution of the topics.

It is derived from the joint probability which can be written for the document $\boldsymbol{d}$ and the cluster k as:

$$\mathbb{P}(\boldsymbol{d}, k|\alpha, \delta, \beta) = \mathbb{P}(\boldsymbol{d}|k, \beta)\mathbb{P}(k|\alpha, \delta) \tag{3}$$

We have:

$$\mathbb{P}(\boldsymbol{d}|k, \beta) = \int \mathbb{P}(\boldsymbol{d}|k, \phi)\mathbb{P}(\phi|\beta)d\phi \tag{4}$$

$\mathbb{P}(\boldsymbol{d}|k, \phi)$ is a multinomial distribution given by [24]:

$$\mathbb{P}(\boldsymbol{d}|k, \phi) = \prod_{k=1}^{K} \prod_{w=1}^{V} \phi_{k,w}^{n_k^{(w)}} \tag{5}$$

and $\mathbb{P}(\phi|\beta)$ is a Dirichlet given by [24]:

$$\mathbb{P}(\phi|\beta) = \frac{\Gamma(\sum_{k=1}^{K} \beta_k)}{\prod_{k=1}^{K} \Gamma(\beta_k)} \prod_{w=1}^{V} \phi_{k,w}^{\beta_k - 1} \tag{6}$$

From (5) and (6) we have:

$$
\begin{aligned}
\mathbb{P}(d|k,\beta) &= \int \prod_{k=1}^{K} \prod_{w=1}^{V} \phi_{k,w}^{n_k^{(w)}} \frac{\Gamma(\sum_{k=1}^{K} \beta_k)}{\prod_{k=1}^{K} \Gamma(\beta_k)} \prod_{w=1}^{V} \phi_{k,w}^{\beta_k-1} d\phi_k \\
&= \frac{\Gamma(\sum_{k=1}^{K} \beta_k)}{\prod_{k=1}^{K} \Gamma(\beta_k)} \int \prod_{k=1}^{K} \prod_{w=1}^{V} \phi_{k,w}^{n_k^{(w)}} \phi_{k,w}^{\beta_k-1} d\phi_k \\
&= \frac{\Gamma(\sum_{k=1}^{K} \beta_k)}{\prod_{k=1}^{K} \Gamma(\beta_k)} \int \prod_{k=1}^{K} \prod_{w=1}^{V} \phi_{k,w}^{n_k^{(w)}+\beta_k-1} d\phi_k
\end{aligned} \tag{7}
$$

We have $\Delta(\beta)$ is the dirichlet integral of the first kind for the summation function given by:

$$
\Delta(\beta) = \frac{\Gamma(\sum_{k=1}^{K} \beta_k)}{\prod_{k=1}^{K} \Gamma(\beta_k)} \tag{8}
$$

We have integrating over the probability density function equals to 1:

$$
\int \frac{1}{\Delta(\beta_k')} \prod_{k=1}^{K} \prod_{w=1}^{V} \phi_{k,w}^{n_k^{(w)}+\beta-1} d\phi_k = 1 \tag{9}
$$

where $\beta_k' = \beta + n_k^{(w)}$

$$
\implies \int_{\phi \in \Phi} \prod_{k=1}^{K} \prod_{w=1}^{V} \phi_{k,w}^{n_k^{(w)}+\beta-1} d\phi_k = \prod_{k=1}^{K} \Delta(\beta_k') \tag{10}
$$

$$
\implies \prod_{k=1}^{K} \Delta(n_k + \beta) = \int \prod_{k=1}^{K} \prod_{w=1}^{V} \phi_{k,w}^{n_k^{(w)}+\beta_k-1} d\phi_k = \prod_{k=1}^{K} \frac{\Gamma(\sum_{w=1}^{V}(n_k^{(w)}+\beta))}{\prod_{w=1}^{V} \Gamma(n_k^{(w)}+\beta)} \tag{11}
$$

From (8) and (10) we have:

$$
\mathbb{P}(d|k,\beta) = \prod_{k=1}^{K} \frac{\Delta(n_k+\beta)}{\Delta(\beta)} \tag{12}
$$

Now we will follow the same procedure for $\mathbb{P}(z|\alpha,\delta)$ where:

$$
\mathbb{P}(k|\alpha,\delta) = \int \mathbb{P}(k|\theta)\mathbb{P}(\theta|\alpha,\delta)d\theta \tag{13}
$$

We have $\mathbb{P}(k|\theta)$ is a multinomial given by:

$$
\mathbb{P}(k|\theta) = \prod_{k=1}^{K} \theta_k^{m_k} \tag{14}
$$

and $\mathbb{P}(\theta|\alpha,\delta)$ is a generalized Dirichlet distribution given by [23]:

$$\mathbb{P}(\theta|\alpha,\delta) = \prod_{k=1}^{K} \frac{\Gamma(\alpha_k + \delta_k)}{\Gamma(\alpha_k)\Gamma(\delta_k)} \theta_k^{\alpha_k - 1}(1 - \sum_{j=1}^{l} \theta_j)^{\gamma_l} \tag{15}$$

From (14) and (15) we have:

$$\begin{aligned}
\mathbb{P}(k|\alpha,\delta) &= \int \mathbb{P}(k|\theta)\mathbb{P}(\theta|\alpha,\delta)d\theta \\
&= \int \prod_{k=1}^{K} \theta_k^{m_k} \prod_{k=1}^{K} \frac{\Gamma(\alpha_k + \delta_k)}{\Gamma(\alpha_k)\Gamma(\delta_k)} \theta_k^{\alpha_k - 1}(1 - \sum_{j=1}^{l} \theta_j)^{\gamma_l}d\theta \\
&= \prod_{k=1}^{K} \frac{\Gamma(\alpha_k + \delta_k)}{\Gamma(\alpha_k)\Gamma(\delta_k)} \int \prod_{k=1}^{K} \theta_k^{\alpha_k - 1 + m_k}(1 - \sum_{j=1}^{l} \theta_j)^{\gamma_l}d\theta \tag{16}
\end{aligned}$$

For the case of generalized Dirichlet we have:

$$\Delta(\alpha,\delta) = \prod_{k=1}^{K} \frac{\Gamma(\alpha_k)\Gamma(\delta_k)}{\Gamma(\alpha_k + \delta_k)} \tag{17}$$

We have:

$$\int \frac{1}{\Delta(\alpha',\delta')} \prod_{k=1}^{K} \theta_k^{\alpha - 1 + m_k}(1 - \sum_{j=1}^{l} \theta_j)^{\gamma_l'}d\theta = 1 \tag{18}$$

where $\alpha' = \alpha + m_k$ and $\delta' = \delta + \sum_{l=k+1}^{K} m_l$

$$\implies \int \prod_{k=1}^{K} \theta_k^{\alpha - 1 + m_k}(1 - \sum_{j=1}^{l} \theta_j)^{\gamma_l'}d\theta = \Delta(\alpha',\delta') = \prod_{k=1}^{K} \frac{\Gamma(\alpha + m_k)\Gamma(\delta + \sum_{l=k+1}^{K} m_l)}{\Gamma(\alpha + m_k + \delta + \sum_{l=k+1}^{K} m_l)} \tag{19}$$

$$\Delta(\alpha + m, \delta + \sum_{l=k+1}^{K} m_l) = \int \prod_{k=1}^{K} \theta_k^{\alpha - 1 + m_k}(1 - \sum_{j=1}^{l} \theta_j)^{\gamma_l'}d\theta = \prod_{k=1}^{K} \frac{\Gamma(m_k + \alpha)\Gamma(\delta + \sum_{l=k+1}^{K} m_l)}{\Gamma(m_k + \alpha + \delta + \sum_{l=k+1}^{K} m_l)} \tag{20}$$

From (17) and (19) we have:

$$\mathbb{P}(d|k,\alpha,\delta) = \frac{\Delta(m + \alpha, \delta + \sum_{l=k+1}^{K} m_l)}{\Delta(\alpha,\delta)} \tag{21}$$

The conditional probability that will give us the hidden cluster will be derived as follows:

$$\mathbb{P}(z_d = k | k_{\neg d}, \boldsymbol{d}) \propto \frac{\mathbb{P}(\boldsymbol{d}, k | \alpha, \beta, \delta)}{\mathbb{P}(\boldsymbol{d}_{\neg d}, k_{\neg d} | \alpha, \beta, \delta)}$$

$$\propto \frac{\mathbb{P}(\boldsymbol{d} | k, \beta) \mathbb{P}(k | \alpha, \delta)}{\mathbb{P}(\boldsymbol{d}_{\neg d} | k_{\neg d}, \beta) \mathbb{P}(k_{\neg d} | \alpha, \delta)} \tag{22}$$

$$\propto \frac{\frac{\Delta(\boldsymbol{m} + \alpha, \delta + \sum_{l=k+1}^{K} m_l)}{\Delta(\alpha, \delta)} \quad \frac{\Delta(\boldsymbol{n}_k + \beta)}{\Delta(\beta)}}{\frac{\Delta(\boldsymbol{m}_{\neg d} + \alpha, \delta + \sum_{l=k+1}^{K} m_{l,\neg d})}{\Delta(\alpha, \delta)} \quad \frac{\Delta(\boldsymbol{n}_{k,\neg d} + \beta)}{\Delta(\beta)}}$$

$$\mathbb{P}(z_d = k | k_{\neg d}, \boldsymbol{d}) \propto \frac{\Delta(\boldsymbol{m} + \alpha, \delta + \sum_{l=k+1}^{K} m_l)}{\Delta(\boldsymbol{m}_{\neg d} + \alpha, \delta + \sum_{l=k+1}^{K} m_{l,\neg d})} \frac{\Delta(\boldsymbol{n}_k + \beta)}{\Delta(\boldsymbol{n}_{k,\neg d} + \beta)}, \tag{23}$$

where $\boldsymbol{n}_k = \{n_k^{(w)}\}_{w=1}^{V}$

To elaborate on this conditional probability, we will rely on three major properties:

1. The property of the Gamma function: $\frac{\Gamma(x+m)}{\Gamma(x)} = \prod_{i=1}^{m}(x + i - 1)$
2. The proposition that: $m_k = m_{k,\neg d} + 1$
3. The assumption that each word can appear at most once in each document

This results as follows:

$$\mathbb{P}(z_d = k | k_{\neg d}, d)$$

$$\propto \frac{\Gamma(\alpha + m_k)\Gamma(\delta + \sum_{l=k+1}^{K} m_{l,\neg d} + \alpha + m_{k,\neg d})\Gamma(\delta + \sum_{l=k+1}^{K} m_l)}{\Gamma(\alpha + \delta + \sum_{l=k+1}^{K} m_l + m_k)\Gamma(m_{k,\neg d} + \alpha)\Gamma(\delta + \sum_{l=k+1}^{K} m_{l,\neg d})} \frac{\prod_{w=1}^{V} \Gamma(n_k^{(w)} + \beta)\Gamma(n_{k,\neg d} + V\beta)}{\prod_{w=1}^{V} \Gamma(n_{k,\neg d}^{(w)} + \beta)\Gamma(n_k + V\beta)}$$

$$\propto \frac{\Gamma(\alpha + m_{k,\neg d} + 1)\Gamma(\delta + \sum_{l=k+1}^{K} m_{l,\neg d} + \alpha + m_{k,\neg d})\Gamma(\delta + \sum_{l=k+1}^{K} m_{l,\neg d} + K - k)}{\Gamma(\alpha + \delta + \sum_{l=k+1}^{K} m_{l,\neg d} + m_{k,\neg d} + 1 + K - k)\Gamma(m_{k,\neg d} + \alpha)\Gamma(\delta + \sum_{l=k+1}^{K} m_{l,\neg d})} \frac{\frac{\prod_{w=1}^{V} \Gamma(n_k^{(w)} + \beta)}{\prod_{w=1}^{V} \Gamma(n_{k,\neg d}^{(w)} + \beta)}}{\frac{\Gamma(n_k + V\beta)}{\Gamma(n_{k,\neg d} + V\beta)}}$$

$$\propto \frac{(m_{k,\neg d} + \alpha) \prod_{i=1}^{K-k} (\delta + \sum_{l=k+1}^{K} m_{l,\neg d} + i - 1)}{\prod_{i=1}^{K-k+1} (m_{k,\neg d} + \alpha + \delta + \sum_{l=k+1}^{K} m_{k,\neg d} + i - 1)} \frac{\prod_{w=1}^{V} (n_{k,\neg d}^{(w)} + \beta)}{\prod_{i=1}^{N_d} (n_{k,\neg d} + V\beta + i - 1)} \tag{24}$$

$$\mathbb{P}(z_d = k | k_{\neg d}, d) \propto \frac{(m_{k,\neg d} + \alpha) \prod_{i=1}^{K-k} (\delta + \sum_{l=k+1}^{K} m_{l,\neg d} + i - 1)}{\prod_{i=1}^{K-k+1} (m_{k,\neg d} + \alpha + \delta + \sum_{l=k+1}^{K} m_{k,\neg d} + i - 1)} \frac{\prod_{w=1}^{V} (n_{k,\neg d}^{(w)} + \beta)}{\prod_{i=1}^{N_d} (n_{k,\neg d} + V\beta + i - 1)} \tag{25}$$

where α, δ are the two parameters of the generalized Dirichlet, β is the parameter of the Dirichlet, V size of vocabulary, $m_{k,\neg d}$ the number of documents in cluster k except for the document d, $n_{k,\neg d}^{(w)}$ number of occurence of word w in the cluster k without considering the document d and $n_{k,\neg d}$ number of words in cluster k without considering the cluster of document d.

From Eq. 25, we can see that the parameters α and δ determine the prior probability of a student choosing a table while the parameter β regulates the factor of sharing the interests in the same table.

4   Experimental Results

4.1   Short-Text Datasets and Preprocessing

The Google News dataset was extracted from the Google News website of November, 27, 2013 where the titles and snippets of 11,109 articles were collected and associated to one of the 152 clusters. This dataset was previously used in [25]. The validity of the dataset was examined manually and was divided to different sets. Our work will focus on the SnippetSet which consists of short texts containing the main information from the articles and on the TitleSnippetSet which contains both the titles and snippets of the short texts.

The data preprocessing of the texts included lowercasing all the words, removing non-latin characters and stop words, using the WordNet Lemmatizer of NLTK to apply the stemming, keeping only sentences ranging between 2 and 15 words and removing words which frequency is less than 2.

4.2   Evaluation Metrics Used

To assess the effectiveness of our contribution to cluster short texts, we used the same metrics as in [19]: Homogeneity (H), Completeness (C), Adjusted Rand Index (ARI), Normalized Mutual Information (NMI) and Adjusted Mutual Information (AMI) [26]. Homogeneity and Completeness are two metrics that give a comparison between the ground truth and the inferred information. The Homogeneity is a cluster-wise metric where it insights if each cluster contains only observations belonging to the same ground truth. Completeness is a data-wise metric where it informs whether all the data points from the same ground truth cluster were assigned to the same cluster. The Normalized Mutual Information (NMI), which gives the same result as the V-measure, is defined as the harmonic mean between the completeness and the homogeneity [27]. The Adjusted Rand Index (ARI) measures the similarity between two data clusterings [28]. The Adjusted Mutual Information (AMI) quantifies the amount of information obtained on one random variable through observing another random variable.

4.3   Comparison of Gibbs Sampling Algorithms

In this subsection, we will show the performance of our approach compared to the GSDMM approach. As given in [19], we set the initial number of clusters to 500, the number of iterations to 30, $\alpha = 0.1$, $\delta = 0.1$ and $\beta = 0.1$ for the working datasets. Figure 1 shows that our approach gives better results than the GSDMM approach. We can see that the GSGDMM approach improved the NMI, H, ARI and AMI metrics while having the completeness quite the same for the SSet dataset. From Table 1, we can see the dataset TSSet for which all the mentioned metrics were increased except for the completeness metric which value slightly decreased. We can also see that both of the GSDMM and GSGDMM models perform better on longer texts. This can open room to many improvements as of giving a better representation to the short texts making it longer through different techniques.

Table 1. Performance of the GSDMM and GSGDMM approaches on the SSet and TitleSnippetSet datasets.

		GSDMM	GSGDMM
TSSet	NMI	0.928	0.933
	H	0.911	0.925
	C	0.945	0.941
	ARI	0.789	0.832
	AMI	0.897	0.913

4.4   Influence of K

In this part, we assess the influence of the initial number of clusters K on the performance of the GSGDMM model. For that, we set $\alpha = 0.1$, $\beta = 0.1$ and the number of iterations to 30. Figure 2 displays the performance of the TitleSnippetSet for different values of K. We can see that with a small number of clusters, it gets easy for the model to assign similar documents to the same cluster which gives a very high completeness. But, this same fact gives a low value of homogeneity as it gets hard for the model to separate between the different documents. As we increase the value of K, we start to reach a certain equilibrium between the value of the homogeneity and the value of the completeness. The latter will start decreasing while the former will increase obviously. As the harmonic mean between the completeness and homogeneity, we can rely on the value of NMI to give us the best number of clusters to start with. The highest value for NMI is given for a value of K equal to 400.

4.5   Influence of the Number of Iterations

In this subsection, we analyze the effect of the number of iterations on the number of clusters found on the two datasets. We set the number of initial clusters K to 400, $\alpha = 0.1$, $\beta = 0.1$ and $\delta = 0.1$. From Fig. 3, we can see that the number of clusters found for both datasets drops quickly from 400 to 182 after only 5 iterations. This observation affirms the initial concept of MGP where the most popular tables will get more popular and the less popular ones will get empty quickly. This is why we see the number of clusters found dropping. We can also see that the final number of clusters found reached is a bit above the actual number of clusters of the Google News dataset. From [19], we can see that the number of clusters found by the GSDMM for the TSSet is very near the actual number of clusters for Google News reaching 161 clusters while the one for the SSet went below reaching 148. In that aspect, GSDMM may seem to be performing better but GSGDMM has a better clustering quality since its homogeneity and completeness are higher. Also, it is predicted that going further 30 iterations will improve the number of clusters found by GSGDMM.

4.6   Performance Given the Parameter δ

In this part, we try to find which value of delta can give us the best results. For that, we set K = 300, $\alpha = 0.1$, $\beta = 0.1$ for the TSSet dataset. We set the number of iterations to 10 and do computations for different values of delta ranging from 0.01 to 0.4. The performance is tracked through the NMI metric as it gives a good idea on how well the model is performing. From Fig. 4, we can see that the highest value for NMI is reached for $\delta = 0.2$.

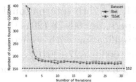

Fig. 1. Performance of the Approaches on the SnippetSet

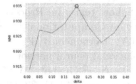

Fig. 2. Performance of GSGDMM with different numbers of K on the TitleSnippetSet

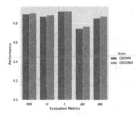

Fig. 3. Number of clusters found by GSGDMM for different number of iterations

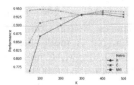

Fig. 4. NMI for different values of delta for TSSet dataset

5   Conclusion

In this paper, we introduced a new approach for short text clustering. We also introduced the background behind the Latent Dirichlet Allocation and presented our approach that introduced the Generalized Dirichlet for short text clustering. Our model improved further the GSDMM approach for all the datasets used. We studied the influence of the number of iterations on the performance of the approach and found the more computations we do the better the results get. Further work should be done using other distributions. As for GSGDMM, it coped very well with the limitations presented by short texts.

References

1. Blei, D.M., Ng, A.Y., Jordan, M.I.: Latent Dirichlet allocation. J. Mach. Learn. Res. **3**(Jan), 993–1022 (2003)
2. Hu, D.J.: Latent Dirichlet allocation for text, images, and music. University of California, San Diego (2009). Accessed 26 Apr 2013
3. Aggarwal, C.C., Zhai, C.: A survey of text clustering algorithms. In: Aggarwal, C., Zhai, C. (eds.) Mining Text Data, pp. 77–128. Springer, Boston (2012). https://doi.org/10.1007/978-1-4614-3223-4_4
4. Frunza, O., Inkpen, D., Tran, T.: A machine learning approach for identifying disease-treatment relations in short texts. IEEE Trans. Knowl. Data Eng. **23**(6), 801–814 (2010)
5. Alsmadi, I., Hoon, G.K.: Term weighting scheme for short-text classification: Twitter corpuses. Neural Comput. Appl. **31**(8), 3819–3831 (2019)
6. Bouguila, N.: A model-based approach for discrete data clustering and feature weighting using MAP and stochastic complexity. IEEE Trans. Knowl. Data Eng. **21**(12), 1649–1664 (2009)
7. Zeng, J., Li, J., Song, Y., Gao, C., Lyu, M.R., King, I.: Topic memory networks for short text classification. arXiv preprint arXiv:1809.03664 (2018)
8. Jin, O., Liu, N.N., Zhao, K., Yu, Y., Yang, Q.: Transferring topical knowledge from auxiliary long texts for short text clustering. In: Proceedings of the 20th ACM International Conference on Information and Knowledge Management, pp. 775–784 (2011)
9. Dos Santos, C., Gatti, M.: Deep convolutional neural networks for sentiment analysis of short texts. In: Proceedings of COLING 2014, the 25th International Conference on Computational Linguistics: Technical Papers, pp. 69–78 (2014)
10. Lee, J.Y., Dernoncourt, F.: Sequential short-text classification with recurrent and convolutional neural networks. arXiv preprint arXiv:1603.03827 (2016)
11. Bouguila, N., ElGuebaly, W.: Discrete data clustering using finite mixture models. Pattern Recogn. **42**(1), 33–42 (2009)
12. Bouguila, N., Amayri, O.: A discrete mixture-based kernel for SVMs: application to spam and image categorization. Inf. Process. Manag. **45**(6), 631–642 (2009)
13. Bouguila, N., Ziou, D.: MML-based approach for finite Dirichlet mixture estimation and selection. In: Perner, P., Imiya, A. (eds.) MLDM 2005. LNCS (LNAI), vol. 3587, pp. 42–51. Springer, Heidelberg (2005). https://doi.org/10.1007/11510888_5
14. Bouguila, N., ElGuebaly, W.: A generative model for spatial color image databases categorization. In: Proceedings of the IEEE International Conference on Acoustics, Speech, and Signal Processing, ICASSP 2008, Caesars Palace, Las Vegas, Nevada, USA, 30 March–4 April 2008, pp. 821–824. IEEE (2008)
15. Bakhtiari, A.S., Bouguila, N.: A variational Bayes model for count data learning and classification. Eng. Appl. Artif. Intell. **35**, 176–186 (2014)
16. Bouguila, N., Ghimire, M.N.: Discrete visual features modeling via leave-one-out likelihood estimation and applications. J. Vis. Commun. Image Represent. **21**(7), 613–626 (2010)
17. Mehdi, M., Bouguila, N., Bentahar, J.: Trustworthy web service selection using probabilistic models. In: Goble, C.A., Chen, P.P., Zhang, J. (eds.) 2012 IEEE 19th International Conference on Web Services, Honolulu, HI, USA, 24–29 June 2012, pp. 17–24. IEEE Computer Society (2012)

18. Zamzami, N., Bouguila, N.: Text modeling using multinomial scaled Dirichlet distributions. In: Mouhoub, M., Sadaoui, S., Ait Mohamed, O., Ali, M. (eds.) IEA/AIE 2018. LNCS (LNAI), vol. 10868, pp. 69–80. Springer, Cham (2018). https://doi.org/10.1007/978-3-319-92058-0_7
19. Yin, J., Wang, J.: A Dirichlet multinomial mixture model-based approach for short text clustering. In: Proceedings of the 20th ACM SIGKDD International Conference on Knowledge Discovery and Data Mining, pp. 233–242 (2014)
20. Carlo, C.M.: Markov chain Monte Carlo and Gibbs sampling. Lecture notes for EEB, 581 (2004)
21. Yildirim, I.: Bayesian inference: Gibbs sampling. Technical Note, University of Rochester (2012)
22. Blei, D.: Cos 597c: Bayesian nonparametrics. Lecture Notes in Priceton University (2007). http://www.cs.princeton.edu/courses/archive/fall07/cos597C/scribe/20070921.pdf
23. Bouguila, N.: Clustering of count data using generalized Dirichlet multinomial distributions. IEEE Trans. Knowl. Data Eng. **20**(4), 462–474 (2008)
24. Heinrich, G.: Parameter estimation for text analysis. Technical report (2005)
25. Banerjee, S., Ramanathan, K., Gupta, A.: Clustering short texts using Wikipedia. In: Proceedings of the 30th Annual International ACM SIGIR Conference on Research and Development in Information Retrieval, pp. 787–788 (2007)
26. Han, J., Pei, J., Kamber, M.: Data Mining: Concepts and Techniques. Elsevier (2011)
27. Becker, H.: Identification and characterization of events in social media. Ph.D. thesis, Columbia University (2011)
28. Zhang, S., Wong, H.-S.: ARImp: a generalized adjusted rand index for cluster ensembles. In: 2010 20th International Conference on Pattern Recognition, pp. 778–781. IEEE (2010)

Comparative Study of Machine Learning Algorithms for Performant Text Analysis in a Real World System

Faraz Islam$^{(\boxtimes)}$ and Doina Logofătu$^{(\boxtimes)}$

Department of Computer Science and Engineering,
Frankfurt University of Applied Sciences, 60318 Frankfurt a.M., Germany
fislam@stud.fra-uas.de, logofatu@fb2.fra-uas.de

Abstract. This work illustrates how the text analysis is done on XYZ GmbH's (company's name was changed due to privacy) ticketing engine "ManageEngine ServiceDesk Plus" (MSP) database using modern technology offered by Azure ML. Here, we will use Azure Machine Learning Studio to implemented Text Analysis techniques and machine learning algorithms on the data set obtained from MSP. This research paper will guide us to the process of data extraction, data processing, creation of a word cloud and keyword extraction. We will compare modern machine Learning algorithm like neural network, averaged perceptron and boosted decision tree available in Azure ML to train and score the model and do a prediction on a ticket's probability of being assigned to a correct department.

Keywords: Text analysis · ManageEngine ServiceDesk Plus · Word cloud · Keyword extraction · Azure Machine Learning Studio · Neural network · Averaged perceptron · Boosted decision tree

1 Introduction

Today, with the rapid growth of population, there is a decent amount of shift towards the electronic media. Whether it is study material, journals, books, novels, newspaper etc. everything is uploaded online and thus creating a massive network of electronic media. The increasing popularity and easy means of literature created a lot more pressure on companies and media firms to opt for unstructured and semi structured information.

The popularity of the Electronic media made a strong foundation base for the use of Machine Learning Techniques in extracting useful knowledge from a heap of unstructured text [1]. The issue starts when a user wants to search something from the heap of unstructured data and he didn't find out a way to extract useful information from it. This problem could be solved if we have a way to find out what information is useful to end user and could be of great value to him.

T.-P. Hong et al. (Eds.): ACIIDS 2021, CCIS 1371, pp. 162–173, 2021.
https://doi.org/10.1007/978-981-16-1685-3_14

Solution to the problem can be answered by Text Mining and Analysis which engages several fields like Information Retrieval (IR), Machine Learning (ML), Natural Language Processing (NLP) and Statistics [2]. One of the emerging research areas in Text Analysis is the document classification and information retrieval. The common classification applications are email categorization, ticketing system filtering, spam filtering and mail routing etc. The solutions to most of the applications are solved by using machine learning algorithms [3].

1.1   Motivation

The use of modern equipment, machines and software to translate professional or technical business documents moves the problem from creation of the document to the correction and evaluation, but it does not solve the problem completely [4]. Therefore, the calculation and evaluation of the professionally translated documentation is a vital step for companies to reduce time and cost as well as to create an efficient way of translating critical and important documents. Additionally, this provides and ensures a certain level of eminence and it is also an efficient and effective way of handling a technical document [5].

Based on the background mentioned earlier, the aim and purpose of this research is to check whether a ticket is in English language or not, remove the other language tickets, process the data, perform text analysis steps and finally, implement a machine learning process that can predict the probability of that ticket being assigned to a correct category or not.

2   Problem Statement

As our company is progressing, developing and expanding each year and our client list is increasing day by day, we are getting a bulk amount of tickets and there is a continues pressure on the First Level Support i.e. Service Desk Unit (SDU) to resolve or forward the ticket to the concerned department under agreed Service Level Agreement (SLA) time limit. Even after so many efforts, there is a great risk of a ticket being violated and not fulfilling the agreed SLAs. Many times, the SDU team is not able to find the right department for the concerned ticket, as they didn't understand the client request, or they are unfamiliar with the information mentioned in the ticket [6]. For example, if an SDU technician is familiar with SQL Server and didn't know anything about Oracle, he will easily forward the tickets related to SQL Server ticket to Database Engineer (DBE) Team, but he will be confused with the tickets related to Oracle. Though both the ticket should be forwarded to DBE team, he could end up with forwarding the request to a wrong department. This could result in waste of time, which is critical for High SLA tickets. The problem faced by "XYZ GmbH" leaves us with the following questions:

- How we can detect the MSP ticket language?
- Do we have enough categories and subcategories for a ticket on MSP?
- How to predict the probability of a ticket being assigned to a correct category?

To answer the above-mentioned problem, we need to find the best Text Analysis approach and an efficient Machine Learning Algorithm by which we can identify the text language, process the data into a high quality data and develop a technique or a way to forecast the probability of a MSP ticket being allocated to a correct and exact category. We also need to check how we can increase our categories and subcategories which can help employees and clients in putting the ticket to a more specific slot than generalizing the raised request.

3   Related Work

A similar type of work "Sentiment Analysis on Twitter Data Using Machine Learning Algorithms in Python" [7] was done by S. Siddharth, R. Darsini and M. Sujithra. The authors showed how they have use python libraries which are quite like the libraries of R in cleansing the raw data they have collected from twitter into MongoDB. They changed their data to lowercase and did tokenization by removal of hash tags, numbers and special characters. They have explained why removal of stops words was important as it is cleaning a lot of garbage values from the raw data.

In [8], the authors studied the performance of over 100 variants of 5 filter feature selection methods using four classifier algorithms (Naive Bayes, Rocchio, K-Nearest Neighbor and Support Vector Machines) and two benchmark collections (Reuters 21578 and part of RCV1). Their study shows the use of filter methods and combining them with Document Frequency (DF) and Information Gain (IG) for eliminating rare words.

In [9], the authors tried to compared Support Vector Machine (SVM) to K-Nearest Neighbors (kNN) and naïve Bayes on binary classification tasks. Their main focus was to compare optimized versions of the above-mentioned algorithms. Their results show that all the classifiers achieved similar performance on most of the problems. But the main result was surprising as the Support Vector Machine, despite providing an overall good accuracy and performance, it was not considered to be perfect and a strong winner.

In [10], the author explained how the noisy information sources can be dealt by data mining. They described that the noisy data could decrease the efficiency and performance of any classification algorithms. Their study was focused on the performance of different classification algorithms and the effect of feature selection algorithms on Logistic Regression Classifier.

4   Data Processing

The data stored in MSP is a time series data and we have extracted relevant data from it by using joins on different tables and merging them all together. Our main data containing 33059 observation with 6 column variables. The Table 1, shows that the columns have different data type, data characters and different values.

Table 1. The table shows the data type, data characteristics, data size and the description of the different columns present in XYZ GmbH's data.

Column name	Data type	Char.	Data size	Description
CreatedTime	datetime2	3	7 Bytes	Date and Time of ticket creation
WorkOrderID	bigint		8 Bytes	A unique number assigned to ticket
Category	nvarchar	100	200 Bytes	Categories for the tickets
SubCategory	nvarchar	100	200 Bytes	Sub level of categories
Subject	nvarchar	250	500 Bytes	Brief description about the problem
Description	nvarchar	10^8	2 GBytes	Reason behind creating a ticket

In this section, we will try to refine our data by passing it through different stages of Data Processing like Data Collection, Data Preprocessing, Word Cloud generation, Keyword Extraction and Data Mapping and Frequency.

4.1 Data Collection

We will connect to the MSP db using R's odbc library dedicated for SQL Server.

```
cn <-odbcDriverConnect('driver={SQL Server};
 ↪  server=localhost\\SQLEXPRESS; Database=MSPTab')
data <- sqlQuery(cn, "select * from [dbo].[tbl_query]")
```

Since our data is split across different tables and we have a common key attribute "ID" among all tables, we will use this column to merge different tables into one table and this table will be our initial data frame.

4.2 Data Preprocessing

When the data is extracted, it then enters the data processing stage. Data processing, often referred to as "pre-processing" is the step by which the raw data is cleaned and prepared for the upcoming stages of data and text mining.

We will select only those rows which have English data. For detecting language, we have used "cld2" library in R as shown by the below mentioned code:

```
data$LanguageT <- detect_language(as.vector(data$Title))
```

4.3 Word Cloud

It is a Text Analysis technique that allow us to expose frequent occurring words in a data containing texts and help us in creating a word cloud, which is basically a visual illustration of text data [11].

The packages "text mining (tm)" and the "word cloud generator (wordcloud)" can be used to get some help in analyzing the data containing text and in visualizing the keywords present in the data set as a word cloud [12] and [13].

The word cloud (shown in Fig. 1) will implement the steps like stemming, removal of stopwords etc. and will produce a data frame with words and their frequency count. Words with low frequency (few occurrence) have no meaning and will be removed from our data frame.

Fig. 1. Top 150 most occurring words of our data frame shown as a Wordcloud.

4.4 Keyword Extraction

In this process all the keywords from "Title" and "Description" column will be extracted and saved as a new column "TitleKeywords" and "DescripKeywords" of the data frame. These words will later be used to train the model and search for similar words present in new MSP tickets. The below mentioned code shows how the keywords were extracted.

```
dt %>% unnest_tokens(word, DESCRIPTION) %>%
  group_by(ID, CATEGORY, SUBCATEGORY) %>%
  summarise(word = f(word)) %>%  ungroup()%>%  data.frame -> dt
```

4.5 Data Mapping and Frequency

In this step, we will check if the words from the column "Category" and "Subcategory" are present in the "TitleKeywords" or "DescriptionKeywords" column. If those words are present in the later columns then we will mark them as "True" else it will be marked as "False" under the column "MappedValue".

We will also check if those words are present then what is the frequency (Occurrence/how many times) of those words. The Table 2, shows one row of our final data frame which will be used for training and scoring the model.

Table 2. The table shows 7 columns and a single row out of 33059 of final data frame that we got after the Data Processing stage.

ID	Category	Sub. Cat	T. Key	D. Key	Freq.	Mapped
2029	Software	Application	Deleted	Details	0	False

5   Two-Class Neural Network

Two-Class Neural networks are used in difficult computer vision tasks, such as letter or digit recognition, image recognition, pattern recognition and document classification. A neural network is a group of interrelated layers where an acyclic graph containing the weighted edges and the nodes are used to connect the inputs layer (First Layer) to an output layer [14].

Recent researches done on deep neural network (DNN) has shown that a DNN with many layers can be very efficient, effective and can perform well with complex tasks such as image, text or speech recognition [15]. The path of the neural network graph starts from the input layer, goes through the hidden layer and then to the output layer. The nodes in one layer is connected to another layer by a weighted edge [16].

5.1   Training the Two-Class Neural Network

In our experiment, we have taken one input node (single parameter), one hidden layer with 100 node, a learning rat of 0.1 to avoid overshooting of local minima and a min-max normalization to rescale the intervals between a range of 0 and 1. We have allowed unknown categorical values to get better prediction for unknown values. The Table 3, show the value of Precision, Accuracy, Recall and F1 score over the Score Bin Range.

Table 3. The experimental results shows that the best accuracy "0.87" is achieved at a fraction above threshold (FAT) of "0.552", best Precision "0.856" at a FAT of "0.502". On a closer look, we can see that the best of both "Accuracy and Precision" could be achieved at a FAT of "0.50".

Score bin	Positive examples	Negative examples	Fraction above threshold	Accuracy	F1 score	Precision	Recall	Negative precision	Negative recall	Cumulative AUC
(0.900,1.000]	2988	503	0.502	0.858	0.858	0.856	0.861	0.861	0.855	0.101
(0.800,0.900]	87	55	0.523	0.863	0.866	0.846	0.886	0.881	0.84	0.115
(0.700,0.800]	23	26	0.53	0.862	0.866	0.841	0.893	0.886	0.832	0.121
(0.600,0.700]	11	7	0.532	0.863	0.867	0.84	0.896	0.889	0.83	0.123
(0.500,0.600]	38	7	0.539	0.867	0.872	0.84	0.907	0.899	0.828	0.125
(0.400,0.500]	18	10	0.543	0.869	0.874	0.839	0.912	0.904	0.825	0.127
(0.300,0.400]	13	9	0.546	0.869	0.875	0.837	0.916	0.907	0.823	0.13
(0.200,0.300]	23	15	0.552	0.87	0.877	0.835	0.922	0.914	0.818	0.134
(0.100,0.200]	20	37	0.56	0.868	0.875	0.828	0.928	0.919	0.808	0.144
(0.000,0.100]	249	2810	1	0.499	0.666	0.499	1	1	0	0.934

6   Two-Class Averaged Perceptron

In this step we will discuss how we can use the "Two-Class Averaged Perceptron algorithm" to create a prediction model based on our data set in Azure Machine

Learning Studio. This classification algorithm requires a tagged data set containing a labelled column because it uses an administered learning method. We can say that the averaged perceptron technique is an initial and very easy form of a neural network. Two-Class Averaged Perceptron can be used with constant training because they are quicker and can handle the cases consecutively [17].

6.1 Training of Two-Class Averaged Perceptron

For Two Class Averaged Perceptron, we have taken one Input layer (Single Parameter) and a maximum number of iterations as 10, which basically allows our algorithm to examine our data set. A learning rate of 1 was chosen which controls the size of each step used in stochastic gradient descent (SGD). No random numbers were chosen as we don't want to check the repetitions across each run. Unknown categorical values were allowed to get better prediction for unknown values. The Table 4, show the value of Precision, Accuracy, Recall and F1 score over the Score Bin.

Table 4. The experimental results shows that the best accuracy "0.86" is achieved at a fraction above threshold (FAT) of "0.554", best Precision "0.852" at a FAT of "0.499" by the Two-Class Average Perceptron Algorithm. On a closer look, we can see that the best of both "Accuracy and Precision" could be achieved at a FAT of "0.50".

Score bin	Positive examples	Negative examples	Fraction above threshold	Accuracy	F1 score	Precision	Recall	Negative precision	Negative recall	Cumulative AUC
(0.900,1.000]	2951	514	0.499	0.851	0.851	0.852	0.85	0.851	0.852	0.103
(0.800,0.900]	87	51	0.518	0.857	0.859	0.843	0.876	0.871	0.838	0.116
(0.700,0.800]	42	29	0.529	0.858	0.862	0.838	0.888	0.881	0.829	0.123
(0.600,0.700]	13	10	0.532	0.859	0.863	0.837	0.891	0.884	0.826	0.126
(0.500,0.600]	10	13	0.535	0.858	0.863	0.834	0.894	0.886	0.823	0.129
(0.400,0.500]	24	19	0.542	0.859	0.865	0.831	0.901	0.892	0.817	0.134
(0.300,0.400]	13	13	0.545	0.859	0.865	0.829	0.905	0.896	0.813	0.137
(0.200,0.300]	31	27	0.554	0.86	0.867	0.824	0.914	0.904	0.806	0.144
(0.100,0.200]	67	72	0.574	0.859	0.869	0.812	0.933	0.922	0.785	0.164
(0.000,0.100]	232	2731	1	0.499	0.666	0.499	1	1	0	0.936

7 Two-Class Boosted Decision Tree

A boosted decision tree is a collective learning method where the errors of the first tree are corrected by the second tree and the errors of the first and the second trees collectively are corrected by the third tree and this process goes further down the line [18]. The predictions done by boosted decision tree are built on forecast done by the complete group of trees all together. To get good performance and results on a vast variations of machine learning tasks, boosted decision tree is the easiest technique when configured properly. The current implementation of boosted decision tree holds everything in memory and therefore, they are more memory intensive learners. Therefore, a very large data set which can be easily handled by linear learners might not be possible for a boosted decision tree model [19].

7.1   Training the Two-Class Boosted Decision Tree

In the starting we will take empty collection of weak learner. Then we will try to get the output of collection of all weak learners. Then we evaluate the gradient for the log loss function for every example. The log-loss (LL) function is used, just like we do in a logistic regression for a binary classification model. Then we use the previously described gradient as the target function to fit a weak learner using the examples. Then we will add that weak learner to a collection with a strength determined by the learning rate. The implementation is grounded on the gradients that we have just calculated using log loss function, the weak learners are the least-squares regression trees. The Table 5, show the value of Precision, Accuracy, Recall and F1 score over the Score Bin Range.

Table 5. The experimental results shows that the best accuracy "0.855" is achieved at a fraction above threshold (FAT) of "0.567", best Precision "0.979" at a FAT of "0.248" by the Two-Class Boosted Decision Tree Algorithm. On a closer look, we can see that the best of both "Accuracy and Precision" could be achieved at a FAT of "0.50".

Score bin	Positive examples	Negative examples	Fraction above threshold	Accuracy	F1 Score	Precision	Recall	Negative precision	Negative Recall	Cumulative AUC
(0.900,1.000]	1687	37	0.248	0.738	0.65	0.979	0.486	0.659	0.989	0.005
(0.800,0.900]	328	86	0.308	0.773	0.719	0.942	0.581	0.698	0.965	0.018
(0.700,0.800]	180	76	0.345	0.788	0.749	0.917	0.633	0.72	0.943	0.031
(0.600,0.700]	828	414	0.523	0.847	0.851	0.831	0.871	0.865	0.824	0.12
(0.500,0.600]	127	94	0.555	0.852	0.86	0.817	0.908	0.897	0.797	0.145
(0.400,0.500]	48	32	0.567	0.855	0.864	0.812	0.922	0.91	0.788	0.153
(0.300,0.400]	45	95	0.587	0.847	0.859	0.795	0.935	0.921	0.76	0.178
(0.200,0.300]	62	150	0.617	0.835	0.852	0.771	0.952	0.938	0.717	0.219
(0.100,0.200]	59	288	0.667	0.802	0.83	0.726	0.969	0.954	0.634	0.299
(0.000,0.100]	106	2207	1	0.499	0.666	0.499	1	1	0	0.928

8   Experimental and Statistical Results

From the Tables 3, 4 and 5, it is clear that we are getting the best precision, accuracy and recall value at approx. 0.50 threshold value. So, by setting threshold to 0.50, we further analyse the results achieved (shown by Table 6) by all the algorithms discussed in this experiment.

Table 6. Comparison of Accuracy, Precision and Recall values at a threshold of 0.5.

Algorithm	True positive	False negative	False positive	True negative	Threshold	Accuracy	F1 score	Precision	Recall	AUC
Two-Class Neural Network	3147	323	598	2881	0.5	0.867	0.872	0.84	0.907	0.934
Two-Class Averaged Perceptron	3103	367	617	2862	0.5	0.858	0.863	0.834	0.894	0.936
Two-Class Boosted Decision Tree	3150	320	707	2772	0.5	0.852	0.86	0.817	0.908	0.928

The Table 7, shows the comparison of the best run-time out of three runs achieved by the algorithms used in this experiment to train and score the model.

Table 7. Comparison of the Run-Time for all the algorithms.

Algorithm	Trained data	Scored data	Run time
Two-Class Neural Network	16214	6949	45 min and 25 s
Two-Class Averaged Perceptron	16214	6949	2 min and 30 s
Two-Class Boosted Decision Tree	16214	6949	7 min and 45 s

8.1   Web Service Deployment and Prediction

Two-Class Averaged Perceptron shows promising results and stand out as a clear winner among the other algorithms discussed in our experiment. We will go ahead with Two-Class Averaged Perceptron Algorithm to deploy our azure web app in order to predict the probability of a ticket being assigned to a category (Fig. 2).

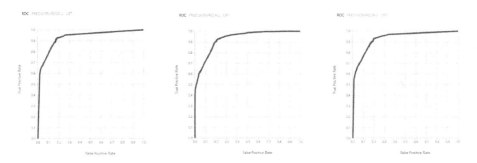

Fig. 2. The figure shows the Area Under Curve (AUC) of Two-Class Neural Network, Two-Class Boosted Decision Tree and Two-Class Averaged Perceptron respectively.

The Table 8 shows the data provided by us to the Web Services. Here we have a "FREQUENCY" of 0, that means there were no words present in the column "TITLEKEYWORDS" or "DESCRIPKEYWORDS" that matched to the "CATEGORY" or "SUBCATEGORY" column.

The Fig. 3, shows the scored label and prediction value at the bottom. It also shows that the data provided to the Web Services have no matching words corresponding to "CATEGORY" or "SUBCATEGORY" column.

Table 8. Table showing the data entered for the web services to predict the probability of a ticket being assigned to a category.

ID	Category	Sub Cat.	Title keywords	Descrip keywords	Frequency
1187	Backup	Backup & Recovery	Management	Access required	0

The prediction result shows that the scored label is "False" and the prediction value is "0.000185822" which is very low and implies that the chances of a ticket being assigned to a correct category is very low based on the data provided to the Web Services.

Fig. 3. The figure at the bottom shows the scored label and the prediction value done on the provided data by the azure web services using the two-class average perceptron.

The Table 9, show the prediction done on different data value provided to the azure Web Services. The table shows the scored label and prediction value of the data provided to the web services. It is clear from the table that the ticket will be assigned to a correct Category if there are matching words present in the "TITLEKEYWORDS" or "DESCRIPKEYWORDS" which corresponds to "CATEGORY" or "SUBCATEGORY" column.

Table 9. Scored label and probability on the data using two-class average perceptron algorithm.

Id	Category	Subcategory	Title keywords	Descrip keywords	Frequency	Scored label	Probability
20330	Softwares	Desktop applications	Printer	Support printer mal- functioned	0	False	2.82516424 476853E−07
20383	Hardware & servers	Local hardware	Replace floor machine	Provide additional information	0	False	1.23329027 701402E−05
39644	Networking & communi- cations	Security	xyz website security	Support network com- munication security	3	True	0.9999999 40395355
59286	It security	Investigations	Noc ids ips alert	Original fault ip	0	False	0.4559161 3650322
38408	Database	Object management	db server	Enter related database	1	True	0.9999980 92651367

9 Conclusion and Future Work

After a thorough and deep inspection it is clear from the Tables 6 and 7, that the neural network performance was the best with an Accuracy of 0.867, Precision of "0.84" and a Recall Rate of "0.907". Though proving the best classification algorithm on the current data set, this algorithm has a long Run-time. The Boosted Decision Tree scored the best Recall Rate of "0.908" but has slightly low Precision Rate "0.817" and Accuracy "0.852" than other algorithms.

The best performant of all the machine learning algorithms used in this experiment came out to be the Two-Class Averaged Perceptron. This classification algorithm was the fastest amongst its competitors and has the second best Precision "0.834" and Accuracy "0.858". Two-Class Averaged Perceptron took only 2 min and 30 s to Train and Score the Data Set.

During the experiment it was found out that the probability of ticket being assigned to a correct category is high when we had a greater "Frequency". This proves that the MSP database needs an updated table containing the new categories and subcategories to allow the employees and clients a vast option to choose from and describe their problem in a more informative manner.

We can do much better in text analysis steps like data processing, cleansing and Keyword extraction by using "Extract N-Gram Features from Text" module and "Feature Hashing" offered by Azure ML Studio, which can help us in data cleansing and extracting keywords from a data set faster and more precisely than by using R as explained by [20] and [21].

Further, we can create a high availability setup for our Azure Web Services. The setup consist of an application gateway with a back end hosting two Azure Web Services. The Azure Web Services are connected to the MSP database hosted on Azure SQL Server via a secure connection string. This whole setup will help us in connecting our Web Services directly to the MSP Database and start predicting the probability of each ticket being assigned to a correct category.

References

1. Baharudin, B., Lee, L.H., Khan, K., Khan, A.: A review of machine learning algorithms for text-documents classification. J. Adv. Inf. Technol. **1**, 4–20 (2010)
2. Kao, A., Poteet, S.: Text mining and natural language processing: introduction for the special issue. ACM SIGKDD Explor. Newsl. - Nat. Lang. Process. Text Min. **7**(1), 1–2 (2005)
3. Nihthiya Althaf, N.A.N., Vijayaraghavan, M.: Development of novel machine learning approach for document classification. Int. J. Comput. Trends Technol. (IJCTT), 62–64 (2017)
4. Luckert, M., Schaefer-Kehnert, M.: Using machine learning methods for evaluating the quality of technical documents. Master thesis (2015)
5. Doherty, S.: The impact of translation technologies on the process and product of translation. Int. J. Commun. 947–969 (2016)
6. Mathenge, J.: Service desk support analyst: roles and responsibilities. https://www.bmc.com/blogs/service-desk-support-analyst/. Accessed 14 Mar 2020
7. Siddharth, S., Darsini, R., Sujithra, M.: Sentiment analysis on twitter data using machine learning algorithms in python. Int. J. Eng. Res. Comput. Sci. Eng. **5**(2), 285–290 (2018)
8. Rogati, M., Yang, Y.: High-performing feature selection for text classification. In: International Conference on Information and Knowledge Management, pp. 659–661 (2002)
9. Colas, F., Brazdil, P.: Comparison of SVM and some older classification algorithms in text classification tasks. In: Bramer, M. (ed.) IFIP AI 2006. IIFIP, vol. 217, pp. 169–178. Springer, Boston, MA (2006). https://doi.org/10.1007/978-0-387-34747-9_18

10. Thota, H., et al.: Performance comparative in classification algorithms using real datasets. J. Comput. Sci. Syst. Biol. **2**, 97–100 (2009)
11. R Studio and Amazon WS: Text Mining using R. https://rstudio-pubs-static.s3. amazonaws.com/539725_9c9396a7f2624b1b8ee12e16fec39362.html. Accessed Mar 2020
12. E. Guides: Text mining and word cloud fundamentals in r: 5 simple steps you should know. https://www.r-bloggers.com/2017/02/text-mining-and-word-cloud-fundamentals-in-r-5-simple-steps-you-should-know/. Accessed Nov 2017
13. STHDA: Text mining and word cloud fundamentals in r: 5 simple steps you should know. http://www.sthda.com/english/wiki/text-mining-and-word-cloud-fundamentals-in-r-5-simple-steps-you-should-know. Accessed 2020
14. Kraipeerapun, P., Fung, C.C., Wong, K.W.: Multiclass classification using neural networks and interval neutrosophic sets. In: International Conference on Computational Intelligence, Man-Machine Systems and Cybernetics - Proceedings, vol. 1, pp. 123–128 (2006)
15. Martens, J., et al.: Two-class neural network. https://github.com/MicrosoftDocs/ azure-reference-other/blob/master/studio-module-reference/two-class-neural-network.md. Accessed 2020
16. Kraipeerapun, P., Fung, C.C., Wong, K.W.: Lithofacies classification from well log data using neural networks, interval neutrosophic sets and quantification of uncertainty. Int. J. Comput. Inf. Eng. **2**, 2941–3937 (2008)
17. Ericson, G., et al.: Two-class averaged perceptron. https://docs.microsoft. com/en-us/azure/machine-learning/studio-module-reference/two-class-averaged-perceptron. Accessed June 2019
18. Veloso, M., Martens, J., Petersen, T., Sharkey, K.: Two-class boosted decision tree Module overview. https://github.com/MicrosoftDocs/azure-reference-other/blob/ master/studio-module-reference/two-class-boosted-decision-tree.md#module-overview. Accessed Sept 2019
19. Lu, P., Martens, J., Sharkey, K., Takaki, J.: Two-class boosted decision tree. https://docs.microsoft.com/en-us/azure/machine-learning/studio-module-reference/two-class-boosted-decision-tree. Accessed June 2019
20. Li, B., Baccam, N., Zhang, X., Jackson, S., Wells, J.: Extract n-gram features from text module reference. https://github.com/MicrosoftDocs/azure-docs/blob/ master/articles/machine-learning/algorithm-module-reference/extract-n-gram-features-from-text.md. Accessed Jan 2019
21. Parente, J., Lu, P.: Extract key phrases from text. https://docs.microsoft.com/en-us/azure/machine-learning/studio-module-reference/extract-key-phrases-from-text. Accessed Aug 2019

CONTOUR: Penalty and Spotlight Mask for Abstractive Summarization

Trang-Phuong N. Nguyen[✉] [iD] and Nhi-Thao Tran [iD]

Vietnam National University, University of Science, Ho Chi Minh City, Vietnam
tttnhi@mso.hcmus.edu.vn
http://en.hcmus.edu.vn

Abstract. The act of transferring accurate information presented as proper nouns or exclusive phrases from the input document to the output summary is the requirement for *Abstractive Summarization* task. To address this problem, we propose CONTOUR to emphasize the most suitable word in the original document that contains the crucial information in each predict step. CONTOUR contains two independent parts: *Penalty* and *Spotlight*. *Penalty* helps to penalize inapplicable words in both training and inference time. *Spotlight* is to increase the potential of important related words. We examined CONTOUR on multiple types of datasets and languages, which are large-scale *(CNN/DailyMail)* for English, medium-scale *(VNTC-Abs)* for Vietnamese, and small-scale *(Livedoor News Corpus)* for Japanese. CONTOUR not only significantly outperforms baselines by all three ROUGE points but also accommodates different datasets.

Keywords: Abstractive summarization · Emphasize words · CONTOUR · Penalty · Spotlight

1 Introduction

Text Summarization task generates the complete sentences that can be understood as a shorter version of the original text while still able to convey the full meaning. Depends on the aim of the output summary, there are two approaches: *Extractive* and *Abstractive*. *Extractive summarizing* is the act of combining verbatim words and phrases from some specific sections from the input. Furthermore, *Abstractive summarizing*, which is not limited to words that appeared in the source text, takes valuable information then interprets it in a new way. While both approaches need to guarantee the readability and grammatical sense, *Abstractive method* is closer to human-like actions [12] and more advanced [12]. This article only focuses on *Abstractive Summarization*.

Introduced in 2014 by Sutskever, Sequence-to-Sequence [13] becomes a sustainable framework for several tasks including Abstractive Summarization. It can be described as an encoder-decoder architecture, where the model would process the whole input before making any prediction. Afterwards, the framework is used by Rush et al. [11], Nallapati et al. [9], and Chopra et al. [3] with *Recurrent Neural Networks* (RNN) association. While being a crucial improvement to

© Springer Nature Singapore Pte Ltd. 2021
T.-P. Hong et al. (Eds.): ACIIDS 2021, CCIS 1371, pp. 174–187, 2021.
https://doi.org/10.1007/978-981-16-1685-3_15

the Text Summarization task, the traditional Sequence-to-Sequence with RNN is restricted because of the *out-of-vocabulary* (OOV) [12] problem.

Pointer Generator Network [12] raises a good approach to resolve the OOV by considering copying one word from the input [14] and selecting one word from the fixed vocabulary set. In this way, the particular words that do not appear in the normal dictionary would not be ignored. Therefore, the problem with OOV is partially solved. On the other hand, Cho et al. proposed SELECTOR [2] with the idea of creating fixed salient areas to be focused along with the input via a trainable binary mask. However, the methods still have an issue of optimizing the word's level of significance, which will be discussed further in Sect. 3.1.

Recent resources for text summarization on languages other than English are in short supply. For instance, languages estimated as complicated like Japanese or non-common spoken such as Vietnamese. By the actual demand in multilingual processing, it is worth expanding the linguistic boundary of summarizing. Moreover, we would like to examine our model on different datasets intensities to analyze the effectiveness. Therefore, apart from large-scale dataset for English as *CNN/DailyMail* [5], we examine the small-scale Japanese - *Livedoor News Corpus* [15] and propose a medium-scale Vietnamese - *VNTC-Abs* (Sect. 4.1).

In this paper, we continue to address the issue of significant level by proposing CONTOUR. The purpose of CONTOUR is to emphasize words containing important information at each specific time by producing distinctive contours of word's potential. In contrast to Pointer Generator Network and SELECTOR, we aim to take advantage of word that has just been selected from the Decoder for the output word, called as the *latest predicted word*. CONTOUR is an association of two independent sub-methods: (1) **Penalty** - builds on the available SELECTOR model by the updated version of SELECTOR that improves the focus areas generation and optimizes the inference time, and (2) **Spotlight** - acts as an extra mask to re-ranking words potential at each predicting step based on the last predicted word. We conduct experiments on a variety of languages and achieve the result for Abstractive Summarization task: (1) **English:** outperforming baselines on the *CNN/DailyMail* dataset; (2) **Japanese:** the first method to work on the original version of *Livedoor News Corpus*, and (3) **Vietnamese** the first to make an end-to-end report on the *VNTC-Abs* dataset.

Our implementation is available online[1].

2   Related Work

Pointer Generator Network. [12] proposed by See et al. to control the *OOV*. While the *fixed vocabulary set* are distinct and built from the most popular words, if w is *OOV*, then the probability of w in the vocabulary distribution $P_{vocab}(w)$ equal 0. Therefore, they try to extend the vocabulary source by examining words in the input. Finally, a *generation probability* p_{gen} work as a propagation gate to copy words from the source text or choose a word in the fixed vocabulary set. Figure 1 visualize our baselines including Pointer Generator Network concept.

---

[1] Our released implementation: github.com/trangnnp/CONTOUR.

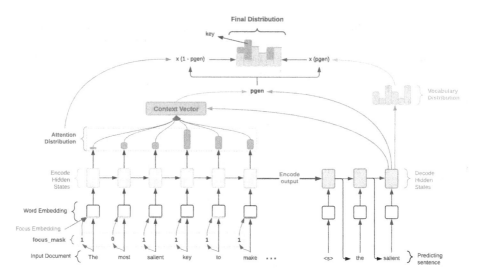

Fig. 1. The baseline sequence-to-sequence with *Pointer Generator Network* with *Attention* and SELECTOR

To train, they took advantage of the Attention outputs which are *attention score* e^t and *attention distribution* a^t to produce a context vector $h_t^\star$ at each time step t. The Attention given by Bahdanau et al. [1] are showed in Eq. 1 and 2, where v, W_h, W_s, and b_{attn} are learnable parameters, S is the input length. Equation 3 shows how the $h_t^\star$ is formed by a^t and encode hidden states h. In the next step, $h_t^\star$ is used to calculate P_{vocab} in Eq. 4. The *generation probability* $p_{gen} \in [0,1]$ provided in Eq. 5 by added learnable parameters w_h, w_s, and w_x. Equation 6 uses p_{gen} to modify words probability and adds the probability of all words in the input document. The output word is which has the highest value in the extended vocabulary distribution $P(w)$.

$$e_i^t = v^T \tanh(W_h h_i + W_s s_t + b_{attn}) \qquad i = 1, 2 \ldots, S \qquad (1)$$

$$a^t = softmax(e^t) \qquad (2)$$

$$h_t^\star = \sum_i a_i^t h_i \qquad (3)$$

$$P_{vocab} = softmax(V'(V[s^t, h_t^\star])) \qquad (4)$$

$$p_{gen} = \sigma(w_h^T h_t^\star + w_s^T s_t + w_x^T x_t) \qquad (5)$$

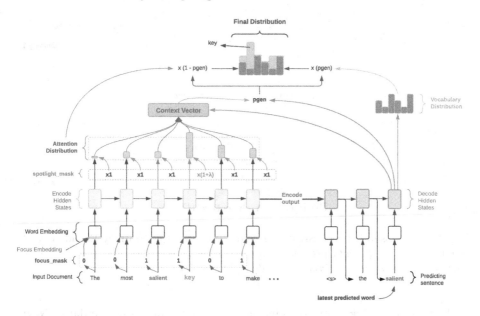

Fig. 2. Our CONTOUR with improved *focus mask* and *spotlight mask*

$$P(w) = p_{gen} \star P_{vocab}(w) + (1 - p_{gen}) \sum_{i:w_i=w} a_i^t \tag{6}$$

SELECTOR. [2] Cho et al.'s SELECTOR idea is to use a BiGRU layer to produce N different *focus masks* to mark phrases that need paying attention along the input. This mask works as the additional information, which is the blue information in Fig. 1, added to the word embedding. Cho et al. reported SELECTOR as a plug-and-play model to be ready to apply to an arbitrary Sequence-to-Sequence model. Equation 7 describes the *focus mask* generation by Cho et al. with the input length S and position t $(1 \le t \le S)$ (Fig. 2).

$$\begin{aligned}(h_1...h_S) &= BiGRU(x) \\ o_t^z &= \sigma(FC([h_t; h_1; h_S; e_z]))\end{aligned} \tag{7}$$

For details, they feed the *BiGRU* hidden states and the embedding e_z $(0 \le z < N)$ to a *Sigmoid* function. Then a Bernoulli distribution transforms the *focus logit* o^z to a binary mask m_z. Those *focus masks* are fed to the encoder-decoder simultaneously to produce N summaries, then choose only one of these summaries for the output. Figure 3a is the overview of how SELECTOR is associated with a general Encoder-Decoder Architecture.

3   CONTOUR: Penalty and Spotlight Mask

The proposed CONTOUR contains two independent modules: *Penalty* and *Spolight*. In this section, we first describe *Penalty* (Sect. 3.1) by reviewing some

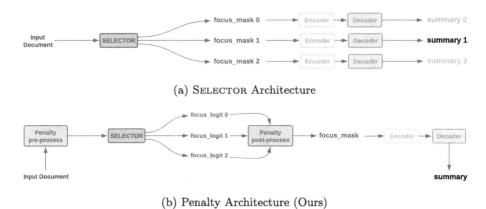

(a) SELECTOR Architecture

(b) Penalty Architecture (Ours)

Fig. 3. Comparison between SELECTOR and *Penalty* on the arbitrary Encode-Decode architecture. Our *Penalty* provides two extra steps: pre-process and post-process to improve the *focus_ mask quality and reduce the test time*

of the inherited works from baselines and then introducing our updates. Next, we describe *Spolight* (Sect. 3.2) by explaining the algorithm in detail. Finally, Sect. 3.3 is the visualization of the whole concept of CONTOUR.

3.1 Penalty

We present *Penalty* as a pre-processed and post-processed updated version of SELECTOR to improve the focus areas' quality. Unlike Cho et al., in the pre-process phase before training SELECTOR, our *Penalty* defines a new rule for the ground-truth to for SELECTOR model. While the main Sequence-to-Sequence model's Encoder needs to focus on informative words. Therefore, we ignore the most popular tokens like stopwords and symbols, with the obvious exceptions of the overlapping between source and target as ground-truth. As a result, our predicted *focus mask* marks meaningful phrases.

$$o_t = \sigma(\sum_{z=1}^{N}(o_t^z))$$

$$p(m_t^{'}|x) = Bernoulli(o_t)$$

(8)

Follow Cho et al. [2, 3.2], at training time, we use the SELECTOR ground-truth for computation. In the inference time, as a post-process step, we sum up all logit candidates together to achieve o_t and $m_t^{'}$. The detailed formula is described in Eq. 8 and the *Penalty* architecture is visualized in Fig. 3b. The sum of logit candidates provides three benefits:

1. *The inference time is optimized*: By producing one *focus mask* $m_t^{'}$ from multiple *focus logit* candidates, the encoder-decoder is executed once.

Algorithm 1. Spotlight mask algorithms

    Input: focus_mask, last_id, et   **Output**: rerank_et   **Constant**: lambda

1: **procedure** RERANKING_WORDS_POTENTIAL
2:    focused = (Binary) focus_mask == 1
3:    is_behind = (Binary) **input_ids** *equal* **last_id**
4:    is_behind = Concatenate([0], is_behind[:–1]))
5:    **is_spotlight = focused * is_behind**
6:    spotlight_mask = is_spotlight * lambda + 1
7:    rerank_et = et * spotlight_mask
8:    Return rerank_et
9: **end procedure**

---

2. *The focus mask accuracy is evaluated naturally*: Independent from the predicted result of the main model, *focus mask* accuracy is calculated by the number of correct positions over input length.
3. *The focus mask accuracy is improved*: Summing up candidates helps to penalize the negative points and enhance good points.

3.2 Spotlight

In the purpose of re-ranking words potential, *Spotlight* generates *spotlight mask* with a fixed-length as the input document. *Spotlight mask* helps to emphasize the most suitable word containing valuable information at a specific time. To construct this mask, we use a static parameter λ ($\lambda > 0$) and the current context of predicting sentence, which is formed by incorporating the latest predicted word and the Attention result from Eq. 2.

There are two primary inclusion criteria for the word w in the input document to become a *spotlight*. First, w is focused by the *focus mask* (from Sect. 3.1). The use of *focus mask* (denoted as m in general) is to narrow the number of word candidates. Secondly, w stands right behind the latest predicted word w_{prev} in the source text. However, *spotlight* just re-raking words and dose not select any word for the output directly. There is only one criterion for the word w to become the next output words that w achieve the highest probability after Eq. 6.

The generating *spotlight mask* process is repeated in each decoder step. Equation 9 presents the *spotlight mask* at time step t, denoted as s^t, and Eq. 10 shows how the *spotlight mask* modify the *attention scores*.

$$s_i^t = \begin{cases} 1 + \lambda, & \text{if } w_{i-1} = w_{prev} \text{ and } m_i = 1 \\ 1, & \text{otherwise} \end{cases} \tag{9}$$

$$e^t = e^t * s^t \tag{10}$$

Algorithm 1 describes the *Spotlight*'s effect on words potential.

Table 1. CONTOUR visualization. Including: highlighted words by *focus mask* , Spotlight word with very high possibility, and normal words. Assume that "new" in red is the latest predicted word

Baseline (truncated Input): the 49-year-old woman showed a staggering response to a new combination of two drugs, to treat skin cancer . doctors at the memorial sloan-kettering cancer center in new york have hailed their patient's recovery ' one of the most astonishing responses' , ever seen . a 49-year-old cancer patient's giant tumour underneath her left breast -lrb- starred left -rrb- was ' completely destroyed' and 'dissolved away' -lrb- right -rrb- after she received a pioneering new drugs cocktail to treat melanoma . they hope the new therapy could save thousands of lives in future .
CONTOUR (truncated Input): the 49-year-old woman showed a staggering response to a new combination of two drugs , to treat skin cancer . doctors at the memorial sloan-kettering cancer center in new york have hailed their patient's recovery ' one of the most astonishing responses', ever seen . a 49-year-old cancer patient's giant tumour underneath her left breast -lrb- starred left -rrb- was ' completely destroyed' and 'dissolved away' -lrb- right -rrb- after she received a pioneering new drugs cocktail to treat melanoma . they hope the new therapy could save thousands of lives in future .
Reference: doctors at memorial sloan-kettering cancer center in new york were trialling a new combination of drugs to treat advanced skin cancer . combined standard drug ipilimumab with new drug nivolumab .

Input: lambda: λ, focus_mask: *focus mask*, latest_id: latest predicted word's index, e_t: *attention scores* from Eq. 1

Output: rerank_et: the modified *attention scores*

Line 2: the *focused* condition reuses the *focus mask* result.

Line 3–4: Generate *is_behind* binary mask for indices next to the *latest_id*.

Line 5: The multiplication produces a binary mask. Indices that match both *focused* and *is_behind* conditions are set to 1.

Line 6: Generate the *spotlight mask*. *Spotlight* positions are set to $1 + \lambda$. All other indices will have value 1, including the padding positions.

Line 7: Re-ranking input words potential. Words marked by *spotlight* will increase itself $1 + \lambda$ times, and normal words will be unchanged.

3.3 CONTOUR

By combining *Penalty* and *Spotlight*, CONTOUR becomes a two stages process that closer to human-like actions in Abstractive Summarizing. Formally, from the human point of view, the *Penalty* stage is similar to skimming and highlighting crucial information such as personal names, locations, dates, or unique phrases. On the other hand, the *Spotlight* stage can be understood as scanning words that stand next to one given word. Looking as a whole, the similarity between CONTOUR and human action promises the output that is closer to human results.

Table 1 visualizes the comparison between the whole concept of CONTOUR to baselines. The table shows that *Penalty* aids in collecting informative words and excluding common words such as "a", "one", "with", "and", "of". On the other hand, in this example, with the latest predicted word new, spotlights combination, york , drugs , therapy are recognized. Moreover, the phrases that formed by word new and these spotlights are in the reference summary at 3 over 4 of them. Therefore, *Spotlight* proves to be useful in emphasizing the most suitable words for the next output. To sum up, CONTOUR would help to reduce noise areas and estimate correctly high potential words.

4 Experiments and Results

4.1 Datasets

CNN/DailyMail [5]. Collected from *CNN News*[2] by Hermann et al. This dataset contains 287,226 training pairs, 13,368 validation pairs, and 11,490 test pairs. We experiment on the same *non-anonymized* version as See et al. [12].

VNTC-Abs. Base on *VNTC* [6] dataset, which collected from a Vietnamese Online News website named *VNExpress*[3] by Hoang et al. for *Webpage categorizing* task, we modify it to be compatiple with *Abstractive Summarization* task and rename to *VNTC-Abs*[4]. When the original dataset has some duplicated articles in analogous categories, we differentiate and achieve 49,289 distinct articles. We extract the available summary as target output and the main content as an input source. Each source and target pair has a length of 309 and 43 on average. We split 70% of the total data for training and 15% for each validate and test set.

Livedoor News Corpus (Livedoor) [15]. Collected from Japanese Online News named ライブドアニュース *Livedoor News*[5] by Vo C. Each source and target pair have 664 and 16 tokens on average, respectively. Because of the small-scale of data, we split 90% of the total 7,376 articles as the train set, 10% for the test set, and ignore the validate set.

4.2 Implementation Details

In experiments on *CNN/DailyMail*, we followed the configuration by See et al. and previous works. We set the maximum input length to 400 tokens, maximum of 100 tokens for the output summary, and 50,000 most popular words for the vocabulary. We set 256 dimensions for hidden states for both Encoder and Decoder. Besides, we trained word embedding from scratch with 128 dimensions.

---

[2] CNN News: edition.cnn.com.
[3] VNExpress: vnexpress.net.
[4] Our released VNTC-Abs: github.com/trangnnp/VNTC-Abs.
[5] Livedoor News: livedoor.com.

As the first reporter for *VNTC-Abs* and *Livedoor* for *Abstractive Summarizing*, we did experiments and analysis for the most suitable configuration. In particular, for *VNTC-Abs*: the vocabulary size is limited to 12,000, maximum input length set to 650, and output length arranges from 20 to 60 tokens, where UETsegmenter [10] is used as tokenizer. On the other hand, for *Livedoor*, we limited the vocabulary size to 15,000, set the maximum input length to 800, and bounded the output length from 10 to 40 tokens, applying Knok's preprocess[6] for the official version of this dataset[7], which uses Mecab [7] for tokenizing.

4.3 Evaluate Metrics

We evaluate our models by ROUGE score [8]. ROUGE calculates the similarity of two paragraphs based on the word-overlapping, bigram-overlapping, and the longest common subsequence, which are presented as ROUGE-1, ROUGE-2, ROUGE-L points, respectively. The higher the ROUGE points are, the closer between reference and predicted summary. We used the *rouge-score*[8] library which is a pure implementation of *Rouge155*[9] that used by See et al. [12].

4.4 Comparison Basis

Baselines. We measure our models with *Pointer Generator Network* by See et al. [12] and SELECTOR by Cho et al. [2] as baselines. We re-implemented and trained the baselines from scratch. For more details, we trained *CNN/DailyMail* by the same configuration reported by See et al. [12]. As the first work on *VNTC-Abs* and *Livedoor*, these datasets are trained under our own settings.

CONTOUR (Ours). We first report the result of using *Penalty* and *Spotlight* separately, then report CONTOUR as the fully proposed method. We also compare the effectiveness of n as the number of *focus logit* candidates to *Penalty*, and λ value as the degree of impact to *Spotlight*.

Recent Researches. Besides baselines on the *CNN/DailyMail*, we compare our models with PEGASUS by Zhang et al. [17] and Bottom-up by Gehrmann et al. [4] as recent researches. In detail, PEGASUS uses the Transformer-based method, while Bottom-up proposes bottom-up attention and likely phrases. We compare directly to the reported results by Zhang et al. and Gehrmann et al.

4.5 Result

Our results are given in Table 2 and 3. Overall, the result is better than baselines for all three ROUGE points of three datasets. For more details, the result are

---

[6] Knok's preprocess: github.com/knok/ldcc-summarize.

[7] The official version download at https://www.rondhuit.com/download.html.

[8] pypi.org/project/rouge-score/0.0.4/.

[9] pypi.org/project/pyrouge/0.1.3/.

Table 2. The comparison of CONTOUR to baselines. Denote: PG for *Pointer Generator*, R for ROUGE, best scores are bolded

Method	CNN/DailyMail			VNTC-Abs			Livedoor		
	R-1	R-2	R-L	R-1	R-2	R-L	R-1	R-2	R-L
Baselines									
PG Network	39.09	17.25	35.57	25.21	9.11	21.70	27.72	12.66	26.28
SELECTOR	41.29	18.66	38.20	25.72	8.23	21.84	29.03	12.27	27.44
Our methods									
$Penalty_{n=3}$	41.43	18.82	38.34	27.00	**9.51**	23.79	31.63	14.03	29.26
$Penalty_{n=5}$	41.51	18.80	38.42	27.19	9.31	24.09	31.01	13.68	**29.34**
$Penalty_{n=10}$	41.39	18.76	38.28	26.99	9.18	23.45	**31.68**	**14.83**	29.26
$Spotlight_{\lambda=0.25}$	41.64	18.72	38.36	27.01	9.25	23.32	30.42	12.67	27.16
$Spotlight_{\lambda=0.5}$	41.79	18.79	38.59	**27.27**	9.10	23.70	30.85	13.01	28.31
$CONTOUR_{n=3,\lambda=0.25}$	42.01	18.98	38.97	26.97	9.35	23.71	30.45	12.92	27.87
$CONTOUR_{n=3,\lambda=0.5}$	41.80	18.86	38.89	26.90	9.24	23.47	31.00	13.37	27.72
$CONTOUR_{n=5,\lambda=0.25}$	**42.08**	**19.11**	**39.10**	26.98	9.19	23.89	31.47	14.03	28.69
$CONTOUR_{n=5,\lambda=0.5}$	41.93	18.87	38.86	26.80	**9.50**	**24.16**	30.85	13.66	28.67

increased by +0.79 R-1, +0.45 R-2, +0.90 R-L for *CNN/DailyMail*, +1.59 R-1, +0.39 R-2, +2.46 R-L for *VNTC-Abs*, and +3.96 R-1, +2.17 R-2, +2.98 R-L for *Livedoor*. The results show that our work is compatible with these datasets. However, each sub-method still has its own characteristics for the different datasets. Table 4 presents one of the sample summary output for each language.

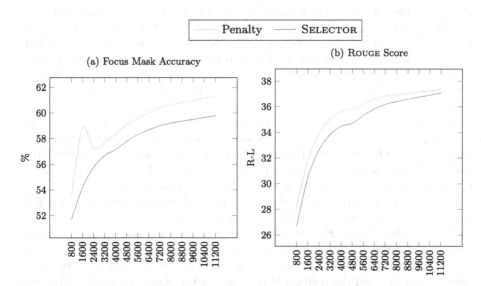

Fig. 4. Penalty effection visualization. Collected from the best result of $Penalty_{n=3}$ and $SELECTOR_{n=3}$. The x-axis is the number of tested data, ascending sorted by R-L. Each point presents the average value from the first to current number of data.

Table 3. The comparison of CONTOUR to related researches on *CNN/DailyMail*

Method	CNN/DailyMail		
	R-1	R-2	R-L
Reported baselines			
PG Network [12]	39.53	17.25	36.38
SELECTOR [2]	41.72	18.71	38.79
Recent researches			
PEGASUS [17]	38.27	15.03	35.48
Bottom-Up [4]	41.22	18.68	38.34
Our implementations			
PG Network	39.09	17.25	35.57
SELECTOR	41.29	18.66	38.20
CONTOUR(*Ours*)	**42.08**	**19.11**	**39.10**

In *CNN/DailyMail* experiments, the parameters analyzing in Table 2 shows that $\lambda = 0.25$ and $n = 5$ are holding the best score for CONTOUR and each of its sub-method. The positive outcome proves that CONTOUR is well suited for *CNN/DailyMail*. By some specific configuration on Cho et al.'s implementation, our R-L is a little less than the reported SELECTOR. However, with the provided settings by Cho et al., our CONTOUR$_{n=3,\lambda=0.25}$ (as well as the sub-methods *Penalty*$_{n=3}$, *Spotlight*$_{\lambda=0.25}$) achieved better results than our re-implement baseline of SELECTOR$_{n=3}$ by +0.72 R-1, +0.32 R-2, +0.77 R-L.

With *VNTC-Abs*, CONTOUR$_{n=5,\lambda=0.5}$ achieves two of the best ROUGE points. In fact, *VNTC-Abs* contains some passages that only serve as an opening without containing any essential words or key information. Therefore, the *focus mask* would become incoherent in some cases. However, all experiments on *VNTC-Abs* are greater than baselines. The predicted summaries are fluent and able to transfer valuable information.

In *Livedoor* experiments, *Penalty* is the most stable when holding the highest ROUGE points. With a small summary size, *Livedoor* does not contain numerous long phrases. Therefore, the dataset is not an ideal condition to utilize CONTOUR's the full potential of copying continuous words. However, our proposed methods still overcome the low-source data and surpassed baselines. Moreover, CONTOUR achieve surprising performance producing the summary that fulfills the readability and grammatical requirements.

In analyzing the relation of *focus mask* accuracy and the output summary, we prove that the improvement of *focus mask* quality helps *Penalty* to generate better summaries. Figure 4 visualizes the effectiveness of *Penalty* to the *focus mask*'s accuracy and the relation of accuracy to the summary quality. Figure 4a proves the positive impact of *Penalty* on the optimization of *focus mask* accuracy with the average differential of 1.64%. Moreover, Fig. 4b shows that higher accu-

Table 4. Sample summaries

CNN/DailyMail dataset
Input (truncated): an oregon man adopted from south korea as a toddler fears that he could be torn away from his wife and three children because his u.s. citizenship was never registered . adam crapser , 39 , of salem , was issued with deportation papers by the department of homeland security in january and a hearing is set for april 2 . the father-of-three , who has a fourth child on the way , arrived in america with the name shin sonh hyuk in 1979 around the age of three along with his biological sister . however , his first set of adoptive parents abandoned him and the second set turned out to be abusive . adding to crapser 's struggles , at no point did his guardians seek the green card or citizenship for him that they should have . deportation fears : korean adoptee adam crapser , left , poses with daughters , christal , one , christina , five , and his wife , anh nguyen , in the family 's living room in oregon . now , after a life battling joblessness because of his lack of immigration papers , homelessness and crime , crapser is facing deportation . ' the state abandoned him when he was a child , ' his attorney , lori walls , said . ' now the us is throwing him out .
Reference: adam crapser , 39 , was issued with deportation papers by the department of homeland security in january and a hearing is set for april 2 . the father-of-three arrived in america with the name shin sonh hyuk in 1979 around the age of three along with his biological sister . however , his first set of adoptive parents abandoned him and the second set turned out to be abusive . adding to crapser 's struggles , at no point did his guardians seek the green card or citizenship for him that they should have .
Predicted: adam crapser , 39 , of salem , oregon , was issued with deportation papers by the department of homeland security in january and a hearing is set for april 2 . the father-of-three , who has a fourth child on the way , arrived in america with the name shin sonh hyuk in 1979 around the age of three with his biological sister . his first set of adoptive parents abandoned him and the second set turned out to be abusive .
VNTC-Abs dataset
Input (truncated): giu nguyen hon nhac trinh tôi là một người yêu nhạc trịnh giống như tất cả mọi người . vấn đề đổi mới nhạc trịnh mà hiện nay một số ca sĩ đã làm thật không thể chấp nhận được . tất nhiên họ có lý khi cho rằng đó là cách hát nhạc trịnh cho phù hợp với thời đại . chúng ta cũng như những người thân của trịnh công sơn sẽ rất lấy làm vui , vì như thế có nghĩa là nhạc trịnh vẫn đầy sức hấp dẫn . song họ phải hiểu thế nào là cái hồn của nhạc trịnh , và nếu họ hát như cách mà họ cho là ' thổi vào đó một hơi thở mới của thời đại ' , e rằng nếu không có phần giới thiệu thì người nghe khó có thể biết đó là nhạc trịnh .
Reference: nếu họ hát như cách mà họ cho là "thổi vào đó một hơi thở mới của thời đại " , e rằng nếu không có phần giới thiệu thì người nghe khó có thể biết đó là nhạc trịnh , nhất là đối với những ca khúc của ông .
Predicted: hiện nay khi là cách hát nhạc trịnh cho phù hợp với thời đại . chúng ta cũng như những người thân của trịnh công sơn sẽ rất vui , vì như thế có nghĩa là nhạc trịnh vẫn vào đó một hơi thở mới của thời đại " , e rằng nếu không có phần giới thiệu thì người nghe khó có thể biết đó là nhạc trịnh .
Livedoor News Corpus
Input (truncated): スマホで絵本を見よう!私も子供の頃にたくさんの絵本を読んだ気がします。家にも幼稚園にもいろいろな絵本があって、今でも内容を思い出せる絵本がたくさんあります。今回紹介するandroid向けアプリ「けいたいえほんライブラリー」は絵本を読み聞かせできるアプリです。では、さっそくアプリを紹介していきましょう。まず読みたい絵本をタップで選びます。絵本は1回の配信で6編が収録されています。絵本はナレーションと同時に文字が表示されるようになっています。音声offで読み聞かせてあげたり、お子様が1人でも楽しめるようになっています。文字は平仮名と片仮名で書かれているので、文字を覚えだしたお子様にも良さそうです。絵本は絵がアニメーションしたり、タップした時に動く絵もあります。デジタルならではの遊びの仕掛けで楽しく絵本を読んでくれそうですね。外出中でも場所を選ばず絵本が読めるので、お子様とのお出かけ時にはインストールしておきたいですね
Reference: 子供といっしょに読もう!音声再生にも対応した「けいたいえほんライブラリー」
Predicted: スマホで読みしよう!スマホで絵本「けいたいえほんライブラリー」

racy of *focus mask* would generate summary with better ROUGE score. In turn, this means focusing on the right words and phrases will produce finer results.

With pre-trained data, recent state-of-the-art on *CNN/DailyMail* are much better than ours, which are 44.31 on R-1, 41.60 on R-L by ERNIE-GEN$_{LARGE}$ of Xiao et al. [16], 21.35 on R-2 by PEGASUS$_{LARGE}$ Zhang et al. [17]. However, compared to models without pre-training, CONTOUR provides a competitive ROUGE scores by exceeding most of the mentioned studies and achieving the highest ROUGE score. As seen from Table 3, we outmaneuver Xiao et al.'s PEGASUS by +3.81 R-1, +4.08 R-2, +3.62 R-L, and Bottom-up of Zhang et al. by +0.86 R-1, +0.43 R-2, +0.76 R-L.

5   Conclusion

In this work, we presented CONTOUR architecture consists of two independent modules: *Penalty* and *Spotlight*. Our model helps to improve the focus areas and highlight containing crucial information at a specific time. CONTOUR achieves the significantly better results when comparing with recent works on

CNN/DailyMail dataset. We proved the adaptability of CONTOUR to other natural languages by examining the *Livedoor News Corpus* for Japanese, moreover, we modified and explored the *VNTC-Abs* dataset for Vietnamese. Future work involves CONTOUR with the automatic adjustment to an arbitrary dataset.

Acknowledgement. We thank the anonymous reviewers for their helpful suggestions on this paper. This research was supported by the Department of Knowledge Engineering, funded by the Faculty of Information Technology under grant number *CNTT 2020-12* and the Advanced Program in Computer Science from Vietnam National University, Ho Chi Minh City University of Science. The corresponding author is Trang-Phuong N. Nguyen.

References

1. Bahdanau, D., Cho, K., Bengio, Y.: Neural machine translation by jointly learning to align and translate. ArXiv 1409, September 2014
2. Cho, J., Seo, M., Hajishirzi, H.: Mixture content selection for diverse sequence generation. In: EMNLP (2019)
3. Chopra, S., Auli, M., Rush, A.M.: Abstractive sentence summarization with attentive recurrent neural networks. In: Proceedings of the 2016 Conference of the North American Chapter of the Association for Computational Linguistics: Human Language Technologies, pp. 93–98. Association for Computational Linguistics, San Diego, June 2016. https://doi.org/10.18653/v1/N16-1012. https://www.aclweb.org/anthology/N16-1012
4. Gehrmann, S., Deng, Y., Rush, A.: Bottom-up abstractive summarization. In: Proceedings of the 2018 Conference on Empirical Methods in Natural Language Processing, pp. 4098–4109. Association for Computational Linguistics, Brussels, October-November 2018. https://doi.org/10.18653/v1/D18-1443. https://www.aclweb.org/anthology/D18-1443
5. Hermann, K.M., et al.: Teaching machines to read and comprehend. In: Advances in Neural Information Processing Systems, pp. 1693–1701 (2015)
6. Hoang, C.D.V.: VNTC (2006). github.com/duyvuleo/VNTC
7. Kudo, T.: MeCab : yet another part-of-speech and morphological analyzer (2005)
8. Lin, C.Y.: ROUGE: a package for automatic evaluation of summaries. In: Text Summarization Branches Out, pp. 74–81. Association for Computational Linguistics, Barcelona, July 2004. https://www.aclweb.org/anthology/W04-1013
9. Nallapati, R., Zhou, B., dos Santos, C., Caglar, G., Xiang, B.: Abstractive text summarization using sequence-to-sequence RNNs and beyond. In: Proceedings of The 20th SIGNLL Conference on Computational Natural Language Learning, pp. 280–290. Association for Computational Linguistics, Berlin, August 2016. https://doi.org/10.18653/v1/K16-1028. https://www.aclweb.org/anthology/K16-1028
10. Nguyen, T.P., Le, A.C.: A hybrid approach to vietnamese word segmentation. In: 2016 IEEE RIVF International Conference on Computing Communication Technologies, Research, Innovation, and Vision for the Future (RIVF), pp. 114–119, November 2016. https://doi.org/10.1109/RIVF.2016.7800279
11. Rush, A.M., Chopra, S., Weston, J.: A neural attention model for abstractive sentence summarization. In: Proceedings of the 2015 Conference on Empirical Methods in Natural Language Processing, pp. 379–389. Association for Computational Linguistics, Lisbon, September 2015. https://doi.org/10.18653/v1/D15-1044. https://www.aclweb.org/anthology/D15-1044

12. See, A., Liu, P.J., Manning, C.D.: Get to the point: summarization with pointer-generator networks. In: Proceedings of the 55th Annual Meeting of the Association for Computational Linguistics (Volume 1: Long Papers), pp. 1073–1083. Association for Computational Linguistics, Vancouver, July 2017. https://doi.org/10.18653/v1/P17-1099. https://www.aclweb.org/anthology/P17-1099
13. Sutskever, I., Vinyals, O., Le, Q.V.: Sequence to sequence learning with neural networks. CoRR abs/1409.3215 (2014). http://arxiv.org/abs/1409.3215
14. Vinyals, O., Fortunato, M., Jaitly, N.: Pointer networks. In: Cortes, C., Lawrence, N.D., Lee, D.D., Sugiyama, M., Garnett, R. (eds.) Advances in Neural Information Processing Systems, vol. 28, pp. 2692–2700. Curran Associates, Inc. (2015). http://papers.nips.cc/paper/5866-pointer-networks.pdf
15. Vo, C.C.: Livedoor news (2020). www.kaggle.com/vochicong/livedoor-news
16. Xiao, D., et al.: ERNIE-GEN: an enhanced multi-flow pre-training and fine-tuning framework for natural language generation, January 2020
17. Zhang, J., Zhao, Y., Saleh, M., Liu, P.J.: PEGASUS: pre-training with extracted gap-sentences for abstractive summarization (2019)

Heterogeneous Information Access System with a Natural Language Interface in the Context of Organization of Events

Piotr Nawrocki[1,2]([envelope]) [iD], Dominik Radziszowski[2,3,4] [iD],
and Bartlomiej Sniezynski[1,2] [iD]

[1] Institute of Computer Science, AGH University of Science and Technology,
Kraków, Poland
`piotr.nawrocki@agh.edu.pl`
[2] Eventory sp. z o.o., Kraków, Poland
[3] PoleForce sp. z o.o., Kraków, Poland
[4] XTRF Management Systems S.A., Krakow, Poland
`https://eventory.cc`

Abstract. The article presents an innovative approach to providing information to event and conference participants using a natural language interface and chatbot. The proposed solution, in the form of chatbots, allows users to communicate with conference/event participants and provide them with personalized event-related information in a natural way. The proposed system architecture facilitates exploitation of various communication agents, including integration with several instant messaging platforms. The heterogeneity and flexibility of the solution have been validated during functional tests conducted in a highly scalable on-demand cloud environment (Google Cloud Platform); moreover the efficiency and scalability of the resulting solution have proven sufficient for handling large conferences/events.

Keywords: Natural language interface · Information access system · Chatbot · Events

1 Introduction

Event marketing is becoming the leading promotion instrument for many B2B applications. Dynamic development of this industry generates a number of challenges which can be – to a lesser or greater extent – solved with the help of the latest technologies. Still, event marketing remains technologically insufficiently developed in relation to other marketing instruments (e.g. social media). Experts deal with increasing the attractiveness of organized events from the point of view

The research presented in this paper was supported by the National Center for Research and Development (NCBR) under Grants No. POIR.01.01.01-00-0878/17 and POIR.01.01.01-00-0327/19.

T.-P. Hong et al. (Eds.): ACIIDS 2021, CCIS 1371, pp. 188–200, 2021.
https://doi.org/10.1007/978-981-16-1685-3_16

of participants, while maintaining efficient allocation of resources and achieving business goals from the point of view of organizers and sponsors.

Many challenges related to organization of events can be identified. Participant groups can be very diverse, as can their expectations and goals. However, they are usually linked by the need to share content, emotions and experience with others; seeking new interaction channels outside of social media; expecting an immediate response/reaction to their queries, and personalizing experience to the maximum extent possible.

Each participant expects a thorough understanding of their needs, building trust and lasting relationships, and communicating in the simple language of benefits. The vast majority also want to exert an active influence on the course of the event. Understanding of these needs differs from participant to participant. The influence of technology on participants' expectations is not without significance. Many expect access to tools that would facilitate their preparation for the event, enrich their experience and adapt the message to their personalized needs. Therefore, they are open and willing to use new solutions, and their positive emotions and experience translate into real benefits for the organizer and sponsor of the event.

At the same time, many modern IT systems are beginning to exploit the capability of communicating with users through a natural language interface. The most popular solutions of this type include voice assistants (such as Apple Siri or Amazon Alexa) and voice control systems in cars. In addition to voice-based solutions, an important area of development involves systems that enable automatic interaction through chatbots.

At the moment, the main methods of engaging and motivating participants are based on mobile applications, social media, virtual reality (stands and presentations) and traditional methods of playing with light, arrangement of stands, catering, gadgets, etc. The solution developed by us on the basis of chatbot concepts enables creation of a heterogeneous system that delivers information as naturally as possible and evokes positive emotions in event participants.

The remainder of this paper is structured as follows: Sect. 2 contains a description of related work, Sect. 3 deals with defining, in detail, the solution's architecture, Sect. 4 describes our evaluation of the heterogeneous information access system, while Sect. 5 contains conclusions and a description of further work.

2   Related Work

In recent years, systems enabling control of devices and exchange of information using speech recognition interfaces have experienced strong evolution. On the one hand, voice assistant solutions such as Apple Siri, Google Assistant or Amazon Alexa, operating primarily on mobile devices, have been devised, while on the other hand, many cars are now equipped with voice communication systems providing control over certain functions of the car. Another interesting area involves chatbots capable of automatically obtaining information using text-based dialogue with the user. Chatbot solutions will become more and more common at large companies or state institutions which, by including such solutions on their websites, may provide users with useful information or assistance. One area where this technology sees great potential (despite the current lack of in-depth research and developed solutions) is the use of chatbots during events.

An important issue related to the design of chatbots is how to develop effective mechanisms of conducting dialogue with the user in such a way that they perceive it in the most natural way possible. In [7] the authors analyzed existing evaluation methods and presented their own methodology for determining the effectiveness and naturalness of the dialogue system. They also presented a prototype methodology for evaluating the LifeLike virtual avatar design. Assessing the naturalness of dialogue may be particularly important for chatbots operating during events due to their duration and potential complexity.

In [4] the authors analyze available chatbot design techniques, focusing on their similarities and differences. They also present factors that influence the design of a chatbot, such as techniques which can be used to enter into dialogue with the user, or the knowledge base used. Among these techniques, apart from pattern matching and genetic algorithms, machine learning algorithms play an important role [9]. The second important aspect is the creation of the knowledge base. In [6] the authors present an interesting approach to automatic extraction of knowledge from discussion forums to support the development of chatbots. They applied the SVM classifier and conducted experiments which confirm the effectiveness of its operation. All of the issues outlined above, especially knowledge extraction, are important in the context of developing conference chatobots. The knowledge bases of various events and conferences can differ strongly from one another, and depend, among other things, on their subject matter. This is why it is so important to prepare an appropriate knowledge base for the chatbot and to develop mechanisms for its use, taking into account the context resulting from the topic or type of the event.

Little research has been performed in the area of creating and applying chatbots for event support. In [10] the author describes the iBot environment which enables creation of chatbots for various application areas (including events). The iBot environment uses the concept of agents and operates as a multi-agent system. The developed environment enables the creation of intelligent chatbots, able to perform complex tasks and maintain dialogue control, while being flexible enough to allow for software development in different domains. The weakness of this solution compared to our solution is that it does not account for the specificity of events in the context of communicating with the chatbot. The proposed solution is universal, but this also means that it is not always well suited to the specific context of events.

Another important article in the area of chatbots is [5] in which the author discusses various issues related to creating chatbots, including possible architectures, NLP issues, creating a knowledge base and dialogue management. Aspects of security in the creation and use of chatbots are also presented. This work also presents an overview of ready-made solutions, including the solution developed by IBM Watson and popular Virtual Personal Assistants (VPAs) such as Apple Siri, Microsoft Cortana, Google Assistant or Amazon Alexa. The article furthermore covers various areas where bots can be applied, such as transportation bots, dating, mediation, fitness bots, weather bots and medical bots. However, there is no description of the use of bots during events.

Commercial aspects of using the chatbot have been analyzed in [8]. The author, in addition to presenting various types of chatbots developed by companies in Poland, proposes several indicators to assess the quality of the chatbot's operation, including its performance and usability. The paper analyzes various aspects of chatbot functionality, including appearance, implementations, speech synthesis unit, built-in knowledge base, knowledge presentation, conversational skills, sensitivity to context and the ability to react to unexpected or emergency situations. However, this work lacks an analysis of the architectures of individual solutions and does not cover the use of chatbots during events and conferences.

Analysis of existing chatbot solutions reveals that few of them deal with event handling and at the same time there is no research on this topic. The most important solutions of this type include EVABot [2], ConfBot [1] and Morph.ai [3]. As chatbots, all these solutions are based on AI mechanisms. They appear to be the most advanced ones in this field – even though they are still at a very early stage of development. In terms of functionality, these tools offer similar services to the solution developed by the authors, i.e. they use AI to personalize communication with users, they can research opinions, collect feedback, analyze sentiment, and send feedback to and from event managers. In these solutions, operators can obtain an up-to-date view of the interactions taking place, which enables them to introduce changes and improvements in real time as well as in subsequent events. Each of the above-mentioned solutions also provides the ability to integrate the chatbot with many event management tools that their clients (organizers) already use – however, they remain platform-independent solutions, costly to integrate and connect data, and they can operate (learn) only in the context of one event.

An important advantage of the solution developed by the authors will be its universality and the ability of the chatbot to learn from a constantly expanding database of participants' questions and preferences, along with the event database available on the Eventory platform. Having a much larger source database upon which the learning loop was based, the developed chatbot will learn faster and more efficiently than potential competing algorithms, ultimately bringing better results in the form of participant satisfaction and ROO (Return on Objective).

3   Architecture

A key part of our project is an environment for information access systems in which it is possible to carry out experiments on various conversation automatizing solutions. The environment should have the following characteristics:

- easy integration with various conversation agents, specializing in selected areas,
- easy integration with various communication systems used to exchange messages with users (e.g. Messenger, WeChat, Slack, Siscord);
- ability to manage conversation context, which means not only reacting to messages received, but also using information collected in previous interactions and knowledge stored in external data sources;
- ability to conduct the conversation in a way which enables collection of all data necessary to reach a predefined goal (such as providing a comprehensive answer, collecting marketing information, etc.);
- fulfilling standard scalability and efficiency requirements.

The research environment in question is therefore a hybrid solution, combining various existing technologies to reduce production costs as much as possible, and fitting into existing (used by customers) infrastructures which support organization of events. Its general architecture is illustrated in Fig. 1.

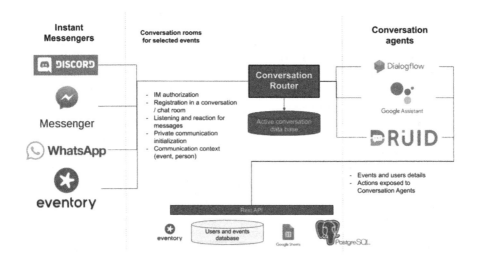

Fig. 1. Architecture of the research environment for information access systems.

The main component of the environment is the conversation router. It can receive messages from different conversation rooms and from different messengers, and forward them to the appropriate conversation agents. It is responsible for selecting the most appropriate agent and for exchanging subsequent messages until the end of the conversation. It also allows the operator to independently connect multiple conversation agents with different specialization areas for different events handled through different channels. On the one hand, it provides a communication and authorization layer in communicators, while on the other hand it unifies access mechanisms for various agents.

A key property of the conversation router is the ability to manage the afore-mentioned context of the conversation. Incoming messages are assigned to a conversation based on three key parameters:

- communication channel – type of messenger used and channel identifier;
- person sending the message – unique identifier of the author of the message (within the communicator);
- timestamp of the message – maintaining context depends on the time elapsed since the last message exchange. Given the application area, the communication is of a short-term nature. Therefore we assume that any message received after a long time (e.g. 1 h) concerns a new thread rather than a continuation of an earlier conversation.

Based on these assumptions, a module managing active conversations was built. The *Database of Active Conversations* contains conversations that are currently being conducted, i.e. those for which the time since the last message exchange is not greater than the time specified in the configuration (1 h) and the conversation has not been explicitly ended (some implementations of conversation agents can recognize the end of a conversation by looking for a closing statement such as "goodbye"). Messages sent to active conversations are always redirected to the conversation agent responsible for the given conversation. The most difficult task is to determine which conversation agent should handle the first message that initiates the conversation. This is done by the conversation router according to the following scheme:

- each new conversation initiation message is forwarded to all registered conversation agents;
- each agent replies to the conversation router with its own message. Apart from the content, these responses contain information about the extent to which the message matches the function performed by the agent. For example, given the following user message: "what is the conference address", a "don't understand" response – considered the default response when the agent does not know what to do with the message – will be assigned a poor match score by the agent, or will be accompanied by a fallback flag. However, the answer "what conference are you asking about" indicates that the agent may be competent to continue the conversation and will be assigned a higher match level;
- all agent responses are compared, whereupon the one with the highest match level is selected and passed to the user. The agent that was its author is also assigned to the conversation and without further selection it will handle subsequent user messages (until the conversation ends);
- if none of the agents "knows" what to do with the message, communication is not performed. This is especially important for communicating in group channels – according to the good practice of "if you don't know what's going on, don't speak up".

In a typical communication tool there are group channels (with many, sometimes thousands of participants) and individual channels used for direct communication with one specific user. Due to its confidentiality, some information should not or even cannot be disclosed in a group channel. Other information may not be confidential, but its public disclosure would increase communication noise – therefore, it should also be communicated using private channels. The same principles are applied to communication with the use of agents.

The way a given agent communicates – in a group or private channel – is determined by the configuration of the conversation router. Conversation with an agent may be initiated when a user asks a question in a group channel (e.g. "what is my password for the conference room?"), but it will be continued by the agent in the individual channel of the user. In this case, the conversation router opens an individual communication channel and redirects all communication to that private channel.

The conversation router supports multiple concurrent conversation agents and multiple communication channels (from different providers). For the purposes of the test environment, it is assumed that each agent is available on every channel.

Conversation agents are defined in the router configuration file. It should be noted that this configuration contains all the data needed to operate in a highly distributed environment. Conversation agents can be created and managed completely independently of the main system. Communication and access security issues are handled by using encrypted connections and authorization based on a public/private key infrastructure.

Integration with the messenger is implemented mainly on the messenger side. Most communication software providers offer the option of creating so-called user bots, i.e. conversation participants which are machines. Creating a user bot consists of determining what kind of events should reach it (new message, reaction to a message, adding/removing a user from a conversation, etc.) and generating a token used for authorization. Depending on the platform, registration of the bot consists of providing the URL of the service (endpoint) to be called in order to process events using a callback mechanism (Slack) or using a library provided for communication, which is initialized with the API token (Discord).

The bot user has access to many channels and can simultaneously participate in many independent conversations (group and private). It is added or removed to a channel by that channel's administrator. In the case of events, it is the creator of the event channel who decides to add/remove the bot. The bot listens on all channels to which it has been added, while maintaining a single connection to the conversation router.

The data needed by the conversation agents to competently conduct the conversation can be provided from a variety of sources. This includes data about conferences (Eventory platform), specific data of organizers (e.g. access codes maintained in a data repository) and other network services. A common feature of these data sources is that they may be accessed through a REST API interface.

The effect of such access to data is the ability for agents to perform specific actions (operations) during conversations. For example, they can not only ask about the details of a pizza order, but also actually place the order itself.

Data access security issues are always individually negotiated between the agent creator and the data owner.

It should be emphasized that for the purposes of determining the communication context, the conversation agent is provided by the conversation router with full information about a specific conversation that can be obtained from the communicator (username, id, channel name, etc.). This data is used by the agent to identify the correct dataset for the purpose of formulating a response (e.g. the channel name identifies the conference so that the agent can retrieve access codes from the correct datasheet).

4   Evaluation

In order to verify the operation of the developed solution, it was subjected to functional and performance tests. In order to perform these experiments, Discord software and the Google Cloud Platform (GCP) environment were used.

4.1   Functional Test

The research environment developed as part of the project was subjected to functional tests. For this purpose, a hybrid test configuration was prepared as a special case of the architecture shown in Fig. 2. The Discord communicator was used for testing, and the conversation agents were prepared in the Dialogflow system. Conversations were handled by the Conversation Router implemented in the Node.js technology. Additionally, a Google spreadsheet was used to support the data warehouse for which the REST service was created to handle calls from Dialogflow. The architecture of the test environment is shown in Fig. 2.

For the tests, two conversation agents were implemented and used: ConferenceInformer and CodeGiver. The former provides information about the event; in particular, the agent can provide information about the event organizer and instruct users how to register for the event. The CodeGiver agent's task is to communicate the code with which the registered user logs into the event streaming server.

A series of sessions were conducted using the test infrastructure, which verified the functionality of the developed solution. The crucial aspect subject to validation was the correctness of context identification. We assumed that user and conference context identification could be performed in the following ways:

– via the bot, using properly programmed dialogue;
– by proper organization of system infrastructure.

Specifically, identification of the user context using Dialogflow was implemented by creating a bot in which two intents play a crucial role: Code and CorrectAdresForCode. The first responds to the user's statement asking for the code and acquires the email parameter, which represents the address used by the user during registration. This intent calls a web service that checks the user database to see if registration has been performed for the address given. If not,

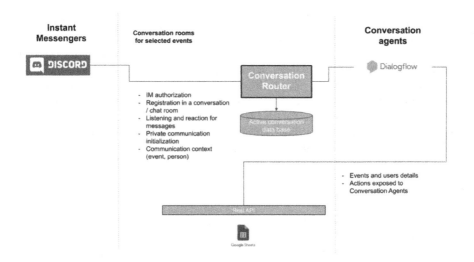

Fig. 2. Test environment architecture.

the bot asks for a valid address and the wrong-code-email context is activated, which allows matching of the CorrectCodeAddress intent, in turn, which acquires the address. The process repeats until a correct address is obtained. The bot worked properly, but it was not immune to interception of the email address by unauthorized persons. In the production version of the system additional security should be introduced, e.g., in the form of a password, the requirement to click on a link delivered to the user in an e-mail message or the requirement to enter a code texted to the user's phone. However, such a solution might prove less convenient for the user.

Proper identification of the user and conference context is handled by the appropriate organization of the system infrastructure. A separate instance of the instant messaging bot represents each conference. Therefore, the user who wishes to carry out a conversation will have to log into the system and connect/interact with the appropriate bot. As a result, identification of the conference context is automatic and results in 100% accuracy. The user, in turn, is identified by the handle used during login and prior registration. We also noticed that, if necessary, a hybrid solution can be used. Identification of the conference context might be supported by the system infrastructure, while user identification is performed by the bot.

Functional tests confirmed that, owing to appropriate design, the selected method allows for correct recognition of the context with the frequency of errors below 1%. The observed limitation is a consequence of the context expiration mechanism built into the router. After a predetermined time (the test threshold was 2 h) the system forgets the conversation and, as a result, the context is lost. However, this behavior reflects the properties of real-life conversations where a considerable interval between statements requires re-contextualization.

4.2   Performance Analysis

A vital element of the verification of the developed research environment was to analyze its performance in the Google cloud computing infrastructure (GCP).

As part of the scalability study of the developed research infrastructure, a scalable system configuration was prepared, consisting of the following (see Fig. 3):

- load balancer mechanism exposing an anycast IP address for external traffic attached to a dedicated test domain, and handling traffic via the https protocol;
- a group of Compute Engine instances on which the conversation router was started: e2-standard-2 instances (2 vCPUs, 8 GB memory), with a configured policy of automatically adding instances upon reaching processor load of 60 %, in the 1–10 instance range;
- a load generating node: e2-standard-2 (2 vCPUs, 8 GB memory).

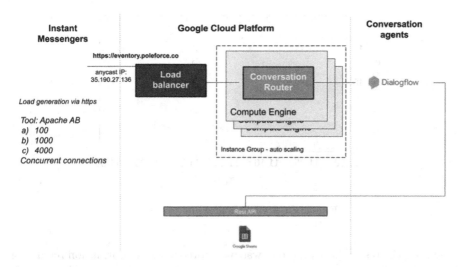

Fig. 3. GCP architecture for performance testing.

The load testing process consisted of generating messages for processing by the system, transmitting them via the web interface (http protocol, GET method) and measuring the system response time.

The tests were carried out using the Apache AB tool which simulates system load by making GET/POST queries to the given http/https address. Key parameters include the -n option, which specifies the number of queries to be executed, and the -c option, which specifies how many queries can be executed in parallel. The final parameter is the http address that is subject to testing.

The invocation queries for the start of the conversation and triggers the execution of the functional logic described in detail in the previous section. It ultimately returns the system's response to the transferred message. Using Apache AB enabled us to obtain statistical information about the processed queries – particularly their duration.

4.3   Tests Performed and Results Obtained

The tests began with generating load which corresponded to 20 clients operating in parallel.

These tests were then repeated for the following configurations respectively:

- 100 queries with 20 parallel connections;
- 200 queries with 100 parallel connections;
- 1,000 queries with 100 parallel connections;
- 4,000 queries with 200 parallel connections.

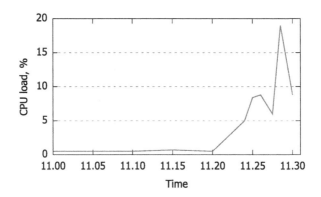

Fig. 4. GCP CPU load while testing.

During the tests, the server load was monitored, as was information regarding the number of open conversations on the conversation router – see Figs. 4, 5, 6.

The obtained performance results are very promising: 4,000 test messages were processed within 10.806 s; the median processing time of one query was 529 ms, while the maximum time was 1.366 s.

At the same time, the observed server load did not exceed 20% – this still leaves a lot of headroom which can be additionally increased by using more efficient machines (vertical scaling).

During tests the number of active conversations on the conversation router side exceeded 5,000, which is an order of magnitude greater than the 500 parallel conversations assumed in the application. These conversations were handled with qualitative parameters which stipulated that the response time must remain below the SLO (Service Level Objective) of 2 s.

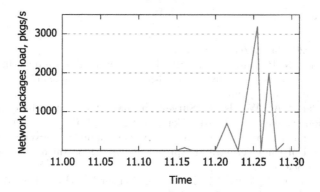

Fig. 5. GCP network packages load while testing.

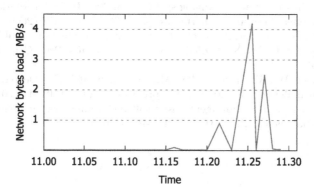

Fig. 6. GCP network bytes load while testing.

5   Conclusions

In this work we present an innovative approach to providing information to event and conference users using a natural language interface and a chatbot. The proposed system architecture facilitates exploitation of different conversation agents as well as integration with various instant messaging platforms. It also permits easy integration with organizer infrastructure and external data sources. An important part of the system is the Conversation Router which manages the flow of conversations and keeps track of their contexts. The heterogeneity and flexibility of the solution have been validated in functional tests conducted in a highly scalable, on-demand cloud environment (Google Cloud Platform). During performance tests, 4,000 messages were sent and processed by the system within 10.806 s. The server-side load of the test infrastructure which hosted the developed solution did not exceed 20%. This allows for further scaling of the amount of messages processed by the system. Further work on the presented system will focus on extending the context of the chatbot and applying advanced machine learning methods to conduct dialogue in an even more natural way.

References

1. ConfBot (2020). http://confbot.ninja/
2. EVABot - the NLP & ml powered HCM conversational assistant (2020). https:// eva.ai/evabot-technology-conversational-assistant-chatbot-hcm/
3. Morph (2020). https://morph.ai/
4. Abdul-Kader, S.A., Woods, J.: Survey on chatbot design techniques in speech conversation systems. Int. J. Adv. Comput. Sci. Appl. **6**(7), 72–80 (2015)
5. Cahn, J.: Chatbot: architecture, design, & development. University of Pennsylvania School of Engineering and Applied Science Department of Computer and Information Science (2017)
6. Huang, J., Zhou, M., Yang, D.: Extracting chatbot knowledge from online discussion forums. In: IJCAI, vol. 7, pp. 423–428 (2007)
7. Hung, V., Elvir, M., Gonzalez, A., DeMara, R.: Towards a method for evaluating naturalness in conversational dialog systems. In: 2009 IEEE International Conference on Systems, Man and Cybernetics, pp. 1236–1241. IEEE (2009)
8. Kuligowska, K.: Commercial chatbot: performance evaluation, usability metrics and quality standards of embodied conversational agents. Prof. Cent. Bus. Res. **2**, 1–16 (2015)
9. Li, J., Monroe, W., Ritter, A., Galley, M., Gao, J., Jurafsky, D.: Deep reinforcement learning for dialogue generation. arXiv preprint arXiv:1606.01541 (2016)
10. Velmovitsky, P.E.: iBot: an agent-based software framework for creating domain conversational agents. Ph.D. thesis, PUC-Rio (2018)

Network Systems and Applications

Single Failure Recovery in Distributed Social Network

Chandan Roy$^{(\boxtimes)}$ (iD), Dhiman Chakraborty$^{(\boxtimes)}$ (iD), Sushabhan Debnath$^{(\boxtimes)}$ (iD), Aradhita Mukherjee$^{(\boxtimes)}$ (iD), and Nabendu Chaki$^{(\boxtimes)}$ (iD)

Department of Computer Science and Engineering, University of Calcutta, Kolkata, West Bengal, India
nabendu@ieee.org

Abstract. This paper aims to propose a single failure recovery for a distributed social network. Single node and multi-node recovery are both non-trivial tasks. We have considered that a user is treated as a single independent server with its own independent storage and is focused only on single-user failure. Here, failure refers to the failure of the server as a whole. With these fundamental assumptions, a new fast recovery strategy for the single failed user is proposed. In this method, partial replication of node data is used for recovery of the node to reduce the storage issue.

Keywords: Distributed graph processing · Distributed social network · Node failure recovery

1 Introduction

A decentralized online social network (DOSN) [3] is an online social network implemented on a distributed social platform. A distributed system is a software system which is a connection of independent computers or nodes where coordination and communication among computers or nodes only happen through message passing, with the intention of working towards a common goal. This system will provide a viewpoint of being a single coherent system to the outside world. So the distributed social network platform will give the flexibility to the user over the control of their own data.

Multiple types of failure may occur in the social network, such as node failure, communication connection channel failure, infrastructure failure for network overload. Here node may be a single user or a group of users. In this paper, node failure detection and recovery of the node is the primary focus. Here actually node failure indicates data failure issues. After analysing the scenario, recovery of a failed user or node is a non-trivial task within a social network. Different solutions are available to solve the problem but graph partitioning [9] plays an

This research is supported by Visvesvaraya Project funded by Digital India Corporation, established by Ministry of Electronics & Information Technology, Government of India.

T.-P. Hong et al. (Eds.): ACIIDS 2021, CCIS 1371, pp. 203–215, 2021.
https://doi.org/10.1007/978-981-16-1685-3_17

important role to solve the problem. The graph partitioning is an NP-hard problem. The most popular solution is to use Secured Scuttlebutt protocol without graph partitioning. But using this method, full replication of every node data is a memory hazard. To improve the situation, checkpoint restart, checkpoint with log and replication-based recovery method with graph partitioning can be chosen. The main issues with these systems are lack of proper partitioning approach and absolute replication of node data (i.e. storage problem).

2   Related Work

There exist quite a few methodologies to solve the problem of failure node recovery in a distributed social network. We present a critical discussion on some of these in this section.

The organization of vertices of social graphs are not the same for every infrastructure of the distributed social graph. These graphs are built up using the distributed graph processing approach. Typically, the graph is partitioned into some subgraphs and the every subgraph is called 'node', where vertices are the members of the node. On the other hand, one vertex is considered as an independent node in some infrastructure.

Based on graph partitioning approach there are many types of failure recovery scheme available, where infrastructure of the social graph is being built on the basis of the recovery scheme. Those recovery schemes are checkpoint restart based - checkpoint with log-based, replication-based, etc.

A checkpoint based system [1, 14] at first takes the input graph and partitions it into multiple servers using edge-cut or vertex-cut method. Subsequently, the replicated vertices are generated in servers where there are edges joining to the origin vertex. This process is also called graph loading. The difference between edge-cut and vertex-cut is whether to replicate edges or not.

It is quite a hectic job for a social graph network system to maintain all execution of processes of a bulky graph without partitioning of the graph. Graph partitioning approaches have been used to get rid of bulky graph problems. Here, replication of vertices denotes replication of vertices data. This data redundancy helps to recover failed vertices easily. Storage of metadata snapshots is located in the servers. This provides easy and speedy access to the metadata and facilitates parallel recovery of many nodes simultaneously.

In this system, proper graph partitioning is too important. Due to improper graph partitioning, vertices in the servers may not be balanced. This imbalance affects the normal execution time severely. Extra (i.e. more than the needed) replication of vertices unnecessarily increases the graph size. This creates graph processing overhead. Over replication may also lead to server failure. If the number of servers and number of vertices within the servers increase then the number and volume of storage every between servers also increases. These, in turn, increase expenses.

A new system, checkpoint with a log for failure node recovery, is introduced [6, 11] to scale out the performance. The primary feature of the checkpoint with log convention is to limit the recomputation to the subgraphs originally residing in failed nodes. Thus it aims to overcome some disadvantages of checkpoint restart based systems that we discussed earlier. In this system, every node stores the copy of their data locally and globally, but the checkpoint restart system stores the copy of data only on globally recognised storage. Iterative node computation helps to get very recent data easily.

Synchronization between local storage and global storage data should be done properly otherwise data mismatch will occur severely. Storage problem and storage expense much greater than checkpoint restart, as a copy of single data store twice, locally and globally. Handling such large data may slow down the system.

Replication based recovery systems can be used to avoid the storage's related problems. In a Replication based recovery [1] system follow social graph system where the input graph is partitioned into some server using any kind of graph partitioning algorithm. There are some types of replica such as 'Mirror Replica', 'Balancing Replica', 'Complementary Replica'.

Cost of storing information on the checkpoint based systems is more expensive than the replication based system. As in the checkpoint based system, every checkpoint stores many information and the number of checkpoints increases with the size of input graph proportionally. It is also time-consuming to update all the checkpoints in the checkpoint based system frequently. It affects the time of normal execution jobs of vertices. There is no headache of checkpoints in the replication based system. No mechanism of balancing the number of vertices is present in servers of checkpoint based systems. For such disparity, a server may crash due to overload of execution jobs of vertices. Synchronization of servers may be delayed for this reason. If a server with a large number of vertices crashes, the system will also take a long time to recover them also. In contrast with this system, the replication based system has a balancing mechanism to balance the number of vertices among servers. So the overhead of execution of each server is low and recovery time of a crushed server is short.

Synchronization messages should be sent in high frequency among nodes otherwise vertex states updation may be affected and also recovery time lengthen. Creation of unnecessary large number of replicas overhead for communication during normal execution, much expensive also for storage purpose. Here all failure recovery processes are done for a server, no specific scope is provided for single vertex failure. A server contains more than one vertex. If a server crashes that means many vertices are crushed. It will be efficient if a single vertex can be treated as a server itself.

In the distributed systems research area replication is mainly used to provide fault tolerance. In active replication, each client request is processed by all the servers. This requires that the process hosted by the servers is deterministic. Deterministic means that, given the same initial state and a request sequence, all processes will produce the same response sequence and end up in the same final state. In order to make all the servers receive the same sequence of operations, an atomic broadcast protocol must be used. An atomic-broadcast [4] protocol guarantees that either all the servers receive a message or none, plus that they all receive messages in the same order. The big disadvantage for active replication is that in practice most of the real-world servers are non-deterministic. In passive replication, there is only one server (called primary) that processes client requests. After processing a request, the primary server updates the state on the other (backup) servers and sends back the response to the client. If the primary server fails, one of the backup servers takes its place. Passive replication may be used even for non-deterministic processes. The disadvantage of passive replication compared to active is that in case of failure the response is delayed.

Still, active replication is the preferable choice when dealing with real-time systems that require a quick response even under the presence of faults or with systems that must handle byzantine faults [5].

One of the most popular projects on the basis of single vertex single node concept is 'Scuttlebutt' and also there is an upgraded version of this project named 'Secure Scuttlebutt'.

Without graph partitioning, one vertex of the graph is considered as one node, and the distributed replication-based system available called Secure Scuttlebutt (SSB) [12,13]. It is a user friendly protocol for building decentralized applications that work well offline and that no one person can manipulate.

Typically, this protocol follows a peer-to-peer network topology. Peers are to be considered here as users. Every user hosts their own content and the content of the peers they follow, which provides easy fault recovery and eventual consistency.

There is another integrated protocol with this called 'Dark Crystal' [2,7], which is used to share a secret(it may be a password or some other) only with a specific number of friends. This dark crystal mechanism uses Shamir's Secret Sharing Algorithm [10], as its message encryption technique.

In another work [7,8], a new 'Neighbor Replication Distribution Technique (NRDT)' technique is introduced to optimize the failure recovery. An index server (IS) and a heartbeat monitoring server (HBM) are introduced, where IS stores replicas (log, service, volume, data) of HBM and vice versa. HBM will monitor and detect the failure occurrence for the application. This fault detection component will continuously monitor the system and update the current status of each application in the IS and provide a space for recovery actions to be taken in a variety of ways such as kill and restart, fail-over, virtual IP and others. This technique is in fact an improvement over the Two Replica Distribution Technique as it uses more than two replicas for failure recovery purposes. However, no solution is provided for the case when HBM and IS fail at the same time. Besides, storage overhead is also another disadvantage as it requires to store full replica of the original data in more than two servers.

3   Problem Statement

This paper focuses on a single failure recovery in a distributed social network. Here it is also considered that the failure means failure of the server and a single user is mapped to a single server in distributed graph processing. In this paper, a new method is introduced to recover a failed user or node in a distributed social network. Along with this, an efficient recovery process is also proposed with a single failure detection mechanism.

4   Proposed Methodology

Using distributed graph processing, recovery methodology of extinct users in a network immensely depends on the infrastructure of the network. Hence, the recovery model will be fabricated after the construction of the network. So, the infrastructure of the network will be explained here first. Then the recovery process will be elaborated.

4.1   Framework

Distributed Social network's configuration is identical to a social graph representation. Here nodes and connections among nodes are alike of vertex and edge of a graph respectively. Also, nodes can be treated as users or user groups and connections among nodes can be treated as communication channels among interested groups or users of the network. Groups are categorized into two types viz. Normal Group and Special Group. Here every group is considered as a single user. Normal Groups are made up of known users or invited friends of group users, where Special Groups assist a new user (i.e. unknown to other users of the network) to make friendship with other users according to the new user's choices/preferences. No user can be alone at least added to any Special group. If any user is found alone then the user must not be a part of the network. Not being a part of the network, that user not only cannot communicate with others but also no process will be accomplished by that user. A new user will come into existence in the network with a highly secured sign up process. Then the user will be provided with credentials like user id, passwords etc. for login. All the existing users of the network are online when the distributed social network is working. If any node is offline, that means the node is considered as dead.

To start communication with other users, a user goes through some phases like Initialization state, Ready state, Running state. There is also an important mechanism to retrieve a crippled user to its previous condition in the network through Recovery state. After logging in of a user, the user enters into Initialization state.

In Initialization state, the user executes some necessary processes such that the network can access the user and set a starting position.

Next is Ready state. In this state, every user is getting ready to connect to each other for communication.

After that, in the Running state, users are connected and they can start communication among them. While they are communicating or connected, every user backs up their conversation into a log file and keeps it to them personally and also shares the log file data chunk by chunk to their neighbour users for backing up the data partially further by a special back up process using Round Robin fashion. Along with, every user publishes and receives heartbeat signals periodically to perceive the neighbour users' existence in the network and they maintain an updated neighbour list. It can be considered a user lost it's existence by neighbour users when the user does not send heartbeat signals to the neighbours. Then the user is in the Dead state. When a user in the Dead state that means it lost its log file too and having the only login credentials like user id, password etc. So, the dead user will log in and after proceeding Initialization state it will enter into a Recovery state.

In the Recovery state, the user publishes request-indication in the network for getting back its log file data with the back up files of neighbour nodes in apple-pie order from the neighbours. After getting back up, it will be recovered and connected to neighbours, the user normally enters into the Running state through Ready state and updates the neighbour list.

4.2   Approach for Simulation

All nodes of the network should be online. If any node is recognised as offline, then the node will be considered as a dead node. It will be firmly established that any other node will not get failure until the last failure one will get recovery. No new outsider node can be added with that no insider node can be removed to or from the intended network except when a single failure of a node will occur. It is also under consideration that communication channels are not noisy and error-free. No unintended messages are allowed in planned communication. There will exist no group (i.e. a group of nodes which is treated as a single node) in the simulated network.

Prerequisites for Simulation

Input Dataset. An intended Dataset will be taken as input for the formation of the graph alike structure of the network system. This Dataset contains Vertex information (i.e. node information) and Edge information (i.e. connection edges information among nodes) of the network. This information will be stored in comma-separated values (CSV) file.

Memory Management. Every node has its own directory to store the Self log file and Backup file(s) for the neighbour(s). A node can read and write only its own log file. Self log files store conversation messages in a line-wise manner. Every new message will append on the next line of the file where the last message was stored. Along with this, another log file is a Backup file. A specific backup file will store messages of a particular neighbour node for partial backup. So, the number of the backup file(s) within a node's directory depends on the number of neighbour(s). Backup files are the append-only encrypted log file and can only be accessed for any recovery purpose of the owning node of original messages. All log files will be stored in the memory as text(.txt) files and the names of the backup files are the same as owning node's id and also the names of directories of the nodes are the same as the owning node's id.

Network Topology. Here peer-to-peer distributed network topology will be selected to develop the communication network. The peer-to-peer (P2P) technologies assure self-organizing and secure-by-design networks in which the final data sovereignty holds to the users. Such networks support end-to-end communication, no need of any centralized server, uncompromising access control, anonymity and resilience against massive data leaks. Here, All peers are suppliers as well as consumers of resources (i.e. data).

Communication Channel. Here the point to be noted, peers of the simulated network established through a virtual local area network (VLAN). So, connectionless communication will be used for this simulation. User datagram protocol(UDP) follows connectionless communication and this protocol will be implemented for message broadcasting purposes. Through transmission control protocol (TCP) peers create a stable connection to begin the proper data transfer process.

These communication channels can be established using Socket programming or Language-specific facilities or a combination of both.

Communication Pattern. One to one connection between nodes is a crucial part of a Social Network. Here, one to one connection will be established using the Push-Pull pattern through the TCP channel. In this pattern, every node acts as a consumer as well as producer of data. For one to one communication, when one node sends data to another node through the communication channel, this is called 'push', i.e. msg pushed in the channel, and when the receiver node receives the message from the sender, this is called 'pull', i.e. message pulled from the channel.

To get rid of the hazards of manual typing of communication messages, and for the sake of simplicity, nodes will use auto-generate messages for communication in the Publish-Subscribe pattern. It will save time and no need to depend on input devices. In the publish-subscribe messaging pattern, senders of messages are called publishers, receivers are called subscribers.

Message Types. Nodes will communicate with each other through Communication messages. Message communication can be completed by using push-pull and pub-sub pattern between or among the nodes. Communication messages will contain a specified size of a string.

On the other hand, there is a special type of message called the status message. There are two types of status messages present, named 'LOST' and 'READY' messages, help to realise the status of a node. At the time of initialization of a node, the 'READY' message will be sent periodically to its peers to inform them that the node is ready to connect with each other. The 'READY' message will be made up of Ready signal, self port number, self-id and other necessary information. It is also used to realise the aliveness of a node. Before the recovery of a lost node, at the time of re-logging in, the node will broadcast the 'LOST' message to its network such that old friend peers(i.e. neighbours of the node when it was functional) can help to recover the lost node. The 'LOST' message will carry Lost signal(i.e. a node wants to recover itself), self port number, self-id and other necessary information.

An Alive Node Existence Sensing Technology, Heartbeat. There is a heartbeat technology to sense a node whether it is alive or not in the network system.

Heartbeat messages generally used for failure detection purposes. To reduce the overhead for generating the heartbeat signal, the 'READY' signal is also used as the heartbeat signal here. Failure detection of a node is fully dependent on heartbeat signal.

Overall Simulation Procedure

Node Creation. At the time of the creation of the social network, social graph information is gathered from the given input data set. Neighbourhood list of a node maintains an order where the first element of the list is the self vertex id and all other neighbour nodes id listed after that. Every node has a unique IP address and one or many port numbers. A Universal port is assigned for broadcast message/s and listening broadcasted message/s in the network. The algorithm of this process is attached to the appendix section, named 'Algorithm 1 Node Creation'.

Node Operation. After collecting all the information about the social graph, every node creates a node process. The directory creation, log file and backup file creation are done before making connections among nodes. If an aforesaid directory and files already exist then no need to create further and previous directories and files will be utilized instead of the new one. There are also some sub-processes related to the main node process. Those are message communication process, backup process, aliveness detection process, heart beating process etc.

After loading all the sub-processes, the main node process will be ready for communication in the network. So after getting ready a node process will broadcast a 'READY' message throughout the network. As per communication pattern, every node acts as a publisher as well as subscriber also. If the 'READY' message will be received for the first time by any node, after extracting the encrypted information of the message, node id will be compared with the neighbour list. If the node id is found in the list, the receiver node will connect to the sender. Inversely when receiver acts as sender and sender acts receiver, and the same previous processes are executed, the inverse connection is also established. So both ways connections are established between to nodes by this process. In the same way, all the friend nodes of the network will be connected.

Four sub-processes of every node process will execute as long as the lifetime of nodes. Those sub-processes are message communication process, heart beating process, aliveness detection process, backup sharing and collection process. Message communication is executed through the pub-sub pattern. Every node sends a periodical heartbeat to their neighbours. Aliveness detection process uses the heartbeat signal to recognise whether the neighbour node(s) is alive or not. If any node detects any neighbour node as dead, the associated processes will be terminated and the communication channel will be dissociated by the detector node. There are backup sharing and collection subprocess to backup node data to its friend nodes in a specific manner.

Recovery of a node only can possible if there is any backup of the data of the node. As this is not a centralized system, so centralized backup is not possible. By the property of a distributed system, storing of the backup can only be possible through replication. There can be many types of replication possible. To remove the superfluousness of the backup data, partial replication is done here. Partial replication is done by round-robin fashion.

A line of the log file will backup with which neighbour node that will rely on a round-robin fashion. This can be done by decreasing one from the total number of neighbour nodes, then the value is modulo divided by Line number of the file and the answer will be increasing by one. The final answer will be the index of the neighbourhood list where the first element is self node id and remaining are the neighbours. Here the zero indexing is followed. Through the index of the neighbourhood list, the target node id can be fetched. round_robin_id = self.neighbourhood[line_no % (len(self.neighbourhood) - 1) + 1]

Here every message line will be encrypted at the time of sending with the original line number and node id, this encrypted message will be sent to a specific node which is specified by the round-robin technique. So unicasting is needed and it is done using the push-pull pattern.

On the other hand, the receiving node receives the messages, here the order of the receiving messages of a specific node is not important for partial backup, as, at the time of appending at receiver's end, the messages will be decrypted for line numbers and messages and stored into the backup file of the owning node. Here the partial data sharing is an instant sharing, the sharing of the line-wise data among neighbours whenever a new line is available in the log file. This procedure will occur for each and every node in the network for backup purposes.

After a node failure, all the data including the neighbour list of the failed node destroyed. So the node is now in the offline state.

Its algorithm is added to the appendix section as 'Algorithm 2 Node Operation'.

Node Recovery: After a single failure, the neighbour nodes of the failed node detect that fault and disconnect the failed node. When the failed node wishes to recover, it will have to just login into the network using the login credentials which is the node-id only. Though the node doesn't have its neighbour list, it only creates a directory for the self-id in the specified path and creates empty log files inside that directory named as the node-id has. The unique IP address and the port will be assigned to it, and background subprocesses are being loaded. After the completion of the background sub-processes, the node process is being ready to execute, but there is no neighbour list to create connection and communication. So the node can recognize itself as a warrior node (i.e. previously failed and just come back to the network), and it has to collect all the backup to recover its previous state.

To collect the backup it will broadcast a 'LOST' message through a universal port, it denotes that the node is failed and wants to be recovered. Other existing nodes present in the network, receive the 'LOST' message and decrypt it and check whether the 'LOST' message is containing id present in their neighbour list or not. If the node id is present in the neighbour list, then the node unicasts the line-wise data that present in the log file in the name of the warrior node using the push-pull pattern. This message also contains the id of the warrior node. The node also unicasts encrypted backup messages with the line number and node id of their own part to store the backup again with the warrior node. If node id is not present in the neighbour list, then the receiving node discards the 'LOST' message.

On the other side, the warrior node receives the encrypted messages from the sender nodes. Then the messages will be decrypted, if it contains the own id, the messages will be stored in the own log file with the specific line number, if it contains other than own id, the id will be added to the neighbour list if not listed, and creates a backup file(s) for the neighbour(s). Also, store their back up messages with line no in the corresponding file. After receiving all the backup files, it will establish all the connections as the neighbour list shows and recovers successfully to the previous full-fledged working condition.

The algorithm of this crucial process is described in the appendix section with the title, 'Algorithm 3 Node Recovery'.

5   Experimental Study

5.1   System Specification

Experimented in a personal laptop equipped with Intel(R) core(^{TM}) i5-8250 CPU @ 1.60GHz 1.80 GHz processor, 8 GB of memory, 1 TB SATA HDD, installed Windows 10 Operating System. It is also tested in the Linux Operating System.

5.2   Result

Every case contains five data sets with five different topological representations and every representation has the same number of nodes. Among representations, only a distinct node remains unchanged with its neighbours in every topology. Every representation has failure detection time and failure recovery time. In every case, the average failure detection time and the average failure recovery time of five representations are the failure detection time and the failure recovery time itself.

After executing the experimental setup several times, it is seen that the recovery of a failed node achieved an accuracy level in between 95% to 100%.

In case of detection of a failed node (Fig. 1 (a)) in the system no matter when the failure has happened (i.e. how long the process is been running, how much

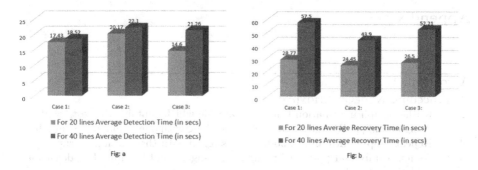

Fig. 1. Representation of average failure detection time and average failure recovery time

data it has appended in its log file). When it fails its neighbour node detects the failure pretty fast in between 15–30 s.

In case of recovery of a failed node (Fig. 1 (b)), for the obvious reason if the failed node had a larger amount of data in its log file it take more time to recover. But there is a good thing that if a failed node's neighbour number is more, the shared data chunk to every neighbour node will be less, then recovery time eventually will be less.

Considering an example, there are n number nodes in a system, and if a node has (n − 1) neighbour and the node fails then recovery time will be utmost less than any other condition. Hence, the recovery time will be less as the number of neighbours of a failed node is high.

6   Conclusion

Different technologies are available to build distributed social networks. It is already discussed how a single node failure can be recovered using different technologies that we have already discussed. Keeping an eagle eye vision with their merits and demerits, new methodology is proposed in this paper. In this methodology, a failed node can be recovered through the co-operative collaboration by getting back partial replication of the failed node data from the friend nodes. The simulation is showing good accuracy and less recovery time to recover a failed node with some predefined benchmark dataset.

Appendix

Algorithm 1. Node Creation

1: **procedure** NODECREATION(*dataset*)
2:      Gather graph information from the data file which contains the test graph.
3:      Assign a unique IP address to every node along with one or many port numbers.
4:      Repeat Step 3 until all the nodes are assigned with their desired ports.
5:      Assign a universal port for broadcast message/s and listening broadcast messages in the network.
6:      All the nodes of the graph are created.

Algorithm 2. Node Operation

1: **procedure** NODEOPERATION(*Nodes*)
2:      Each node creates a node process.
3:      The directory creation, log file and backup file creation are carried out.
4:      If step 3 is completed then no more new directories will be created for the same node.
5:      Sub-processes related to each node start execution in the background.
6:      The primary node process for each node starts connecting with their neighbour nodes by broadcasting a 'READY' message.
7:      When a 'READY' message is received by any node, node id will be compared with the receiver's neighbour list.
8:      If neighbour list contents of Step 7 are matched, the nodes get connected else they do not get connected.
9:      Step 6 and step 7 will be executed until all the friend nodes of the network get connected.
10:      Entire network gets established.
11:      Communication among the nodes starts and messages are logged into the log file of every node.
12:      A heartbeat process is running in the background as an aliveness indicator, which broadcasts a periodic heartbeat signal in the network.
13:      A partial backup process is also running in the background of every node process, by which the log file will be backed up among the neighbours using round-robin fashion.

Algorithm 3. Node Recovery

1: **procedure** NODERECOVERY(*SocialNetwork*)
2:     The failed node logs into the network using login credentials which is nodeID.
3:     The node creates a directory for the selfID and an empty log file.
4:     Unique IP address and port are assigned to it and background subprocesses are being loaded.
5:     The node recognizes itself as 'Warrior' node as it has to collect all backup to recover its previous state.
6:     The 'Warrior' node broadcasts a 'LOST' message through a universal port.
7:     Others nodes receive a 'LOST' message and decrypt it to check whether its containing id is present in their neighbour list or not.
8:     If Step 7 is fulfilled then the node/s sends the backup data to the warrior node, else the 'LOST' message is discarded.
9:     The 'Warrior' node gets recovered by receiving backup from all the other nodes.

References

1. Chen, R., Shi, J., Chen, Y., Zang, B., Guan, H., Chen, H.: PowerLyra: differentiated graph computation and partitioning on skewed graphs. ACM Trans. Parallel Comput. (TOPC) **5**(3), 1–39 (2019)
2. M Collective: Dark crystal. Accessed 30 July 2020
3. Datta, A., Buchegger, S., Vu, L.H., Strufe, T., Rzadca, K.: Decentralized online social networks. In: Furht, B. (ed.) Handbook of Social Network Technologies and Applications, pp. 349–378. Springer, Boston (2010). https://doi.org/10.1007/978-1-4419-7142-5_17
4. Défago, X., Schiper, A., Urbán, P.: Total order broadcast and multicast algorithms: taxonomy and survey. ACM Comput. Surv. (CSUR) **36**(4), 372–421 (2004)
5. Lamport, L., Shostak, R., Pease, M.: The Byzantine generals problem ACM transactions on programming languages and systems, vol. 4, no. 3, pp. 382–401 (1982)
6. Lu, W., Shen, Y., Wang, T., Zhang, M., Jagadish, H.V., Du, X.: Fast failure recovery in vertex-centric distributed graph processing systems. IEEE Trans. Knowl. Data Eng. **31**(4), 733–746 (2018)
7. Noor, A.S.M., Deris, M.M.: Failure recovery mechanism in neighbor replica distribution architecture. In: Zhu, R., Zhang, Y., Liu, B., Liu, C. (eds.) ICICA 2010. LNCS, vol. 6377, pp. 41–48. Springer, Heidelberg (2010). https://doi.org/10.1007/978-3-642-16167-4_6
8. Noor, A.S.M., Deris, M.M.: Fail-stop failure recovery in neighbor replica environment. Procedia Comput. Sci. **19**, 1040–1045 (2013)
9. Onizuka, M., Fujimori, T., Shiokawa, H.: Graph partitioning for distributed graph processing. Data Sci. Eng. **2**(1), 94–105 (2017)
10. Shamir, A.: How to share a secret. Commun. ACM **22**(11), 612–613 (1979)
11. Shen, Y., Chen, G., Jagadish, H., Lu, W., Ooi, B.C., Tudor, B.M.: Fast failure recovery in distributed graph processing systems. Proc. VLDB Endow. **8**(4), 437–448 (2014)
12. Tarr, D.: Secure Scuttlebutt. Accessed 29 July 2020
13. Tarr, D., Lavoie, E., Meyer, A., Tschudin, C.: Secure scuttlebutt: an identity-centric protocol for subjective and decentralized applications. In: Proceedings of the 6th ACM Conference on Information-Centric Networking, pp. 1–11 (2019)
14. Yan, D., Cheng, J., Yang, F.: Lightweight fault tolerance in large-scale distributed graph processing. arXiv preprint arXiv:1601.06496 (2016)

Improving Reading Accuracy in Personal Information Distribution Systems Using Smartwatches and Wireless Headphones

Takuya Ogawa$^{(\boxtimes)}$, Keiichi Endo, Hisayasu Kuroda, and Shinya Kobayashi

Graduate School of Science and Engineering, Ehime University, Matsuyama, Japan
ogawa@koblab.cs.ehime-u.ac.jp

Abstract. There is a huge amount of news information on the Internet, and users have the advantage of being able to access a large amount of information. However, users are not interested in all the news that exists on the Internet. In our previous research, we developed a smartphone application that enables users to provide information via a smartwatch in order to solve the information overload in news information on the Internet. However, considering the screen size of the smartwatch, it was necessary to limit the information that could be obtained, and we could not provide the information that users wanted. Therefore, we propose an application that provides news information by voice and allows users to read out the news from a smartwatch. By learning the user's interests, this application automatically selects articles and provides news that the user is interested in with priority. In addition, by providing information by voice, news information can be obtained in a crowded train or while driving a car, which is expected to improve the convenience of the system. As a result of the evaluation experiment, we were able to provide more information to the users than the conventional applications. However, there was a problem of misreading. In this paper, we refer to misreadings and discuss countermeasures.

Keywords: Information distribution system · User-aware · Smartwatch · Wireless headphone

1 Introduction

The Internet is filled with a vast amount of news information, but not all of it is of interest to users. Therefore, when users read the news, they need to find the information they are interested in among the vast amount of news information. When the amount of information increases and it becomes difficult for users to sort the information, it is called information overload. We developed PINOT (Personalized INformation On Television screen), an information distribution system for individuals, to solve the problem of information overload [1]. PINOT displays news information in ticker format, and the user can pause, fast-forward, and rewind on the displayed information. PINOT inferred the user's interests

© Springer Nature Singapore Pte Ltd. 2021
T.-P. Hong et al. (Eds.): ACIIDS 2021, CCIS 1371, pp. 216–226, 2021.
https://doi.org/10.1007/978-981-16-1685-3_18

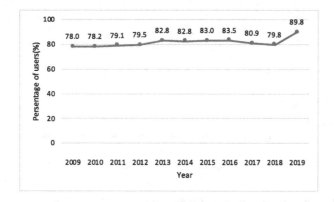

Fig. 1. Internet usage rate

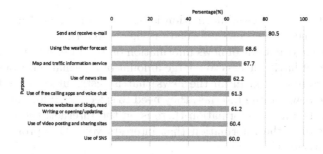

Fig. 2. Purpose of Internet use

from their actions and provided news that the user was interested in. After that, we developed a smartphone application, PINOT, to meet the widespread use of smartphones [1]. However, this application had the problem of not being able to learn the user's interest in news delivered during the period when the user was not able to use smartphone. To solve this problem, we proposed a smartwatch cooperation PINOT [2]. However, it was necessary to limit the information to be displayed in consideration of the screen size of the smartwatch. As a result, it did not become an app that users wanted to use.

2   Research Backgrounds

2.1   Spread of the Internet

In recent years, many people have been using the Internet. The percentage of people who used the Internet at least once during the past year is shown in the Fig. 1 [3]. This figure shows that 89.8% of people use the Internet in 2019.

One of the Internet services that people use is to obtain Internet news. Figure 2 shows a selection of Internet users' purposes and uses of the Internet, the percentage of which exceeds 60% [4]. The percentage of people who use

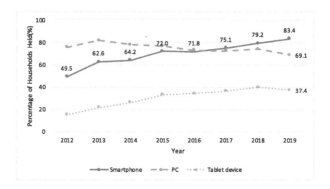

Fig. 3. Ownership rate of information terminals (household)

the Internet to obtain news information is 62.2%. From this figure, we can see that many people obtain news information via the Internet.

There is a huge amount of news information on the Internet. This gives users the advantage of having access to a large amount of information. However, users are not always interested in all the news that is distributed. Therefore, when users read the news on the Internet, they need to select the information they are interested in. The more information there is on the Internet, the more difficult it is to select only the information that interests you. This problem is called information overload.

2.2 Personalized Information Distribution System PINOT

To solve the problem of information overload caused by the increase in the amount of information on the Internet, a TV-based personalized information distribution system, PINOT, was proposed. PINOT is used by connecting the set-top box to the TV and the Internet. The set-top box acquires news information from the Internet, filters it according to the user's interests, and displays it on the TV screen. The user selects the news information by skipping or pausing the character information displayed on the TV. PINOT updates the filter by learning and analogizing the user's interests from the user's operations. PINOT automatically provides the news information that users want to know, which helps to solve the problem of information overload. In addition, since the system is based on a television, users who are not used to operating information terminals can also obtain information.

2.3 Smartphone Penetration

In recent years, smartphones have become popular among the general public, and many people are using them as information terminals. Figure 3 shows the changes in household ownership of smartphones, PCs, and tablet devices. As

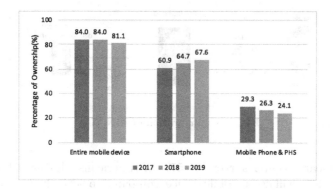

Fig. 4. Mobile device ownership (individual)

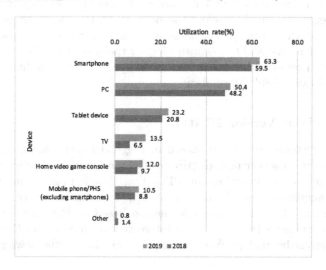

Fig. 5. Percentage of terminals using the Internet

Fig. 3 shows, as of 2019, 83.4% of households in Japan owned a smartphone, which is more than the percentage of households that owned a computer [3].

Figure 4 shows the individual ownership rate of mobile devices. As of 2019, 67.6% of individuals own a smartphone, indicating that many individuals own a smartphone.

Many people also use Internet services via smartphones. Figure 5 shows a graph showing which terminal is used to access the Internet. This figure shows that 63.3% of people use the internet using their smartphones as of 2019. It can be said that smartphones are used in a very high percentage of cases compared to other devices.

Since many people use the Internet on their smartphones, it is expected that many people will use their smartphones to get the news. In fact, according

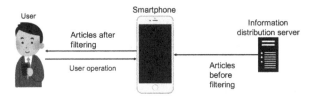

Fig. 6. Configuration of PINOT

to a Pew Research center survey [5], 57% of Americans who get Internet news frequently in 2019 will use a mobile device, compared to 30% who use a computer.

The amount of news available in news apps is enormous, and not all of it is useful to users. Since each user has different interests and targets of interest, the users themselves need to select the news when they want to get the news information. Sorting through the huge amount of news to find something of interest is burdensome and time-consuming for the user. There is a need to solve the information overload in smartphone news applications to reduce the burden on users to obtain useful information.

2.4   Smartphone Version PINOT

The PINOT system structure is shown in Fig. 6. PINOT consists of an information distribution server that distributes the news and a smartphone used by the user. When a user launches the PINOT application on a smartphone, the smartphone acquires news information from the information distribution server. PINOT filters the retrieved news information based on the user's interest and provides only the news that the user is interested in. In addition, PINOT learns the user's interest by analogy from the user's operation on the news provided to the user.

2.5   Problem of Smartphone Version PINOT

In order to provide articles that accurately match the user's interests, we have to ask users to read the article headlines and select the articles they are interested in from them. However, the articles provided by PINOT to users are limited to those that are available at the time the user launches the PINOT application. Therefore, articles that were distributed during the period when the user did not use the app will not be provided to the user, even if the content is relevant to the user's interests. In such a situation, the opportunity for the app to learn the user's interests is lost. As learning opportunities decrease, the articles provided to users may not take their interests into account exactly. In other words, users are unable to obtain the information they are interested in. In order to reduce the need for users to select information, we need to prevent the loss of learning opportunities.

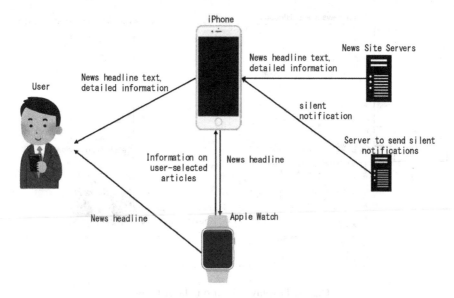

Fig. 7. Configuration of Smartwatch Cooperation PINOT

2.6 Widespread Use of Smartwatches

The size of the smartwatch market was $12,412 million in 2019 and is expected to increase thereafter [6]. As a means of providing users with information from their smartphones, smartwatches can be used by many people. However, there are few applications that can provide news information to users using smartwatches. Even fewer of these apps provide information that takes into account the interests of the user.

One of the features of a smartwatch is that it is a device worn by the user. Wearing it allows us to use it in more situations than with devices such as smartphones. For example, in a crowded train, it may be difficult to use a smartphone because the user is forced to be in a different position than usual. In addition, it is difficult to pick up the device if it falls, so it is not easy to use a smartphone in a crowded train. On the other hand, smartwatches are designed to be operated by the other hand only, so they are relatively easy to operate and there is no risk of falling. Thus, under certain conditions, smartwatches can be more useful devices than smartphones.

2.7 Smartwatch Cooperation PINOT

One of the reasons why users do not launch the smartphone version PINOT is that they are unable to operate their smartphones. We proposed a smartwatch cooperation PINOT, which is a smartphone version PINOT that uses a smartwatch so that news information can be confirmed even in such a situation [2]. The smartwatch cooperation PINOT adds news provision function on smartwatch to

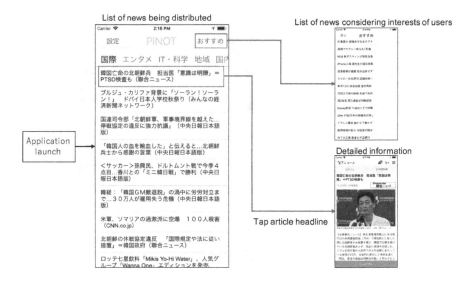

Fig. 8. Display screen on the iPhone

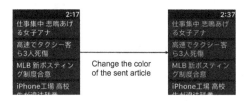

Fig. 9. Display screen on Apple Watch

the existing smartphone version PINOT. However, it is only the headline of the news that can be checked on the smartwatch, and to read the details, launch the app on the smartphone. When the news headline provided on the smartwatch is news information of interest, the user launches a smartphone app to learn more about the news. When launching the smartphone app, we expected users to look through other news items, which we believed would lead to an increase in the number of times the app would learn about the user's interests.

Figure 7 shows the system configuration of the smartwatch cooperation PINOT. The smartwatch used in this paper is the Apple Watch, which currently has the largest market share [7]. Since Apple Watch is a smart device for iPhone, the smartphone used is iPhone, and smartwatch cooperation PINOT is an iOS application.

The smartwatch cooperation PINOT consists of a server that sends silent notifications, a news site server, an iPhone and an Apple Watch used by the user.

The display screen on the iPhone is shown in Fig. 8. When the user launches the iPhone app, a list of current news items is displayed. When the user taps on the headline, a web page with detailed information about the news article is displayed. When the user taps the "Recommendations" button at the top right of the list screen, the system filters the current news based on the user's interest and displays only the news that is determined to be of interest to the user.

The display screen on the Apple Watch is shown in Fig. 9. Apple Watch displays the articles sent from the iPhone. By tapping an article that the user wants to read, the user can designate it as an "Articles read later".

The information on the articles designated as "Articles read later" is sent to the iPhone. "Articles read later" is displayed when the iPhone app is launched by the user. By tapping on this headline, users can get detailed information about the article. To find out more about "Articles read later", users need to use the iPhone application. That's what triggers the user to launch the application. When users launch the app, they will see other news as well. If the news was of interest to the users, they would read the news, and thus the number of times they learned would increase.

2.8   Problem of Smartwatch Cooperation PINOT

The results of the evaluation experiment of smartwatch cooperation PINOT showed no change in the learning frequency of the user's interest. One of the reasons for this was the limited information provided by the Apple Watch. In the smartwatch cooperation PINOT, the information provided to users from the Apple Watch was limited to the news article headline text, taking into account the screen size of the Apple Watch. However, the information that users want is not the headline but the detailed information, and the use of smartphones is indispensable to obtain the information they want by using smartwatch cooperation PINOT. Therefore, we believe that the benefits of using smartwatches were diminished and the app did not become something that users wanted to use.

In order to learn user's interests in situations where users cannot use their smartphones, we believe that an app with a mechanism to provide information to users without limiting them is required.

3   Proposed System

To solve the problem of smartwatch cooperation PINOT, we propose NEAR (News EAR), an application that provides audio news information to users using a smartwatch and wireless Headphones [8]. In the case of providing information by voice, there is no need to consider the screen size and therefore no need to limit the information. In addition, we expect to increase the number of situations in which users can use the system to obtain information without gazing at the screen, and thus increase the opportunities for learning based on their interests.

The configuration of NEAR is shown in Fig. 10.

NEAR consists of a server that provides site information, a server for news sites, an iPhone used by users, an Apple Watch, and wireless headphones. The server that provides the site information has the names and URLs of the RSS sites, and the user selects a news site from them and registers the news site with the application. The server of the news site provides news information to the iPhone. The iPhone provides news information to the user, and learns the user's interest from the user's behavior toward the provided news information. Users can operate voice reading function on Apple Watch. By allowing users to operate the voice reading function from Apple Watch, they will be able to retrieve information without having to operate their iPhones.

By learning the user's interests even in the provision of information by voice, the application will be able to learn more and it will be possible to select articles accurately.

By making it possible to operate the device from the Apple Watch, users will be able to obtain detailed information on the news they are interested in without touching their iPhones at all.

4   Evaluation and Improvement of the Proposed System

In order to evaluate the usefulness of NEAR, we conducted an evaluation experiment. In this experiment, we evaluate how the use of the Apple Watch or not makes a difference when using NEAR.

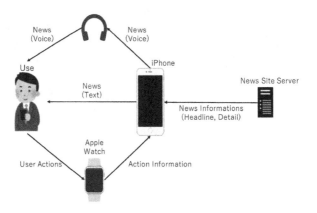

Fig. 10. Configuration of NEAR

Table 1. Providing information through voice information

	Number of times to read	Number of learning interests
With Apple Watch	13.3	5.9
Without Apple Watch	1.0	0.3

In our evaluation experiment, we ask multiple users to use the app, and record the number of times the app is launched, the number of times the news is provided by text and voice information, and the number of times the user

Table 2. Providing information through textual information

	Number of app launches	Number of articles displayed	Number of learning interests
With Apple Watch	1.4	40.0	2.1
Without Apple Watch	1.0	27.9	0.8

learns about their interests. In this experiment, we investigate the usage of the app among 9 users who use the Apple Watch and 12 users who do not use the AppleWatch. By comparing app usage between users who use the Apple Watch and those who do not use the Apple Watch, we evaluate whether the presence or absence of the Apple Watch has an impact.

4.1   Experimental Results and Discussion

Table 1 shows the number of times information was provided by voice and the number of times interest was learned by speech function. For the items in the table, we give the average per user per day.

From Table 1, we can see that users who use Apple Watch are more likely to acquire information by voice information. Users who are not using the Apple Watch can get the voice information, but they need to use their iPhone to get the voice information. On the other hand, this result may have been obtained because users using the Apple Watch can obtain information without operating their iPhones by performing the reading operation from the Apple Watch. In addition, a lot of information is provided by voice, which is used for learning of interest. Therefore, it is useful to provide audio information using smartwatches as a way to provide news information.

Next, Table 2 shows the number of times the smartphone application was started, the number of times the article information was provided based on the text information, and the number of times the learning was performed with interest based on the text information. For the items in this table, we also give the average per user, per day.

Table 2 shows that users who used the Apple Watch ran the app more often than those who used the Apple Watch. Since users using Apple Watch can obtain information by voice information without operating their iPhones, we believe that providing voice information to users could encourage them to start the app. As a result, we were able to provide more information to the users and to learn more about their interests.

These results show that smartwatches and voice information can provide users with more information and can learn more of their interests.

4.2   Improvement of Misreadings

While using the app, we found that reading mistakes occurred frequently when reading Japanese voice aloud. What is occurring is a misreading of the abbreviations of the alphabet and a misreading of the Chinese Character(kanji). In the case of alphabetic abbreviations, there are cases where a reading exists or where the word is read directly from the alphabet, but if the word is the same as the abbreviation, the word is read in the same way as the abbreviation. In addition, special abbreviations may not be read aloud correctly. To correct mispronunciations of the alphabetic abbreviations, we will attempt to correct them as follows.

1. Extraction of words consisting of uppercase alphabets only from the Japanese news text
2. Refer to the dictionary and replace corresponding abbreviations with readings, if they are present. If not, move on to the next step.
3. A space is inserted in the uppercase alphabet and the letters are read out one by one.

Since many abbreviations consist only of upper-case letters, words consisting of upper-case alphabets are covered. The dictionary in step 2 contains the abbreviations of the alphabet and their readings in Japanese. By referring to a dictionary, we can deal with abbreviations that have a special reading. In step 3, a space is inserted between the letters of the alphabet so that they are read out one letter at a time. We will evaluate and improve this method in the future. We will also propose an improvement method for misreading kanji.

References

1. Ono, S., Inamoto, T., Higami, Y., Kobayashi, S.: Development of the smartphone application to choose and display news for individuals based on an operation history. In: Multimedia, Distributed, Cooperative and Mobile Symposium 2016 Proceedings in Japanese (2016)
2. Ogawa, T., Fujihashi, T., Endo, K., Kobayashi, S.: Increasing the chance of interest learning in the user-aware information distribution system using a smart watch. In: International Symposium on Affective Science and Engineering 2019 (ISASE 2019) (2019)
3. Ministry of Internal Affairs and Communications, Japan: 2020 White Paper on Information and Communications in Japan
4. Ministry of Internal Affairs and Communications, Japan: 2019 White Paper on Information and Communications in Japan
5. Pew Research Center, Americans favor mobile devices over desktops and laptops for getting news. https://www.pewresearch.org/fact-tank/2019/11/19/americans-favor-mobile-devices-over-desktops-and-laptops-for-getting-news/. Accessed 26 Oct 2020
6. Gartner, Worldwide Wearable Devices End-User Spending by Type. https://www.gartner.com/en/newsroom/press-releases/2019-10-30-gartner-says-global-end-user-spending-on-wearable-dev. Accessed 27 Oct 2020
7. T4's Market Research, Smartwatch Market Share. https://www.t4.ai/industry/smartwatch-market-share. Accessed 27 Oct 2020
8. Ogawa, T., Fujihashi, T., Endo, K., Kobayashi, S.: A proposal of personalized information distribution system using smart watch and wireless headphones. In: Asian Conference on Intelligent Information and Database Systems (ACIIDS 2020) (2020)

Cyber Security Aspects of Digital Services Using IoT Appliances

Zbigniew Hulicki[1]([⊠]) and Maciej Hulicki[2]

[1] AGH University of Science and Technology, Kraków, Poland
[2] Cardinal Stefan Wyszyński University, Warsaw, Poland
hulicki@kt.agh.edu.pl

Abstract. In the modern digital world, we have observed the emergence of end-less potential for electronic communication using diverse forms of data transmission between subscriber devices. As such, the necessity of a proper networking infrastructure to sustain this communication was a crucial factor for the further development. But now, the successful advancement of networks and systems is dependent on the appropriate status of cyber security in networking and services.

This article describes threats to the customer premises network, and explores both the security mechanisms and the network vulnerability to attacks resulting from the use of devices associated with the IoT applications.

Keywords: Cyber security · Communication networks · Digital services · Internet of Things · Security threats · Blockchain

1 Introduction

Nowadays communication possibilities of different subscriber devices are practically endless. They can transfer information to each other in various forms accordingly to user requirements. Hence, in the modern world the performance of communication networks and systems continues to grow. Moreover, it is an increasing need for alternative, in relation to the classical and already known, forms of digital services. On the other hand, the security of transmitted information causes a number of problems [9]. Without proper knowledge related to cyber security mechanisms, end users of modern, electronic devices can be exposed to a number of threats, e.g. known or new attacks carried out by means of inexpensive equipment available on the market.

Digitalization and new ICT technologies exert more and more influence on the development of communication networks and systems, including the Internet and in particular the Internet of Things (IoT) applications, and hence increases the scope of threats to the ICT infrastructure. Therefore, ensuring the security of communication networks and services is a big challenge for network operators and customers/users of modern, electronic devices [4, 6].

The aim of this contribution is an analysis of threats to the customer premises network that includes devices using the IoT applications, overview and examining of the security

© Springer Nature Singapore Pte Ltd. 2021
T.-P. Hong et al. (Eds.): ACIIDS 2021, CCIS 1371, pp. 227–237, 2021.
https://doi.org/10.1007/978-981-16-1685-3_19

mechanisms, and in particular the vulnerability to attacks resulting from the use of such devices in the network.

The rest of the paper is organized as follows. In the next section, security threats to communication networks are explored taking into account botnets and ransomware, followed by an analysis of security aspects dealing with the customer premises network those concern communication protocols and modeling of threats and incidents. Then, the security testing and validation is discussed and a tentative proposal of the blockchain database in a sample IoT application is outlined. Conclusions and final remarks are given in the last section.

2   Security Threats to Communication Networks

For more than two decades portable subscriber devices evolved from simple appliances, used to set up voice calls, to the sophisticated computer equipment which combines the advantages of cheap multimedia communication with the convenience offered by mobile phones. Similar tendency does also concern household equipment. At the same time, we observe rapid development of the Internet and diverse communication services using Wi-Fi access and Machine-to-Machine (M2M) communication. Inexpensive systems attached to new appliances provide wireless connectivity and monitoring with a low energy consumption. On the other hand, the broadcast nature of wireless networks operating near each other may be the source of many issues including degraded performance and security threats [2]. If the number of interconnected devices (IoT) increase, there is a real need to simultaneously increase the awareness of end users regarding the security of equipment, networks and ICT services. This is very important especially in IoT area where except of communication and information processing, such functions as identification, detection and service provision should be performed simultaneously. Due to the frequent deficiency of cyber security mechanisms (which are standard in modern telecommunications) in many applications, the IoT security is a big challenge for manufacturers and service providers [1]. Therefore, there is a need for secure protocols and cryptographic algorithms that are efficient enough to ensure the confidentiality, integrity and authenticity of data in various IoT devices.

2.1   Botnets and Ransomware

Any application is exposed to the danger of attack and the number of potential threats increases with time. Moreover, it is possible that applications already available on the market do not have adequate security against new attacks, e.g. in 2016, one of the largest attacks carried out with the help of infected devices (IoT) was Distributed Denial of Service (DDoS) attack [5]. The source code of the Mirai malware allowed a search and remote assault on IoT devices such as web cameras, recording appliances and other equipment with Linux operating system. Using known, default passwords and selected ports, the bot software automatically logged to unsecured devices. The scale of Mirai attacks was huge. In spite of that, new versions of the Mirai bot were created to disable various services in subsequent attacks. In recent years new groups of malware software (e.g. Hajime, Persirai, Petya and WireX botnets) for DDoS attacks have been identified

[5]. Significant incidents were also spam and phishing. Some problems have become rare, accidental because security awareness among people involved in software development has increased. On the other hand, attacks can be carried out using the malware encryption software such as Wanna Cry. In 2017 such ransomware launched a remote code with the use of the MS SMBv1 protocol and several disclosed tools of the American National Security Agency. The attack had international coverage, i.e. over 100 countries, state institutions, railways, banks and large companies were assaulted, and although it was not directly targeted at IoT, the lack of regular updates of the equipment can significantly increase the level of security threats in the future.

2.2 Threats to Applications of IoT

It has been already mentioned that the development of the Internet and telecommunications market increases the scope of attacks on the ICT infrastructure. Because threats to IoT applications concern network functionalities, increasing number of interconnected devices should result in rising of user awareness with regard to the security of appliances and services.

In order to counteract security threats, developers of IoT equipment and applications/services must pay attention to both small resources of a device and such functionalities as identification and location of the IoT facility, object authentication and authorization, user privacy, security protocols and encryption methods, including software vulnerabilities to known attacks. Therefore, a general security model for a specific implementation of IoT takes into account security mechanisms, ensuring privacy and high reputation [3].

3 Security Aspects of Customer Premises Network

Cyber security may have different meaning depending on the entity concerned. It will be important for system customers because they need to know whether the software ensures secure use of the application features as well as confidential data processing and storage in a proper manner to preserve no unauthorized access.

From the network operator or service provider point of view, it is important that the confidential data be inaccessible to an ordinary user/customer. An important aspect here is a correctly configured security system, i.e. proper network architecture, access rights to resources, etc.), as well as appropriate employee training in cyber security.

In turn, the software developers will struggle with technical problems, i.e. how to implement the security mechanisms and which components should be used in the application because they directly affect the security level of that application. The solutions used by the programmers determine the extent to which the software will be resistant to attacks by cybercriminals.

According to the report [4], the three most important threats related to private user practice are leakage and loss of data, downloading of dangerous applications and infecting equipment with malware. In many cases the customer premises network does not have sufficient security mechanisms to ensure the appropriate level of protection for network, devices or privacy of users [10].

3.1 Communication Protocols' Aspects

Some of the threats to customer premises network are related to vulnerabilities known from working networks and applications, e.g. the Bluetooth standard, Constrained Application Protocol (CoAP) and Datagram Transport Layer Security (DTLS) protocol used in IoT applications. As regards the Bluetooth standard, the "Blueborne" vulnerability was found in all versions of the standard, i.e. one can read address of the device in the access code field of the header in the Bluetooth package. Besides, when the standard L2CAP (Logical Link Control and Application Layer Protocol) connection is made in Bluetooth, another vulnerability related to buffer overflow in the lowest layer L2CAP protocol was found [8].

IoT applications utilize the CoAP protocol to communicate with each other and with the Web, i.e. communication is based on the HTTP REST (REpresentational Web Transfer) protocol functions which ensure the integrity and confidentiality of message exchange using the DTLS protocol. Another open and easy to implement protocol used for IoT applications is the Message Queue Telemetry Transport (MQTT) protocol, operating in the "publish-subscribe" model with an intermediary node. In the above mentioned protocols the introduction of protection is already possible at the stage of defining the assumptions of the protocol.

3.2 Modeling of Threats and Incidents

In order to secure the customer premises network, one has to create a model of security threats. A sample process of threats' modeling has been proposed in [7]. It specifies a list of resources to be protected, determinates vulnerabilities and attack vectors, arranges possible security breach scenarios and set priorities related to those scenarios. To model the threats, the proces defines susceptibility categories related to the acquisition of the user or device identity, unauthorized modification of equipment, disclosure of private information, blocking access to the service or taking over administrative access.

4   Security Testing and Validation

In order to check the vulnerability of IoT networks and devices to malicious activities, such as access control, remote code execution and interception of communications, a proprietary test plan should be formulated, followed by design of a test environment for customer premises network with IoT devices to enable threat analysis and security verification.

4.1 Testing Environment

As part of the work, a software and hardware environment was designed with IoT devices, allowing the use and testing of (the Internet) services and IoT appliances (cf. Fig. 1).

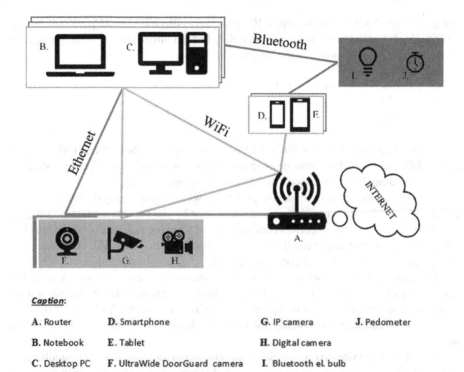

Caption:

A. Router   **D.** Smartphone   **G.** IP camera   **J.** Pedometer

B. Notebook   **E.** Tablet   **H.** Digital camera

C. Desktop PC   **F.** UltraWide DoorGuard camera   **I.** Bluetooth el. bulb

Fig. 1. Software and hardware environment for testing.

Tests of IoT devices can be divided into 4 categories:

- collecting information about the device
- testing (locally) of known vulnerabilities (with authentication)
- remote testing of known vulnerabilities (without authentication)
- testing of web applications.

Proposed categories include the most common vulnerabilities of IoT, such as:

- unprotected interface in the www application
- insufficient authentication and authorization process
- unsecured network services
- no traffic encryption and integrity verification
- insufficient configuration of security
- unprotected software.

Besides, a number of tests have been done also with the IoT devices, i.e. testing of web application was done using *Burp Suite Professional* 1.7.27 software. The program scans scripts in the device at the given IP address and port number with respect to the application security, e.g. for one of the IoT device in Fig. 1 the following addresses and ports have been scanned: https://10.255.255.1; https://10.255.255.1:8088; https://10.255.255.1. For the flash cross-domain policy (a high threat's level), a sample request to the server and its response to the program look like:

request:

GET /crossdomain.xml HTTP/1.1

Host: 10.255.255.1

Connection: close

answer:

<?xml version="1.0"?><!DOCTYPE cross-domain-policy SYSTEM "https://www.adobe.com/xml/dtds/cross-domain-policy.dtd"><cross-domain-policy><allow-access-from domain="*" secure="true" /></cross-domain-policy>

In such case, the application access policy allows access from any domain. It creates a high risk of other domains that can be taken over by the attacker and, in accordance with the policy used, enable correct communication allowing full control by the attacker. Some other tests (that have been done) include cross-site scripting, SSL certificate, link manipulation and unencrypted communications.

As a result of these tests, for the model of customer premises network with IoT devices depicted in Fig. 1, one can propose additional intermediary device (as a main protection) between the access router and IoT equipment, with the VPN server running. In such configuration, communication with IoT appliances within the local network will be safer.

IoT solutions can involve a complex network of smart devices (e.g. home appliances) and are being successfully adopted in many different areas [10]. The opportunity to develop new services based on cloud-enabled connected physical devices provides also huge opportunities for businesses and consumers. In spite of this, current centralized, cloud-based IoT solutions may not scale and meet the security challenges faced by large-scale enterprises [11]. The use of blockchain as a distributed database of transactions and peer-to-peer communication among participating devices can solve such problems.

Considering vulnerabilities and threats present in contemporary networks and systems, and those described above, the next section provides an outline of blockchain-enabled IoT solution and discusses how to use the blockchain platform and smart contracts for an IoT application in a multi-partner environment.

4.2 Blockchain Database in a Sample IoT Application

Considering vulnerabilities and threats present in contemporary networks and systems, and those described above, the blockchain database can be proposed in a sample IoT application for smart home (cf. Fig. 2) to counter attacks against IoT devices and limit their possible effects on houshold appliances.

The smart home concept depicted in Fig. 2 can serve as a basis for implementation of various IoT applications using houshold instalations/devices. An example of such solution can be IoT application for the photovoltaic or solar power system designed to generate electrical power, i.e. supply electricity for houshold equipment. It consists of an arrangement of several components, including solar panels (to absorb and convert sunlight into electricity), a solar inverter (to change the electric current from DC to AC) as well as other electrical accessories to control the flow of energy bought and sold. It may also use a tracking system to improve the system's overall performance and include an integrated battery solution. Surplus of renewable energy produced, i.e. excess beyond what is used or required, can be sold both to the local micro-grid and to the operator of regional grid. Sample functionalities of the IoT application (designed and already implemented in the smart home) have been shown in a few screenshoots (cf. Fig. 3 a, b, c).

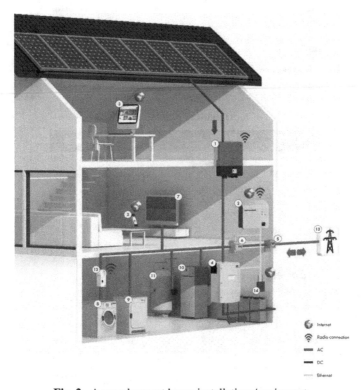

Fig. 2. A sample smart home installations/equipment.

Omitting the above mentioned vulnerability of IoT networks and devices to malicious activities, such as access control, remote code execution and, the main problem of such applications deals with interception of communications and/or security of transactions among participating entities/IoT appliances. Therefore, it seems to be obvious that the use of blockchain as a distributed database (or ledger) of transactions can help to solve that problem in a multi-node environment.

Blockchain refers to a distributed database where a list of transactions is stored in multiple participating servers rather than on a central transaction server [11]. While the key blockchain usage scenarios are in the financial domain [10], blockchain can significantly help in facilitating transactions and coordination among interacting IoT devices. Each IoT appliance participating in the blockchain network is granted access to an up-to-date copy of this encrypted database so it can read, write, and validate

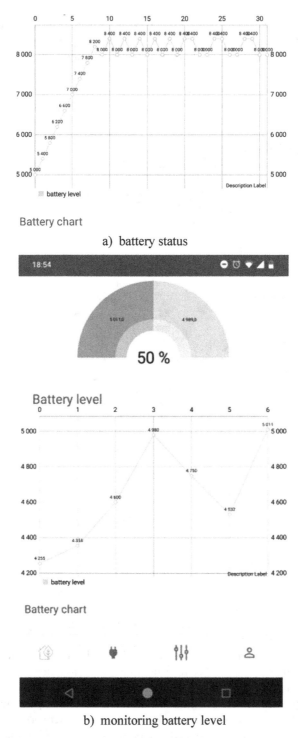

a) battery status

b) monitoring battery level

Fig. 3. Sample functionalities of the application for the smart home.

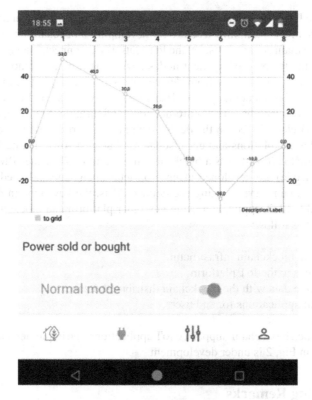

c) monitoring of energy sold to the grid

Fig. 3. (*continued*)

transactions. Hence, blockchain-supported IoT networks of energy grids enable peer-to-peer transactions of energy. In one application, excess rooftop solar energy is sold to other users who need it—all of which is paid for and recorded through a blockchain. The concept of using smart contracts to secure transactions has been shown in Fig. 4.

Fig. 4. The concept of using smart contracts to secure transactions.

In order to implement such blockchain-supported IoT application, its architecture and all components should be defined first. Although this is not a trivial task, one can develop it using (as a basis) the high-level architecture of IoT applications that exploit

IBM cloud-based hyperledger services described in [11] and [12]. That architecture provides the private blockchain infrastructure for developing blockchain-enabled solutions and includes two main components, i.e. the IoT platform to connect, manage and analyze data, and the module comprising client nodes/devices as well as submodule for application/blockchain services. Data from devices are sent to IoT platform using MQTT protocol (M2M/IoT connectivity protocol) and blockchain proxy in this platform sends the data to the chaincode based on a pre-defined configuration. Smart transactions are executed in IBM cloud based on the device data [11]. Smart contracts create the core of blockchain-based solutions and encapsulate the business logic, i.e. each invocation of a smart contract is recorded as a blockchain transaction. While the IBM blockchain contracts (chaincode) are developed using proprietary language and need to be registered with blockchain services using pre-defined APIs, the key steps in developing a blockchain-enabled IoT application using any other platform and blockchain database can be specified as follow:

- set up a private blockchain infrastructure
- connect devices to the IoT platform
- integrate device data with the blockchain distributed database
- develop client applications for end users.

Currently, such blockchain-supported IoT application to be implemented in the smart home depicted in Fig. 2 is under development.

5   Concluding Remarks

Threats to the customer premises network that includes devices using the IoT applications have been described in this paper. Security mechanisms and the network vulnerability to attacks resulting from the use of devices associated with the IoT applications have been also discussed. Investigation of security threats took into account botnets and ransomware. The analysis of aspects those concern communication protocols and modeling of threats and incidents has been done as well. Then, the security testing and validation was discussed and a tentative proposal of the blockchain database in a sample IoT application has been outlined.

References

1. Al-Fuqaha, A., Guizani, M., Mohammadi, M., Aledhari, M., Ayyash, M.: Internet of things: a survey on enabling technologies, protocols, and applications. IEEE Commun. Surv. Tutor. 17(4), 2347–2376 (2015)
2. An Internet of Things blueprint for a smarter world. Capitalizing on M2M, big data and the cloud. Nokia. Espoo (2016). https://resources.ext.nokia.com/asset/190140
3. Babar, S., Mahalle, P., Stango, A., Prasad, N., Prasad, R.: Proposed security model and threat taxonomy for the Internet of Things (IoT). In: Meghanathan, N., Boumerdassi, S., Chaki, N., Nagamalai, D. (eds.) CNSA 2010. CCIS, vol. 89, pp. 420–429. Springer, Heidelberg (2010). https://doi.org/10.1007/978-3-642-14478-3_42

4. Cybersecurity Trends. 2017 spotlight report: Information Security LinkedIn Group Partner, Crowd Research Partners, (ISC)2 (2017)
5. IoT devices being increasingly used for DDoS attacks: Symantec Security Response. Symantec Corporation (2016)
6. ISO/IEC 27000: ISO/IEC Information Security Management Systems (ISMS) standards
7. Ryoo, J., Kim, S., Cho, J., Kim, H., Tjoa, S., DeRobertis, C.V.: IoE security threats and you. Penn State Altoona, Sungkyunkwan University, JRZ TARGET, St. Pölten University of Applied Sciences, IBM, (2017)
8. Seri, B., Vishnepolsky, G.: BlueBorne: the dangers of bluetooth implementations: unveiling zero day vulnerabilities and security flaws in modern Bluetooth stacks. Armis Labs. Armis (2017)
9. The Network and Information Systems Directive 2016/1148 of the EU; updated in August 2018
10. Capgemini: Blockchain: A fundamental shift for financial services institutions. White Paper, November 2015. https://www.capgemini.com/resources/blockchain-a-fundamental-shift-for-financial-services-institutions/
11. Gantait, A., Patra, J., Mukherjee, A.: Integrate device data with smart contracts in IBM Blockchain. IBM, June 2017
12. IBM ADEPT Practictioner Perspective - Pre Publication Draft - 7 Jan 2015. https://pl.scribd.com/doc/252917347/IBM-ADEPT-Practictioner-Perspective-Pre-Publication-Draft-7-Jan-2015

Computational Imaging and Vision

Simple Methodology for Eye Gaze Direction Estimation

Suood Al Mazrouei[1] and Andrzej Śluzek[2]([⊠])[iD]

[1] Khalifa University, Abu Dhabi, UAE
[2] WULS-SGGW, ul. Nowoursynowska 159 bud. 34, 02-776 Warszawa, Poland
andrzej_sluzek@sggw.edu.pl

Abstract. A simple methodology (based on standard detection algorithms and a shallow neural network) is developed for estimating the eye gaze direction in natural (mostly indoor) environments. The methodology is primarily intended for evaluating human interests or preferences (e.g. in interaction with smart systems). First, techniques for detection faces, facial landmarks, eyes and irises are employed. The results are converted into numbers representing resolution-normalized locations of those landmarks. Eventually, six numbers (comprehensively combining the detection results) are used as NN inputs. The NN outputs represent 17 different directions of the eye gaze, where the neutral direction is defined by the virtual line linking the monitoring camera with the face center. Thus, the identified gaze direction results from combined motions of eyeballs and heads. For a feasibility study, a small dataset was created (showing diversified configurations of head and eyeballs orientations of 10 people looking in the specific directions). Data augmentation was used to increase the dataset volume. Using 67% for training (in 3-fold validation) we have reached accuracy of 91.35%.

Keywords: Gaze tracking · Direction estimation · Face detection · Neural network

1 Introduction

Appearance-based eye gaze detection/tracking has been attracting interest of researchers and practitioners (e.g. [10,20]) for several reasons. First, eye contact can be the only communication channel for severely handicapped people (strokes, cerebral palsy, muscular dystrophy, etc.) and a number of experimental (e.g. [14, 21]) and commercial (e.g. [5,9]) works have been reported in this area. In such applications, both the camera module and (unfortunately) the user's head remain motionless and high precision of gaze direction estimates plays an important role (e.g. for eye typing [14]).

Recently, eye gaze direction estimation is attempted in more unconstrained scenarios. While monitoring the driver behavior/attention is one of the flagship

Supported by KUCARS (Khalifa University).

T.-P. Hong et al. (Eds.): ACIIDS 2021, CCIS 1371, pp. 241–253, 2021.
https://doi.org/10.1007/978-981-16-1685-3_20

topics in this area (e.g. [17,19]) the problem is often addressed from a more general perspective, where the direction of visual attention should be estimated (e.g. [7,8,23]) for other reasons. In such problems, the perfect accuracy of estimates (even if measured in degrees for the sake of performance comparison, e.g. in [13,24]) is less critical, as long as the general direction of eye gaze is correctly identified.

The presented paper follows the same philosophy. The objective is to identify the approximate gaze direction of people captured by a monitoring camera (typically positioned at the average height of a human). The gaze direction (which is estimated relatively to the location of camera) may indicate preferences, interests or intentions of individuals, i.e. the prospective applications are either commercial or security-related.

No specific distance between the camera and detected people is required, but it is generally assumed that the resolution of captured human faces is sufficient to localize the important facial features (i.e. mainly irises, but also noses, mouths, etc.). Apart from that assumption, the approach is scale-invariant.

The methodology is built upon publicly available algorithms for detection of faces and localization of facial features. These algorithms are briefly summarized in Sect. 2. In Sect. 3, the algorithmic aspects of the presented work are discussed. Finally, in Sect. 4, experimental results (including description of the developed dataset) are presented. Concluding remarks are given in Sect. 5.

2   Detection of Facial Features

In general, the pre-processing phase of the proposed methodology consists of the following three steps, for which efficient algorithmic solutions are publicly available. These steps are:

1. Face detection.
2. Detection of facial landmarks.
3. Detection of irises.

In Step 1, we apply standard methods which can be traced back to the pioneer Viola-Jones algorithm [18]. Generally, we follow principles of [16], with the actual solution built around the implementation available in Matlab Computer Vision System Toolbox^{TM}.

For detection of significant facial landmarks (Step 2), we have combined Google Mobile Vision API [1,2] and Matlab file exchange resources (including [3]).

Eventually, the following facial landmarks are identified and memorized for the subsequent operations of the method:

- center of the face (CoF),
- tip of the nose (ToN),
- center of the mouth (CoM),
- centers of eyes (CoEs),
- tips (corners) of eyes (ToEs),

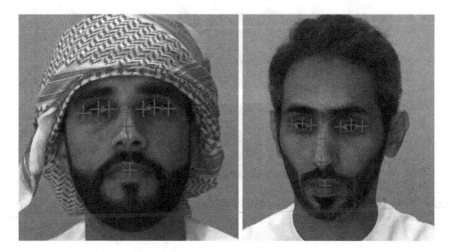

Fig. 1. Exemplary results of facial landmarks detection.

Results of localization of these landmarks in exemplary images are given in Fig. 1.

Additionally, in Step 3, detection of irises is performed within bounding boxes of eyes (those bounding boxes are obtained from ToE landmarks extracted in Step 2). A simple method based on Matlab built-in circular Hough transform is used (following, e.g. [4]). The results are sometimes incorrect, but a straightforward verification mechanism is applied to reject unrealistic results (i.e. too large, too small or obviously incorrectly located irises). Therefore, if the iris detection results are accepted, the location is always at least approximately correct, even in poor-quality and/or low-resolution images. Selected examples (including difficult cases) are given in Fig. 2.

3   Methodology

3.1   Eye Gaze Direction Model

Although in some applications (e.g. eye typing) highly accurate orientation of eye gaze is needed, the tasks we prospectively consider need only a limited number of generally defined directions. Therefore, taking into account the psycho-physiological limitation of the human vision (e.g. [15]) and typical intended applications, we propose to use 17 directions. Those directions (symbolically shown in Fig. 3) correspond to an idealized scenario where the head remains motionless and the monitoring camera is located perfectly frontally at the level of human eyes. Then, the eye gaze directions are categorized into the following types:

(i)   *horizontal* (with *central, left/right* and *far left/right* sub-types);
(ii)   *up* (with *central, left/right* and *far left/right* sub-types);
(iii)   *down* (with *central, left/right* and *far left/right* sub-types);
(iv)   *extreme up* (no sub-types);
(v)   *extreme down* (no sub-types).

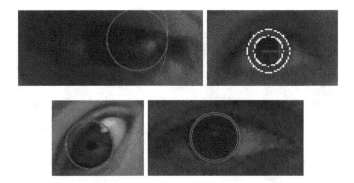

Fig. 2. Examples of iris detection in eye bounding boxes extracted for images of diversified resolution and quality.

It can be noticed that for *extreme up/down* types no directional subtypes exist. This is motivated by physiology and anatomy of eyeballs (e.g. [22]) which can hardly move sideways when maximally lowered or raised. It can be also noticed

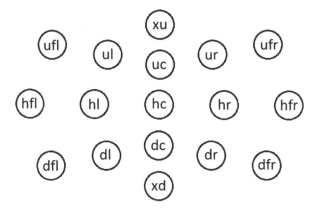

Fig. 3. Symbolic representations of proposed eye gaze directions. The same template is used for database building (see Sect. 4).

that the proposed gaze directions are similar to calibration patterns used in various works where ground-truth directions of gaze orientation are defined (e.g. [11,14]).

When the head is not motionless, the same gaze direction (in terms of its orientation in the real-world coordinates) can be achieved by various combinations of eyeball and head movements. Therefore, we always define the *horizontal-central* direction by the virtual line joining the camera and the center of the eye area in the face (and the other directions are correspondingly defined relatively to this line).

In Fig. 4, several examples illustrate how various gaze directions are obtained by combined movements of eyeballs and heads.

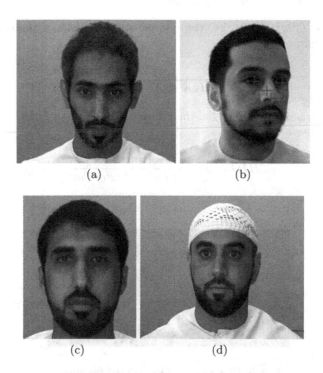

(a)   (b)

(c)   (d)

Fig. 4. In (a), (b) and (c) *horizontal central* gaze direction is present. In (a), it is obtained by lowering the head and upward motion of eyeballs. In (b), the head rotation is compensated by the eyeball motions in the opposite direction. In (c) the head is slightly lifted and eyeballs slightly lowered. In (d), *up left* direction is obtained primarily by movements of eyeballs with only a slight motion of the head.

3.2   Features for Gaze Direction Estimates

As discussed in Subsect. 3.1, the gaze direction is a combined result of head and eyeball movements. Therefore, we assume that both factors should be used in conjunction (rather than separately) in the gaze direction estimates. Similar approach was presented, for example, in [23] where a multimodal convolutional neural network is used to analyze normalized sub-images of eyes and data representing head orientation. However, that was preceded by more tradition image processing techniques which provided parameters needed for the normalization.

In this paper, we argue that similar parameters (features) can be directly fed into a feed-forward (shallow) neural network to produce the gaze direction estimates (within the proposed 17 classes). Eventually, we propose the following six features which are computed from the facial landmark localization data returned by standard detection algorithms (as explained in Sect. 2).

The first four features (two for each eye) define position of the iris center within the eye (as shown in Fig. 5). These are:

- Vertical iris offset (VIO), i.e. the relative displacement of the centre of iris from the main eye axis.

$$VIO = \frac{\pm\|CoI - X\|}{\|InnerToE - OuterToE\|} \tag{1}$$

- Horizontal iris offset (HIO), i.e. the relative displacement of the centre of iris from the inner eye corner.

$$HIO = \frac{\|X - InnerToE\|}{\|InnerToE - OuterToE\|} \tag{2}$$

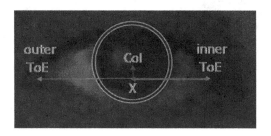

Fig. 5. Illustration of iris-related features.

The remaining two features characterize the face orientation. They approximate the horizontal and vertical head rotations by relative displacements of ToN (*tip of the nose*) landmark from the line joining CoF (*center of face*) and CoM (*center of mouth*) landmarks, as shown in Fig. 6. Thus, we have:

- Horizontal head offset (HHO), i.e. the relative displacement of the nose tip from the main face axis.

$$HHO = \frac{\pm\|ToN - Y\|}{\|CoF - CoM\|} \tag{3}$$

- Vertical head offset (VHO), i.e. the relative displacement of the nose tip from the face center.

$$VHO = \frac{\|Y - CoF\|}{\|CoF - CoM\|} \tag{4}$$

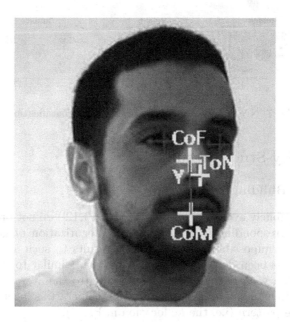

Fig. 6. Illustration of face orientation related features.

It should be highlighted that all these features are scale-invariant (although for high-resolution images they can be computed more accurately). Nevertheless, the resolution should be obviously sufficiently high to detect the facial landmarks needed in Eqs. 1–4.

The features are also rotation invariant, but this property is of secondary significance because in the intended applications people would be generally in upright positions (i.e. either standing or walking).

3.3 Neural Network for Gaze Direction

The features derived in Subsect. 3.2 are fed into a feed-forward neural network developed for gaze direction classification (i.e. there are 17 output nodes). Various architecture have been tested (with the number of hidden layer ranging from 1 to 5) but eventually the structure with two hidden layers (and 15 nodes in each layer) was found the most effective, see Fig. 7. Details of the dataset collection and NN training are discussed in Sect. 4.

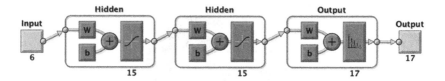

Fig. 7. NN architecture for gaze direction classification.

4 Feasibility-Study Results

4.1 Dataset Building

Since popular publicly available datasets (e.g. [6,10,12]) do not provide ground-truth results corresponding to the proposed categorization of gaze directions (and it would be impossible to convert their results to such a categorization) a new dataset has been developed. First, a pattern (similar to Fig. 3 diagram) was attached to a wall at the level corresponding to the average height of the experiment participants. The image-acquisition camera was located at the central point of the pattern (i.e. the hc location in Fig. 3).

Participants were standing at the shortest distance from which they can see the extreme pattern points (i.e. xu, xd, hfl and hfr in Fig. 3) without moving their heads, i.e. a gaze redirection from one point to another can be achieved through the eyeball movements only.

First, participants were asked to focus on individual points of the pattern by moving only eyeballs (with motionless heads) so that 17 face images were captured from each participant. Then, the participants had to rotate their heads (both vertically and horizontally) with the eye gaze fixed on the selected points, i.e. a combination of head and eyeball movements was involved to keep the gaze direction fixed. In this way, around 5 to 10 images were acquired for each pattern point (i.e. approx. 100–120 images altogether from each participant).

Therefore, with the total number of 10 participants, 1085 images (each annotated by one of the 17 gaze directions) have been collected. To further increase the size of dataset, data augmentation was performed. In general, for each original image, three down-scaled copies were created, two copies with non-linear illumination changes, and two copies with minor random distortions (one barrel and one pincushion). Those images inherit annotations of their corresponding originals. Eventually, almost 8000 usable images were obtained (with the ground-truth annotation provided) since some images from the augmented dataset are rejected, e.g. because of undetectable facial landmarks (too low resolution or excessively distorted image intensities).

Because of certain legal restrictions, the dataset cannot be fully disclosed (please contact the corresponding author if you are interested in more details). Nevertheless, some examples of acquired images are given in Fig. 8.

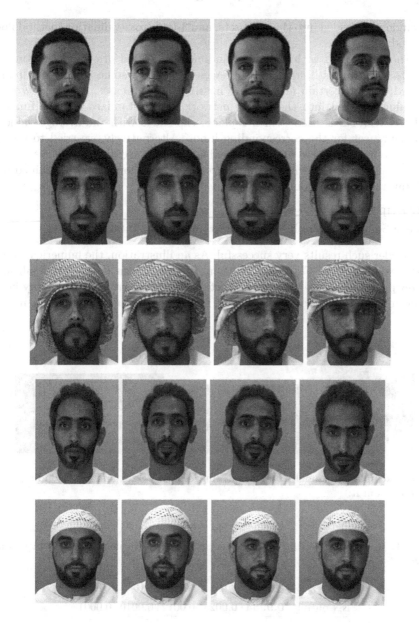

Fig. 8. Exemplary images from the collected dataset.

4.2 Neural Network Training

The proposed architectures of neural networks were trained on 2/3 of the dataset, while the remaining 1/3 was used for validation and testing. Actually, the dataset was randomly divided into three subsets, and 3-fold cross-validation methodology was used.

Although images with no facial landmarks detected are removed from the dataset, there are still numerous images with only some of the landmarks detected, so that certain features are missing. In such cases, NaN (*not a number*) values are fed into the neural network and a standard Matlab function FIXUNKNOWNS is used, which replaces the NaNs with the average value in the training set for the corresponding input. Alternatively, we tried to replace NaNs be zeros, but there are no noticeable performance differences.

Eventually, the following performances have been found for the trained network (average results from 3-fold cross-validation).

– ACCURACY on the validation data: **91.35%**,
– Mean Squared Error (MSE) on the training data: **0.0032**.

We consider such results very successful. As an illustration, the numerical values for the example from Fig. 9 are given in Table 1. It can be seen that all non-zero outputs indicate *up*-type gaze direction, with *up-central* sub-category being the clear winner. Out of curiosity, we tested the method on random images

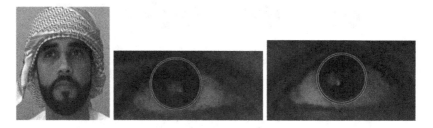

Fig. 9. Exemplary (manually annotated) image of UC class, and eye bounding boxes extracted from it.

Table 1. Results for Fig. 9. Only classes (see Fig. 3) with non-zero NN outputs are shown.

Class	UC	UL	UFL	UR	UFR
NN output	**0.9954**	0.0023	0.0012	0.0010	0.0001

from internet and personal collections (for which, obviously, no ground-truth annotation was available). The results always (if the required facial landmarks can be detected) look plausible. A spectacular example is given in Fig. 10 with the NN outputs provided in Table 2.

Fig. 10. A random image from personal collections.

Table 2. Results for Fig. 10. Only classes (see Fig. 3) with non-zero NN outputs shown.

Class	DFR	HFR	DR
NN output	**0.9949**	0.0050	0.0001

5 Summary

In this paper, we present a methodology for approximate estimation of the eye gaze (using 17 pre-defined directions). Directions are defined relatively to the position of the camera, i.e. the reference direction (*horizontal central* in the proposed terminology) is determined by the virtual line linking the camera and the eye area of the observed person. In case of a perfect frontal view, this direction actually corresponds to the neutral position of eyeballs, but in general the gaze direction is a combined result of head rotation and eyeballs motion.

Because no dataset suitable for the quantitative verification of the proposed methodology is publicly available, we developed a collection of images which adequately address the needs of this task. The dataset has been additionally augmented using simple tools of image degradation/distortion.

The method requires face detection, followed by facial landmark detection. These operation can be, in general, performed by any public-domain or commercially available software. Therefore, we do not focus on this aspect (even though a simple technique for iris localization has been implemented in the project).

From the provided locations of the facial landmarks (we use the following ones: left and right irises, corners of both eyes, tip of the nose, center of the mouth and the overall face center) six numerical features are obtained, which jointly approximate the head orientation and eyeball motions. The features are fed into a shallow neural network which outputs 17 values representing confidence of the proposed 17 directions of eye gaze.

The evaluation on the developed dataset (3-fold cross-validation was used) indicates the average accuracy at 91.35% level (note that only the testing subsets are used in the accuracy estimation).

As mentioned above, the popular publicly available datasets were not used for quantitative evaluation because they do not provide the ground-truth corresponding to the proposed direction categorization (and such a conversion would be very difficult, if not impossible).

In general, the proposed technique is meant for monitoring various aspects of human behavior in public places, both from commercial and non-commercial perspective. These include, for examples, identification of customer interests, estimating attractiveness of ads and promotional campaigns, monitoring pedestrians entering zebra crossing, detecting prospectively malicious intentions, etc.

Even though our implementation study was conducted in selected indoor environments, the whole methodology is easily transferable to many typical indoor or outdoor environment, as long as quality of acquired visual data is sufficient to locate the facial landmarks discussed in the paper.

Acknowledgment. Support of Khalifa University Center for Autonomous Robotic Systems is gratefully acknowledged.

References

1. Detect facial features in photos. https://developers.google.com/vision/ios/detect-faces-tutorial. Accessed 20 May 2020
2. Face detection concepts overview. https://developers.google.com/vision/face-detection-concepts. Accessed 15 June 2020
3. Silva, C.P.E.: Facial landmarks (2020). https://www.mathworks.com/matlabcentral/fileexchange/47713-facial-landmarks. Accessed 20 Sept 2020
4. Dong, C., Wang, X., Pei-hua, C., Pu-Liang, Y.: Eye detection based on integral projection and hough round transform. In: IEEE 5th International Conference on Big Data and Cloud Computing, pp. 252–255 (2015)
5. Eyegaze Inc.: How does the eyegaze edge® work?. https://eyegaze.com/products/eyegaze-edge/. Accessed 28 Feb 2020
6. Funes Mora, K.A., Monay, F., Odobez, J.M.: EYEDIAP: a database for the development and evaluation of gaze estimation algorithms from RGB and RGB-D cameras. In: Proceedings of the ACM Symposium on Eye Tracking Research and Applications. ACM (2014)
7. Guo, Z., Zhou, Q., Liu, Z.: Appearance-based gaze estimation under slight head motion. Multimed. Tools Appl. **76**, 2203–2222 (2017). https://doi.org/10.1007/s11042-015-3182-4
8. Huang, Q., Veeraraghavan, A., Sabharwal, A.: TabletGaze: dataset and analysis for unconstrained appearance-based gaze estimation in mobile tablets. Mach. Vis. Appl. **28**, 445–461 (2017). https://doi.org/10.1007/s00138-017-0852-4
9. Imotions: Eye tracking: Track eye position and movement to access visual attention. https://imotions.com/eye-tracking/. Accessed 28 Apr 2020
10. Jiang, J., Zhou, X., Chan, S., Chen, S.: Appearance-based gaze tracking: a brief review. In: Yu, H., Liu, J., Liu, L., Ju, Z., Liu, Y., Zhou, D. (eds.) ICIRA 2019. LNCS (LNAI), vol. 11745, pp. 629–640. Springer, Cham (2019). https://doi.org/10.1007/978-3-030-27529-7_53
11. Kar, A., Corcoran, P.: A review and analysis of eye-gaze estimation systems, algorithms and performance evaluation methods in consumer platforms. IEEE Access **5**, 16495–16519 (2017)

12. Krafka, K., et al.: Eye tracking for everyone. In: 2016 IEEE Conference on Computer Vision and Pattern Recognition (CVPR), pp. 2176–2184 (2016)
13. Lemley, J., Kar, A., Drimbarean, A., Corcoran, P.: Efficient CNN implementation for eye-gaze estimation on low-power/low-quality consumer imaging systems (2018). http://arxiv.org/abs/1806.10890
14. Liu, Y., Lee, B., Sluzek, A., Rajan, D., Mckeown, M.: CamType: assistive text entry using gaze with an off-the-shelf webcam. Mach. Vis. Appl. **30**, 407–421 (2019). https://doi.org/10.1007/s00138-018-00997-4
15. Loomis, J.M., Kelly, J.W., Pusch, M., Bailenson, J.N., Beall, A.C.: Psychophysics of perceiving eye-gaze and head direction with peripheral vision: implications for the dynamics of eye-gaze behavior. Perception **37**(9), 1443–1457 (2008)
16. Luh, G.: Face detection using combination of skin color pixel detection and viola-jones face detector. In: 2014 International Conference on Machine Learning and Cybernetics, vol. 1, pp. 364–370 (2014)
17. Vicente, F., Huang, Z., Xiong, X., De la Torre, F., Zhang, W., Levi, D.: Driver gaze tracking and eyes off the road detection system. IEEE Trans. Intell. Transp. Syst. **16**(4), 2014–2027 (2015)
18. Viola, P., Jones, M.J.: Robust real-time face detection. Int. J. Comput. Vis. **57**, 137–154 (2004). https://doi.org/10.1023/B:VISI.0000013087.49260.fb
19. Wang, Y., Yuan, G., Mi, Z., Peng, J., Ding, X., Fu, X.: Continuous driver's gaze zone estimation using RGB-D camera. Sensors (Basel) **19**(6), 1287 (2019)
20. Wang, Y., Zhao, T., Ding, X., Peng, J., Bian, J., Fu, X.: Learning a gaze estimator with neighbor selection from large-scale synthetic eye images. Knowl.-Based Syst. **139**(1), 41–49 (2017)
21. Wirawan, C., Qingyao, H., Yi, L., Yean, S., Lee, B., Ran, F.: Pholder: an eye-gaze assisted reading application on android. In: 13th International Conference on Signal-Image Technology Internet-Based Systems (SITIS), pp. 350–353 (2017)
22. Wright, K.W.: Anatomy and physiology of eye movements. In: Wright, K., Spiegal, P. (eds.) Pediatric Ophthalmology and Strabismus, pp. 125–143. Springer, New York (2003). https://doi.org/10.1007/978-0-387-21753-6_8
23. Zhang, X., Sugano, Y., Fritz, M., Bulling, A.: Appearance-based gaze estimation in the wild. In: 2015 IEEE Conference on Computer Vision and Pattern Recognition (CVPR), pp. 4511–4520 (2015)
24. Zhang, X., Sugano, Y., Fritz, M., Bulling, A.: It's written all over your face: full-face appearance-based gaze estimation. In: 2017 IEEE Conference on Computer Vision and Pattern Recognition Workshops (CVPRW), pp. 2299–2308. IEEE (2017)

Dropout Regularization for Automatic Segmented Dental Images

Vincent Majanga and Serestina Viriri[✉]

School of Mathematics, Statistics and Computer Science, University KwaZulu-Natal, Durban, South Africa
viriris@ukzn.ac.za

Abstract. Deep neural networks are those networks that have a large number of parameters, thus the core of deep learning systems. There is a challenge that arises, from these systems as to how they perform against training data, and/or validation datasets. Due to the number of parameters involved the networks tend to consume a lot of time and this brings about a condition referred as over-fitting. This approach proposes the introduction of a dropout layer between the input and the first hidden layer in a model. This is quite specific and different from the traditional dropout used in other fields which introduce the dropout in each and every hidden layer of the network model to deal with over-fitting. Our approach involves a pre-processing step that deals with data augmentation to take care of the limited number of dental images and erosion morphology to remove noise from the images. Additionally, segmentation is done to extract edge-based features using the canny edge detection method. Further, the neural network used employs the sequential model from Keras, and this is for combining iterations from the edge segmentation step into one model. Parallel evaluations to the model are carried out, first without dropout, and the other with a dropout input layer of size 0.3. The introduction of dropout in the model as a weight regularization technique, improved the accuracy of evaluation results, 89.0% for the model without dropout, to 91.3% for model with dropout, for both precision and recall values.

Keywords: Deep learning · Over-fitting · Regularization techniques · Dropout

1 Introduction

Over-fitting is a common problem in various deep learning systems. It usually happens when the model is trained too well to the training dataset, but not as well on the testing dataset. Alternatively, underfitting is when we a model does not perform well on both the training and testing set.

These two conditions can therefore be dealt with through several ways refered to as weight regularization techniques. These include early stopping, L1, L2 regularization and Dropout. In our approach, we used dropout which, consists

© Springer Nature Singapore Pte Ltd. 2021
T.-P. Hong et al. (Eds.): ACIIDS 2021, CCIS 1371, pp. 254–265, 2021.
https://doi.org/10.1007/978-981-16-1685-3_21

of dropping out hidden and visible units in the neural network model. This is done by ignoring units during the training phase of certain set of neurons which are chosen at random [1]. Technically, at each training stage, individual units are either dropped out of the network with probability $1 - p$ or kept with probability p, so that what is left is a reduced network [2].

The key idea behind dropout is to randomly drop units along with their connections, from the neural network during training, to prevent units from co-adapting too much [3]. After dropping units of different network models during training, this makes it easier at testing to approximate the effect of averaging predictions of the networks. From the dropout process, there is reduction of over-fitting, and further gives major improvements over other regularization methods.

In other studies such as [4], show how dropout is applied in deep learning systems. Dropout can be introduced in a network model in several ways. It can be introduced as a layer between the input and the first hidden layer. Secondly, it can be applied between two hidden layers and between the last hidden layer and the output layer.

The approach that we are proposing uses the first method with dropout introduced between the input layer and the first hidden layer. Dropout is very helpful with a large network, with various constraints like learning rate, decay and momentum to improve evaluation performance.

2 Related Work

Dropout has become an essential unit in most deep neural networks that are in existence. In another approach [5], introduce "shakeout" that randomly introduces regularization of weights that chooses to enhance each units of one layer to the next layer in the model. Neural networks have also achieved impeccable results on various computer vision tasks. The introduction of the use of random dropout procedures, improves the overall error by omitting a fraction of hidden units in most or all layers. Dropout helps avoid over-fitting in neural nets, and has also been seen to be successful on small-scale phone recognition tasks using larger neural nets [6].

In recent developments on deep learning, Recurrent Neural Networks(RNNs) seem to be most popular, but the introduction of dropout as a form of regularization, tends to fail them when applied to recurrent layers in the network.

Another approach is to introduce a dropout solution for RNNs that approximates bayesian inference to improve their results. The inference based dropout, is applied in Long Short Term Memory(LSTM) and Gated Recurring Units(GRU) models. These models are used to only access language modelling and sentiment analysis [5].

The method used in [7], show that the performance of Recurrent neural networks (RNNs) with LSTM models can be improved by the introduction of dropout as a regularization method for deep networks.

Dropout has achieved superb results with convolution networks, and this has led to it being applied in recurrent networks without affecting their connections, thus preserving the modelling sequences. Most experiments were conducted on most of the handwritten databases, and confirmed the effectiveness that came with the introduction of dropout on these recurrent networks.

In the method proposed by [8], segmentation algorithm is introduced for brain tumors using deep convolution neural networks (DCNN). They introduce dropout layers in the model to deal with overfitting, brought by the huge number of data parameters in deep neural networks. Dropout is a technique for regularization of neural network parameters, which works by adding noise to the input of each layer of the neural network during optimization. A variational dropout is introduced, with more flexible parameters in the model, for a more generalized result from several conducted experiments [9].

Some researchers proposed a novel way of employing dropouts to multiple additive regression trees(MART) to come up with (DART) dropout additive regression trees algorithm. The algorithm is evaluated using publicly, large scale available datasets on classification tasks. The evaluation results from (DART) seem to be far better than those from (MART) [10]. Another method, (sDropout) split dropout and a rotational convolution was proposed by [11], as a technique to improve the performance of convolution networks on image classification. The approach prevents over-fitting of deep neural networks by splitting data into two, and keeps both rather than discard one, as seen by the standard dropout.

Some networks introduce a fast dropout approach where each parameter adapts by rewarding large weights with under-fitting, and drops them for over-confident predictions. The approach derivatives are based on the training error, and therefore the absence of a global weight, thus the regularizer is not biased towards a certain weight [12].

3   Methods and Techniques

The proposed framework proposes pre-processing and segmentation of the dental images be done using the following steps:

3.1   Dataset Preparation - Augmentation

Lack or inaccessibility of dental images for research, it is impossible to derive intelligent decisions by neural networks. Deep learning networks need a huge amount of data for training and testing purposes to achieve good evaluation performance. Image augmentation is highly prescribed, and involves artificially creating training images through different ways of processing, or a combination of multiple processing ways, for instance data augmentation.

Data augmentation was initially popularized by [4], in order to make systems' simulation easier. Other than the traditional rotation, flips and shifts methods of augmenting data, other techniques have also been witnessed. Some methods like [14], propose generating new images from white noise for skin lesion classification. Datasets can also be made public through challenges organized by various research forums [15], done for melanoma detection for skin lesions.

Furthermore, [16], gathered dermatological images available and used these to train TensorFlow Inception version-3. Training was done for those images with augmentation done, and those without and statistical analysis done to compare results. Data augmentation methods used in this research include horizontally flipping images, random crops, and rotations.

Basically, image shifts via the width shift, range and height shift, range arguments. Image flips through the horizontal and vertical flip arguments. This is done to the few images available for several iterations to come up with 8361 images after augmentation. The Data Augmentation process [17] can be described through the below steps:

Flips: horizontal and vertical flips for each image in the training set. Horizontal flips are commonly used in natural images, and vertical flips too to capture in-variance to vertical reflection in medical images.

Scaling: We scale each I in either the x or y direction; specifically, we apply an affine transformation, $A = (sx00sy)$.

Rotations: The following affine transformation,

$$A = (\cos\theta\sin\theta - \sin\theta\cos\theta) \tag{1}$$

where θ is between 10 and 175°, is applied.

3.2   Pre-processing

After dataset preparation, erosion and dilation are used for removing noise and image enhancement. Canny edge detection is then introduced for edge detection on the images (Fig. 1).

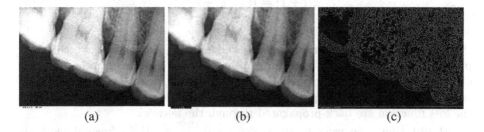

(a)   (b)   (c)

Fig. 1. a. Original image b. Image after erosion c. Canny edge detection

3.3   Model Description

The proposed framework model of this neural network uses the sequential model. This introduces visible input layers, hidden layers and output layers. The model consists of one visible layer, with dropout of size 0.3 introduced. Additionally, there are two hidden layers, both having the RELU activation functions. Lastly, one visible output layer with softmax activation function as seen in Table 1.

Table 1. Dropout sequential model

Layers	Activation function	Parameters
Input Layer (Dropout 0.3)	(RELU) function	0
Hidden Layer 1	(RELU) function	524800
Hidden Layer 2	(RELU) function	262656
Output Layer	Softmax function	1539

3.4 Dropout on a Neural Network

Dropout for a neural network is considered as a network with L hidden layers. Let $l \in \{ 1....L\}$ index the hidden layers of the network [3]. Let $z(l)$ denote the vector of inputs into layer $l, y(l)$ denote the vector of outputs from layer $l (y(0) = x$ is the input) $W(l)$ and $b(l)$ are the weights and biases at layer l. The feed forward operation of a neural network can be described as (for $l0;::::;L1$)

$$z_i^{(l+1)} = w_i^{(l+1)} y^l + b_i^{(l+1)}, \tag{2}$$

$$y_i^{l+1} = f(z_i^{l+1}) \tag{3}$$

Here f is any activation function. For example, $f(x) = 1 = /(1 + exp(x))$. With dropout, the feed forward operation becomes:

$$r_i^l Bernoulli(p) \tag{4}$$

$$\tilde{y}^l = r^l y^l \tag{5}$$

$$z(l + 1) = W_i^{l+1} \tilde{y}^l + b_i{}^{l+1} \tag{6}$$

$$y(l + 1) = f(z_i^{l+1}) \tag{7}$$

Here $r(l)$ is a vector of *Bernoulli* random variables each of which has a probability p of being 1. This vector is sampled for each layer and multiplied element by element with the outputs of that layer, $y(l)$, to create outputs $\tilde{y}^l$. The outputs are then used as input to the next layer. For learning, the derivatives of the loss function are back-propagated through the network.

At test time, the weights are scaled as $W_{test}^{(l)} = pW^{(l)}$. Several methods have also been used to improve stochastic gradient descent with standard neural networks. This include momentum, decaying learning rates and L2 weight decay, being very useful with dropout enhancement in neural networks.

3.5 Rectified Linear Unit Activation Function-RELU

Some techniques [6], acknowledge dropout discouraging co-adaptations of hidden units and their feature detectors, by adding a particular type of noise to the hidden unit activations, during training. At test time dropout does not occur and there may not be additional training, thus multiplying the net input from the layer, by a factor of $\frac{1}{1-r}$, where r is the dropout probability for units in the layer. Specifically, to compute the activation y_t of the tth layer of the network during forward propagation use:

$$y_t = f(\frac{1}{1-r}y_{t-1} * mW + b) \tag{8}$$

Here f is the activation function for the tth layer, W and b are respectively the weights and biases for the layers. * denotes element by element multiplication, and m is a binary mask with entries drawn from $Bernoulli(1-r)$ indicating which activations not to be dropped out.

4   Experimental Results and Discussion

A segmentation method [18], was used to extract teeth contours from images. Image database has two types of dental radio-graphs, those of alive persons and those of dead people. Images are randomly selected and a teeth contour extraction algorithm is executed and results manually compared. Another comparison is done for the same images, this time using the snake method. From the results of the two methods, a comparison is done. The contour extraction algorithm gathers efficiently the boundary of segmentation of teeth, and this can be shown by the below images (Fig. 2):

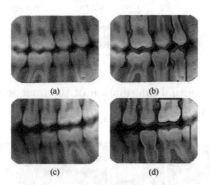

Fig. 2. (a) and (c) Snake method, (b) and (d) Active contour without edges. [4]

Performance evaluation of a CNN was done on a test dataset then compared with that of three expert dentists [19]. The experiments done by the experts were matched with the ground truth data to calculate results for precision, recall and intersection over unit (IOU). Results found a higher accuracy for the more experienced expert dentists compared to the other dentists. After the human exercise of evaluation, the previous fast neural network was trained with the same train and validation datasets. These results from the human experts were compared to those of the automatic system, which in this case the R-CNN.

The introduction of post-processing of images to the neural network, helped in the evaluation performance of the automatic system. The results were close to those of the dentists. The experiment in [13] was to train a VGG-16 net(neural network) on 8 augmented datasets. Each experiment having the learning rate set to 1e-3, L2-regularization set to 1e-7, and the dropout parameter p to 0.5. VGG-16 network was then trained over its corresponding augmented training set for several iterations.

Visualization of the augmentation effect on the training set, presents the mean images result before augmentation, and after augmentation. This helps to understand the relationship that exists between the mean image results from an augmented training set training, and validation results of the VGG-16 network trained on that augmented set. Thus augmentation of original set affects training and validation accuracy (Table 2).

Table 2. Mutual information between mean images + average accuracy.

Augmentation type	MI	Training acc	Validation acc
Noise	2.27	0.625	0.660
Gaussian Filter	2.60	0.870	0.881
Jitter	2.59	0.832	0.813
Scaling	2.67	0.887	0.874
Powers	2.33	0.661	0.737
Rotate	2.20	0.884	0.880
Shear	2.06	0.891	0.879
Flips	2.70	0.830	0.842

Therefore, a higher mutual information of the images correlates to a higher classification accuracy (Fig. 3).

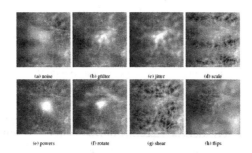

Fig. 3. Images after augmentation process [13].

Dental image segmentation helps to find regions of interest namely, the gap valley and individual teeth isolation. These poses a challenge caused by noise

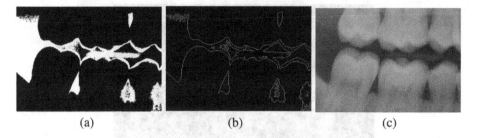

(a)   (b)   (c)

Fig. 4. a. Region growing image, b. Canny edge detected image, c. Identified tooth isolation and gap valley from binary integrated edge intensity curves image

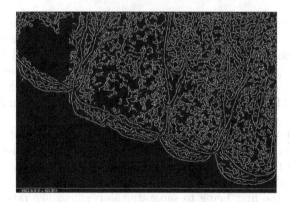

Fig. 5. Original image

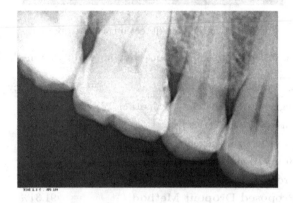

Fig. 6. Canny edge detection without dropout

and intensity variations. The use of gray and binary intensity integral curves is important. A novel method [20], is proposed to find a region of interest for gap valley and tooth isolation using binary edge intensity integral curves. They use the region growing approach, then canny edge detection (Fig. 4).

Fig. 7. Canny edge detection with dropout

Table 3. Comparison of various dropout evaluation methods.

Methods	Accuracy
Segmentation methods evaluation [21]	77.23%
Unsupervised Caries Detection [22]	87.5%
Dropout in Recurrent Neural Networks [5]	73.4%
Brain Tumor Segmentation using DCNN [8]	85.0%
Dropout [3]	79.0%
Proposed Dropout Method	91.31%

Another model is implemented for sentiment analysis where labelled data is not sufficient. From the results from the recurrent neural networks (RNN) trained, have been found to be impressive with the variatio-nal dropout introduced [5]. A dropout neural networks [3], was trained on five image data sets that were used to evaluate the dropout effect. The datasets were, MNIST, SVHN, CIFAR-10, CIFAR-100 and ImageNet, showed improved performance compared to those without dropout (Table 3).

Some experiments were performed on BRATS 2013 dataset and only real patient data used for evaluating the model [8]. Some of the segmentation results generated using the trained neural networks, and the proposed algorithm performed well in specified tumor regions. Another approach [22], show results from two approaches, the first being to separate the performance rates of the upper and lower jaw results, and assess them individually. The information shows how well the algorithm performs on different jaw regions. The other approach is grouping two jaw regions, and finding an average of the two rates to represent the said algorithms overall performance (Figs. 5, 6 and 7).

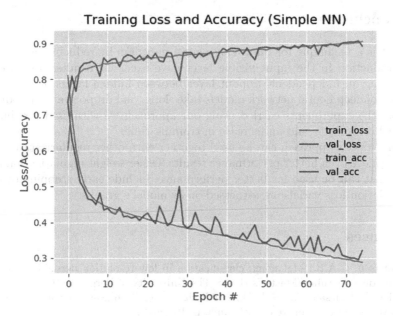

Fig. 8. Graph without dropout.

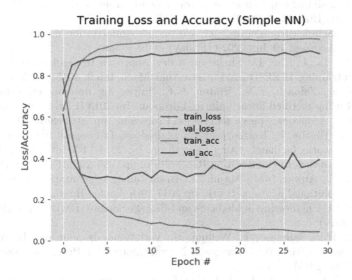

Fig. 9. Graph with dropout.

Researchers in [21], review, compare and summarize some existing dental segmentation methods, that include thresholding, active contours, level set, clustering. These methods were reviewed according to accuracy, precision, and also the speed of computation (Figs. 8 and 9).

5   Conclusion

Dropout is a technique for improving neural networks by reducing over-fitting. Its introduction in the input visible layer gave excellent results, compared to introducing an independent dropout layer between hidden layers of the model. Training of deep neural network models takes long, and dropout aids in reducing it. From the experiments carried using our approach we witnessed a reduction in model complexity and an increase in training time.

Our dropout approach can be used together with other regularization methods to achieve even better performance results. Other weight regularization techniques that can be used for better performance, include early stopping to solve the validation loss variations witnessed on or model's graphs.

References

1. Andersen, C.A.R.: Melanoma classification in low resolution dermoscopy images using deep learning, Master's thesis, The University of Bergen (2019)
2. Miloš, E., Kolesau, A., Šešok, D.: Traffic sign recognition using convolutional neural networks/Science-Future of Lithuania, p. 10 (2018)
3. Srivastava, N., Hinton, G., Krizhevsky, A., Sutskever, I., Salakhutdinov, R.: Dropout: a simple way to prevent neural networks from overfitting. J. Mach. Learn. Res. **15**(1), 1929–1958 (2014)
4. https://machinelearningmastery.com/dropout-regularization-deep-learning-models-keras/. Accessed 19 June 2020 (Friday) at 10.46 am
5. Kang, G., Li, J., Tao, D.: Shakeout: a new approach to regularized deep neural network training. IEEE Trans. Pattern Anal. Mach. Intell. **40**(5), 1245–1258 (2017)
6. Dahl, G.E., Sainath, T.N., Hinton, G.E.: Improving deep neural networks for LVCSR using rectified linear units and dropout. In: 2013 IEEE International Conference on Acoustics, Speech and Signal Processing, pp. 8609–8613. IEEE (2013)
7. Pham, V., Bluche, T., Kermorvant, C., Louradour, J.: Dropout improves recurrent neural networks for handwriting recognition. In: 2014 14th International Conference on Frontiers in Handwriting Recognition, pp. 285–290. IEEE (2014)
8. Hussain, S., Anwar, S.M., Majid, M.: Brain tumor segmentation using cascaded deep convolutional neural network. In: 2017 39th Annual International Conference of the IEEE Engineering in Medicine and Biology Society (EMBC), pp. 1998–2001. IEEE (2017)
9. Kingma, D.P., Salimans, T., Welling, M.: Variational dropout and the local reparameterization trick. In: Advances in Neural Information Processing Systems, pp. 2575–2583 (2015)
10. Rashmi, K.V., Gilad-Bachrach, R.: DART: dropouts meet multiple additive regression trees. In: AISTATS, pp. 489–497 (2015)
11. Wu, F., Hu, P., Kong, D.: Flip-rotate-pooling convolution and split dropout on convolution neural networks for image classification. arXiv preprint arXiv:1507.08754 (2015)
12. He, K., Zhang, X., Ren, S., Sun, J.: Spatial pyramid pooling in deep convolutional networks for visual recognition. IEEE Trans. Pattern Anal. Mach. Intell. **37**(9), 1904–1916 (2015)

13. Hussain, Z., Gimenez, F., Yi, D., Rubin, D.: Differential data augmentation techniques for medical imaging classification tasks. In: AMIA Annual Symposium Proceedings, vol. 2017, p. 979. American Medical Informatics Association (2017)
14. Nyíri, T., Kiss, A.: Style transfer for dermatological data augmentation. In: Bi, Y., Bhatia, R., Kapoor, S. (eds.) IntelliSys 2019. AISC, vol. 1038, pp. 915–923. Springer, Cham (2020). https://doi.org/10.1007/978-3-030-29513-4_67
15. Nyíri, T., Kiss, A.: Novel ensembling methods for dermatological image classification. In: Fagan, D., Martín-Vide, C., O'Neill, M., Vega-Rodríguez, M.A. (eds.) TPNC 2018. LNCS, vol. 11324, pp. 438–448. Springer, Cham (2018). https://doi.org/10.1007/978-3-030-04070-3_34
16. Aggarwal, L.P.: Data augmentation in dermatology image recognition using machine learning. Skin Res. Technol. **25**(6), 815–820 (2019)
17. Van der Walt, S., et al.: scikit-image: image processing in Python. PeerJ **2**, e453 (2014)
18. Shah, S., Abaza, A., Ross, A., Ammar, H.: Automatic tooth segmentation using active contour without edges. In: 2006 Biometrics Symposium: Special Session on Research at the Biometric Consortium Conference, pp. 1–6. IEEE (2006)
19. Chen, H., et al.: A deep learning approach to automatic teeth detection and numbering based on object detection in dental periapical films. Sci. Rep. **9**(1), 1–11 (2019)
20. Modi, C.K., Desai, N.P.: A simple and novel algorithm for automatic selection of ROI for dental radiograph segmentation. In: 2011 24th Canadian Conference on Electrical and Computer Engineering (CCECE), pp. 000504–000507. IEEE 2011
21. Rad, A.E., Rahim, M.S.M., Norouzi, A.: Digital dental X-ray image segmentation and feature extraction. TELKOMNIKA Indones. J. Electr. Eng. **11**(6), 3109–3114 (2013)
22. Osterloh, D., Viriri, S.: Unsupervised caries detection in non-standardized bitewing dental x-rays. In: Bebis, G., et al. (eds.) ISVC 2016. LNCS, vol. 10072, pp. 649–658. Springer, Cham (2016). https://doi.org/10.1007/978-3-319-50835-1_58

Towards Exploiting Convolutional Features for Remote Sensing Images Scene Classification

Ronald Tombe and Serestina Viriri[✉]

School of Mathematics, Statistics and Computer Science,
University of KwaZulu-Natal, Durban, South Africa
viriris@ukzn.ac.za

Abstract. The developments in deep learning have shifting the research directions of remote sensing image scene understanding and classification from pixel-to-pixel handcrafted methods to scene-level image semantics analysis for scene classification tasks. The pixel-level methods rely on handcrafted methods in feature extraction, which yield low accuracy when these extracted features are fed to Support Vector machine for scene classification task. This paper proposes a generic extraction technique that is based on convolutional features in the context of remote sensing scene classification images. The experimental evaluation results with convolutional features on public datasets, Whu-RS, Ucmerced, and Resisc45 attain a scene classification accuracy of 92.4%, 88.78%, and 75.65% respectively. This demonstrate that convolutional features are powerful in feature extraction, therefore achieving superior classification results compared to the low-level and mid-level feature extraction methods on the same datasets.

Keywords: Remote sensing · Image scene classification · Neural networks · Deep learning

1 Introduction

Currently, satellite images facilitate the determination of Earth's surface using detailed information as they are a reliable data source with high significance in earth observations [1,2]. The continuous exponential increase in the number of satellite scene images contain varying spatial and complex structurally geometrical shapes, therefore, understanding their semantic content effectively is of specific importance because of needs for various applications in the remote sensing groups [3–5]. Consequently, research in remote sensing image classification is shifting from traditional hand-crafted pixel-level explanations to scene level understanding [2]. This aim at classifying scene images to sets of semantic land cover and land use classes basing on the image content.

In computer vision research, deep learning algorithms, particularly convolutional neural networks (CNNs) are gaining popularity owing to successes they

© Springer Nature Singapore Pte Ltd. 2021
T.-P. Hong et al. (Eds.): ACIIDS 2021, CCIS 1371, pp. 266–277, 2021.
https://doi.org/10.1007/978-981-16-1685-3_22

have shown for stronger feature representations [21–23]. Motivated by these successes, this paper proposes a novel high-level feature representation technique based on convolutional features in the context of remote sensing images scene classification. The major contributions of this paper are:

1. Empirical validation on the effectiveness and robustness of the convolutional features representation in the context of remote sensing images scene classification.
2. A generic image features representation technique that is based on deep learning convolutional features.

Different from the methods based on Bag-of-visual-Words [15,16] and fisher vectors [20] that model features based on handcrafted methods. These aforementioned methods attain lower accuracy in remote sensing image scene classification. This paper proposes a generic convolutional feature representation technique which extract features directly from images using filters. As the experimental results shows, convolutional features are powerful in feature representation therefore attain superior scene classification results of remote sensing images.

The rest of this paper is organized as follows, in Sect. 2, the a brief review is given on the various scene image feature representation techniques. Section 3 sets the context upon which this work is built, that is it discusses concepts machine learning, neural networks and deep learning while showing their interlinks. In Sect. 4, the proposed technique is presented. Section 5 discusses how experimentation is conducted and the metrics used in results validation. Section 6 provides discussions and finally Sect. 7 provides conclusions and insights for the paper.

2   Related Work

In scene classification it is appropriate to model a scene by creating a holistic representation of the satellite acquired image [6–20]. The underlying motivation in remote sensing image scene classification is that same scene image type should share particular statistically holistic visual features. Therefore, the work of remote sensing images mostly focus on calculating holistic and statistical scene image features for classification. In this regard, remote sensing images scene understanding is grouped into three major categories [1], that is, methods relying on: 1. low-level features, 2. mid-level features and 3. high-level vision feature representation methods which are also referred to as deep learning techniques.

In low-level features, aerial scenes are differentiated by, texture, structural and spectral features. Therefore, an aerial scene is described with feature vectors that are obtained from low-feature characteristics that are either local or global [11]. The classification feature representation vector for local features in this category are formed by combining or pooling the local descriptors of image segments. The Scale Invariant Feature Transform (SIFT)[12] is a good example in the local feature representation category. For representing global spatial order of aerial scenes, Local binary pattern (LBP) for texture [14] and color histogram(CH) [13] descriptors have been applied for scene classification.

The mid-level feature representation methods aim to create a holistic scene representation by representing the high-order statistical patterns developed from the extracted low-level features. A general procedure is to first extract scene image feature using low-level pixel based methods such as LBP, SIFT, color histograms e.t.c, then encode these features for developing a holistic mid-level feature representation of aerial scenes. The Bag-of-Visual-Words (BoVW) [15] is a popular mid-level approach. The BoVW describe low-level features using the SIFT [12], then it learns the vocabulary of visual words using the k-means clustering algorithm which result to an encoded vocabulary. This generates a global feature vector quantization of the image which is the frequency count occurrences of every visual word in the image. BoVW is widely adopted in computing mid-level feature representation for remote sensing images scene classification [15–17,19]. Tombe et al. [20] developed a feature characterization method which combine features generated by low-level feature representation methods, i.e. local ternary patterns and hu-moments to develop a mid-level feature representation method. This method uses fisher vectors to encode the low-level features of local ternary patterns and Hu moments, the method is applied in remote sensing images scene classification [20].

Deep learning techniques have been applied in object and scene recognition tasks [26–28] and they have shown impressive results in scene classification compared to low-level handcrafted and medium-level feature representation methods [1]. Literature exists on the various architectures of deep learning models [21–23]. These architectures have been adopted in the context of remote sensing for feature extraction and image scene classification: The bag of visual words convolutional features(BoCF) [18] comprise four steps: Convolutional feature extraction, creation of codebooks, feature encoding and scene classification, this technique is ambiguous on how the features are engineered from the input images, instead the method BoCF focuses more on feature encoding. The deep learning metrics via learning discriminative CNNs technique [25] that minimize the within class-similarity of same scene images and widen the between class differences with learning metrics regularizer.

3   Foundational Concepts

This section presents foundational concepts of convolutional neural networks, or ConvNets which is a specific type of deep learning technique. ConVnets interconnected systems of processing units(neurons) that compute values from the input layers resulting to outputs that are used on further layer units. Neurons work together to solve a specific problem while learning by example, that is, a network is created for a specific application such as ImageNet Visual Recognition Challenge (ILSVRC) classification for networks [22,23]. ConVnets have several advantages: (i) learn and extract features automatically from inputs, (ii) implement the weight sharing principle which significantly reduces the number of parameters and hence their generalization capacity, (iii) they generate invariant features. Next, the concepts employed in Convolutional neural networks are presented.

3.1   Neurons and Neural Networks

The motivation behind neural networks is the concept that a neuron's computation entails weighted sum of input values [24]. Figure 1 (a) depicts a computation neural network. The input layer neurons receive values which it propagates to neurons in the next layer of the neural-network, commonly referred to as "hidden layer". The weighted sums across hidden layers are finally propagated to the output-layer. Figure 1 (a) is an example of the computations at every layer and this is expressed by Eq. (1)

$$y_k = f(\sum_{i=1}^{3} W_{ik} \times x_i + b) \tag{1}$$

The terms $x_i, W_{ik}, y_k\ b$ are the input activations, weights, output activations and the bias respectively, while $f(.)$ is a non-linear function.

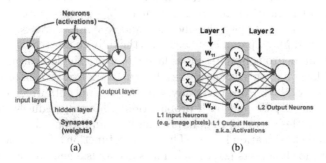

(a)   (b)

Fig. 1. Neural network example. (adopted from [24].) (a) weights and neurons. (b) Calculate weighted sum for every layer

3.2   ConvNet Components

Convolutional Neural Networks, i.e. convNets, are trainable network architectures composed of several stages. The network is built on hierarchical convolutional building blocks called layers. Generally for all convolutional neural networks, the units of initial layers correspond to some section of the input which are then organized into feature representations. The layers close to the input (lower layers) form correlated units which concentrate at local regions and there are less and less number of patches over bigger image regions. ConvNets architectures comprise of various layers that perform different functions:

1. Convolutional layers: Composed of neurons which connects subregions of input image or the outputs of preceding layer. Features are learned at this layers based on the weights applied to the image region by the filters which scans input images. This layer uses filters and strides to generate feature maps.

2. Nonlinear Layers: They immediately follow the convolutional layers and they comprise of different activation functions such as ReLU [29] tanh [30], which perform threshold operations on every element. For instance a ReLU activation function applied to an element in an input value generates an output based on Eq. (2). The activation function does not alter the input size.

$$f(x) = \begin{cases} x, \geq 0 \\ 0, x < 0 \end{cases} \qquad (2)$$

3. Pooling layers: This layer follows the convolutional layers for down-sampling, thus, reduces the number of parameters which are to be learned in the next layers. Further they reduce overfitting [21]. No learning takes place in this layer

4. Fully Connected Layer: Neurons for a fully connected layer join the neurons of previous layer. In this layer, all features(local information) that is learned by and from previous layers of the image combine to identify large patterns. For object identification and recognition problems, this layer combine the features for object category classification and recognition tasks.

5. Dropout Layer: Many parameters occupies the fully connected layer because it is connects all neurons from previous layers to every single neuron of the fully connected layer and this can easily result to network over-fitting problem. The dropout method [31] drops randomly some neuron outputs which do not have effect in forward pass and back propagation. This step reduces the number of neurons in the network, thus improving its training speed.

6. SoftMax and classification Layers: The softmax function adapts and fine-tunes the network for label sets [21,22,33]. The softmax is a generalization function that approximate samples from categorical distribution in the range(0,1) for multi-class classification problems through generation of $n - dimensional$ vectors. Equation (3) depicts a softmax function that can predict probability of the ith class given a sample input vector(X).

$$y_i(X) = \frac{exp((log(\pi_i) + g_i)/T)}{\sum_{j=1}^{k} exp((log(\pi_j) + g_j)/T)} \; for \; i = 1, ..., k. \qquad (3)$$

4   Strategies for Exploiting ConvNets

The main strategies for employing convNets in different scenarios are grouped to three categories, i.e. Full-trained networks, fine-tuned network and convNets as feature extractors [33].

4.1   Full-Trained Networks

This approach trains a network from scratch(using random filter weights for initialization). This approach is beneficial with large datasets because it leads to network convergence; other specific advantages of this approach include: (i) feature extractors are tuned based on specific dataset, this may result to improved accuracy and (ii) better control of the network parameters. A major shortcoming of this approach is that it's expensive due to high computational resources required to train a network from scratch. The two options [33] for training deep neural networks are: (i) A complete designing and training of a new architecture that entail different layers, neurons, type of activations, weight decay, the number of iterations, learning rate among others; (ii) Use the existing architecture and fully train its weights with the target dataset of a given size. In this case, the network architecture and its parameter weights(learning rate, weight decay, e.t.c) are not changed.

4.2   Fine-Tuned Network

With reasonably large datasets which are not sufficient to fully-train a new network, network-fine-tuning is a viable option to extract the effectiveness from pre-trained deep neural network given that they can substantially enhance performance of the classification layer this is demonstrated with examples and results in literature [1,2]. Furthermore, basing on results of ImageNet Visual Recognition Challenge (ILSVRC) networks [22,23] which fine-tuned and adopted for remote sensing scene classification images [18,25], show that earlier layers learn general features independent of the train data. This imply that earlier layers contain generic features that can be useful for various tasks, however later layers are progressively more specific to details of the original dataset classes. With this property, the network initial layers can be retained while the last ones should be modified to suit datasets of interest.

4.3   Convolutional Neural Neworks as Feature Extractors

Pre-trained CNNs can be used for feature extraction from any image, because the CNNs learn generic features(in the initial layers of the network) and they generalize well. For instance, the features obtained with CNNs trained on imageNets datasets have shown to generalize well when used in other contexts which include: flower categorization [34], human attribute detection [35]. In remote sensing CNNs have been applied [9,18] as feature extractors, however, given the enormous amount of images generated periodically by satellites that are useful on earth observations applications for intelligent decision making. CNNs feature extractors frameworks are required to drive development of innovative applications in remote sensing contexts, a gap that this research seek to address.

5   The Proposed Method

The proposed generic convNet feature extraction technique is based on the first four steps as outlined in Sect. 3.2. i.e. i) Convolution layers, ii) Nonlinear layers, iii) Pooling layers and iv) Fully connected layers. note that, the convolutional step and the nonlinear functions steps occur at the convolutional layer. It has been shown that Support vector machines(SVM) achieve superior classification accuracy results [33] compared to the softmax and classification layers of the deep learning neural networks. The following Steps on convolutional feature extraction and classification are depicted in Fig. 2.

1. *Convolutional Feature Engineering*: This step performed at the convolutional-layers, the nonlinear functions RELU [32] are applied in convolving the image for feature engineering. Filters are a set of weights and biases applied to image regions which are connected to receptive fields in the feature-maps of previous kernels. Each feature map contain different weight sets and biases.
2. *Pooling-layers:* are between convolutional-layers to reduce dimensions of the feature maps thereby creating invariance based on movements and distortions of neighboring pixels.
3. *Fully Connected layers*: are the last network layers extract the information relayed by lower layers hence forming a high-level-feature representation for the different class categories. For illustrations, Fig. 2 has six classes, note the size of a fully connected layer vector has six circle elements.
4. *The Support vector machine classifier:* trains a classier using the extracted CNNs features for remote sensing image scene classification.

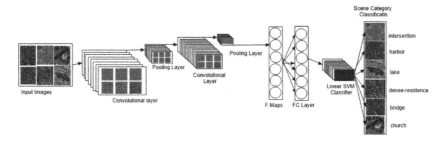

Fig. 2. Mechanics of the generic feature engineering and representation technique for scene category classification tasks

6   Experimentation

6.1   Datasets

Three remote sensing datasets, that is: UCMerced [15], WHU-RS Dataset [36], and RESISC45 [2] are used to evaluate the effectiveness and robustness of con-Vnet features(referred in here as high-level-features). The UCMerced dataset contain 21 classes with 100 images each, WHU-RS dataset has 19 classes with 950 total number of images and the RESISC45 dataset has 45 classes with 700 images per class. For more details on the datasets, see the cited papers. Figure 3 shows a sample of images from the RESISC45 dataset.

Fig. 3. Sample images from the NWPU-RESISC45 dataset [2]

6.2 Experiment Setting

In experimentation, the datasets are set to the train, validation, and test ratios as follows: Whu-RS (160:30:10), Ucmerced (70:20:10), Resisc45(20:75:5). The deep convolutional features are extracted with aid of the state-of-the art model AlexNet [21] which is pre-trained with ImageNet dataset. The AlexNet model is downloaded as a third-party add-on deep learning package for Math-lab software. A pretrained AlexNet network is selected because this research utilize transfer learning to demonstrate the power of convolutional features. Due to space constraints, discussions of other deep learning networks such as VGG16, VGG19 and GoogleNet are omitted in this work. A multi-class binary support Vector Machine(SVM) learners model (*fitcecoc*) provided by matlab 2017b software is applied to learn the convolutional features.

6.3 Evaluation Metrics for High-Level Feature Representation on Effectiveness and Robustness

The overall accuracy [2] is used to evaluate the effectiveness and robustness of the high-level feature convolutional representation. Average accuracy is the number of correctly classified image samples regardless of the class which they belong divided by total number of image samples. The evaluation results of, low-level, mid-level and high-level features representation methods on three datasets(Whu-RS, Ucmerced and Resisc45) are given in Table 1, and Fig. 4 (a, b, and c) respectively on scene classification tasks.

7   Results and Discussion

Table 1 and Fig. 4. (a, b and c) shows the accuracy results of three feature representation methods, i.e. Low-level, Mid-level and High-level methods on Whu-RS19, UCMerced and RESISC45 dataset respectively.

Table 1. Classification Performances on Resisc45 and UCmerced datasets

Feature Extraction method	Whu-RS OA%	Ucmerced OA%	Resisc45 OA%
Local Binary patterns	39.75	35.15	23
LBP and Hu-Moments(Fisher Vectors)	54.85	50.12	31.5
Convolutional Features (AlexNet)	92.4	88.78	75. 65

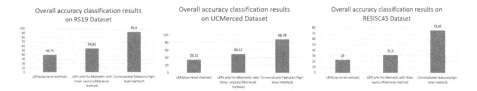

Fig. 4. Overall accuracy classification results on three datasets i.e. a) RS19, b) UCMerced and c) NWPU-RESISC45 with Low-level, mid-level and high-level feature characterization methods respectively.

The accuracy results for the three image feature descriptor techniques: LBPs(low-level), LBPs and Hu-Moments with fisher vectors (mid-level) and the convolutional features (high-level) are depicted on the bar graphs in Fig. 4(a, b and c) for three datasets. Convolutional features demonstrate superior discriminative capacity, hence achieving better classification accuracy results as compared to the low-level and mid-level feature representation methods. The superior accuracy results obtained with convolutional features demonstrate their effectiveness in image feature characterization power. Further, there is a consistent trend with convolutional features attaining the highest classification accuracy results in all the three datasets as compared to low-level and mid-level feature representation techniques. This imply that convolutional neural networks are robust in characterizing image features. Furthermore, from Fig. 4 it is observed that there is a decrease in classification accuracy results with larger image dataset, this can be attributed to two factors: (1.) There is need for Convolutional Neural Network architectures that are specific for remote sensing image scene classification context and (2.) The convolutional neural networks needs to be fine-tuned inorder to extract features that are more discriminatory. The findings of this work are consistent with the literature works [2,33].

8   Conclusions

This paper present a high-level feature representation technique that is based on convolutional features. Given the huge volumes of images generated by satellite technologies periodically on earth observations, there is need for superior scene images feature characterization techniques for reliable remote sensing applications to be developed, this is possible with the advent of deep learning techniques. Experiments in this paper demonstrate that the trend for image feature characterization is moving towards convolutional feature extraction. Our future works and indeed a challenge to the computer scientists is to develop deep-learning architectures for transfer feature learning in the context of remote sensing images so as to spur developments of intelligent applications in the area.

References

1. Xia, G.S., et al.: AID: a benchmark data set for performance evaluation of aerial scene classification. IEEE Trans. Geosci. Remote Sens. **55**(7), 3965–3981 (2017)
2. Cheng, G., Han, J., Lu, X.: Remote sensing image scene classification: benchmark and state of the art. Proc. IEEE **105**(10), 1865–1883 (2017)
3. Du, B., Zhang, L.: A discriminative metric learning based anomaly detection method. IEEE Trans. Geosci. Remote Sens. **52**(11), 6844–6857 (2014)
4. Han, J., Zhang, D., Cheng, G., Guo, L., Ren, J.: Object detection in optical remote sensing images based on weakly supervised learning and high-level feature learning. IEEE Trans. Geosci. Remote Sens. **53**(6), 3325–3337 (2014)
5. Han, J., et al.: Efficient, simultaneous detection of multi-class geospatial targets based on visual saliency modeling and discriminative learning of sparse coding. ISPRS J. Photogramm. Remote. Sens. **89**, 37–48 (2014)
6. Cheng, G., Han, J., Guo, L., Liu, Z., Bu, S., Ren, J.: Effective and efficient midlevel visual elements-oriented land-use classification using VHR remote sensing images. IEEE Trans. Geosci. Remote Sens. **53**(8), 4238–4249 (2015)
7. Zhang, F., Du, B., Zhang, L.: Saliency-guided unsupervised feature learning for scene classification. IEEE Trans. Geosci. Remote Sens. **53**(4), 2175–2184 (2014)
8. Zhu, Q., Zhong, Y., Zhao, B., Xia, G.S., Zhang, L.: Bag-of-visual-words scene classifier with local and global features for high spatial resolution remote sensing imagery. IEEE Geosci. Remote Sens. Lett. **13**(6), 747–751 (2016)
9. Hu, F., Xia, G.S., Hu, J., Zhang, L.: Transferring deep convolutional neural networks for the scene classification of high-resolution remote sensing imagery. Remote Sens. **7**(11), 14680–14707 (2015)
10. Oliva, A., Torralba, A.: Modeling the shape of the scene: a holistic representation of the spatial envelope. Int. J. Comput. Vis. **42**(3), 145–175 (2001)
11. dos Santos, J.A., Penatti, O.A.B., da Silva Torres, R.: Evaluating the potential of texture and color descriptors for remote sensing image retrieval and classification. In: VISAPP (2), pp. 203–208, May 2010
12. Lowe, D.G.: Distinctive image features from scale-invariant keypoints. Int. J. Comput. Vis. **60**(2), 91–110 (2004)
13. Swain, M.J., Ballard, D.H.: Color indexing. Int. J. Comput. Vis. **7**(1), 11–32 (1991)
14. Ojala, T., Pietikäinen, M., Mäenpä, T.: Multiresolution gray-scale and rotation invariant texture classification with local binary patterns. IEEE Trans. Pattern Anal. Mach. Intell. (7), 971–987 (2002)

15. Yang, Y., Newsam, S.: Bag-of-visual-words and spatial extensions for land-use classification. In: Proceedings of the 18th SIGSPATIAL International Conference on Advances in Geographic Information Systems, pp. 270–279. ACM, November 2010

16. Yang, Y., Newsam, S.: Spatial pyramid co-occurrence for image classification. In: 2011 International Conference on Computer Vision, pp. 1465–1472. IEEE, November 2011

17. Shao, W., Yang, W., Xia, G.-S., Liu, G.: A hierarchical scheme of multiple feature fusion for high-resolution satellite scene categorization. In: Chen, M., Leibe, B., Neumann, B. (eds.) ICVS 2013. LNCS, vol. 7963, pp. 324–333. Springer, Heidelberg (2013). https://doi.org/10.1007/978-3-642-39402-7_33

18. Cheng, G., Li, Z., Yao, X., Guo, L., Wei, Z.: Remote sensing image scene classification using bag of convolutional features. IEEE Geosci. Remote Sens. Lett. **14**(10), 1735–1739 (2017)

19. Negrel, R., Picard, D., Gosselin, P.H.: Evaluation of second-order visual features for land-use classification. In: 2014 12th International Workshop on Content-Based Multimedia Indexing (CBMI), pp. 1–5. IEEE, June 2014

20. Tombe, R., Viriri, S.: Local descriptors parameter characterization with fisher vectors for remote sensing images. In: 2019 Conference on Information Communications Technology and Society (ICTAS), pp. 1–5. IEEE, March 2019

21. Krizhevsky, A., Sutskever, I., Hinton, G.E.: ImageNet classification with deep convolutional neural networks. In: Advances in Neural Information Processing Systems, pp. 1097–1105 (2012)

22. Szegedy, C., et al.: Going deeper with convolutions. In: Proceedings of the IEEE Conference on Computer Vision and Pattern Recognition, pp. 1–9 (2015)

23. Simonyan, K., Zisserman, A.: Very deep convolutional networks for large-scale image recognition. arXiv preprint arXiv:1409.1556 (2014)

24. Sze, V., Chen, Y.H., Yang, T.J., Emer, J.S.: Efficient processing of deep neural networks: a tutorial and survey. Proc. IEEE **105**(12), 2295–2329 (2017)

25. Cheng, G., Yang, C., Yao, X., Guo, L., Han, J.: When deep learning meets metric learning: remote sensing image scene classification via learning discriminative CNNs. IEEE Trans. Geosci. Remote Sens. **56**(5), 2811–2821 (2018)

26. Zou, Q., Ni, L., Zhang, T., Wang, Q.: Deep learning based feature selection for remote sensing scene classification. IEEE Geosci. Remote Sens. Lett. **12**(11), 2321–2325 (2015)

27. Zhou, B., Lapedriza, A., Xiao, J., Torralba, A., Oliva, A.: Learning deep features for scene recognition using places database. In: Advances in Neural Information Processing Systems, pp. 487–495 (2014)

28. Yuan, Y., Mou, L., Lu, X.: Scene recognition by manifold regularized deep learning architecture. IEEE Trans. Neural Netw. Learn. Syst. **26**(10), 2222–2233 (2015)

29. Li, Y., Yuan, Y.: Convergence analysis of two-layer neural networks with ReLU activation. In: Advances in Neural Information Processing Systems, pp. 597–607 (2017)

30. Li, M.B., Huang, G.B., Saratchandran, P., Sundararajan, N.: Fully complex extreme learning machine. Neurocomputing **68**, 306–314 (2005)

31. Srivastava, N., Hinton, G., Krizhevsky, A., Sutskever, I., Salakhutdinov, R.: Dropout: a simple way to prevent neural networks from overfitting. J. Mach. Learn. Res. **15**(1), 1929–1958 (2014)

32. Nair, V., Hinton, G.E.: Rectified linear units improve restricted Boltzmann machines. In: ICML, January 2010

33. Nogueira, K., Penatti, O.A., dos Santos, J.A.: Towards better exploiting convolutional neural networks for remote sensing scene classification. Pattern Recogn. **61**, 539–556 (2017)
34. Sünderhauf, N., McCool, C., Upcroft, B., Perez, T.: Fine-grained plant classification using convolutional neural networks for feature extraction. In: CLEF (Working Notes), pp. 756–762 (2014)
35. Hara, K., Jagadeesh, V., Piramuthu, R.: Fashion apparel detection: the role of deep convolutional neural network and pose-dependent priors. In: 2016 IEEE Winter Conference on Applications of Computer Vision (WACV), pp. 1–9. IEEE, March 2016
36. Xia, G.S., Yang, W., Delon, J., Gousseau, Y., Sun, H., Maître, H.: Structural high-resolution satellite image indexing. In: ISPRS TC VII Symposium-100 Years ISPRS, vol. 38, pp. 298–303, July 2010

Contrast Enhancement in Deep Convolutional Neural Networks for Segmentation of Retinal Blood Vessels

Olubunmi Sule, Serestina Viriri$^{(\boxtimes)}$, and Mandlenkosi Gwetu

School of Mathematics, Statistics and Computer Science,
University of KwaZulu-Natal, Durban, South Africa
viriris@ukzn.ac.za

Abstract. The segmentation of blood vessels from the retinal fundus image is known to be complicated. This difficulty results from visual complexities associated with retinal fundus images such as low contrast, uneven illumination, and noise. These visual complexities can hamper the optimal performance of Deep Convolutional Neural Networks (DCNN) based methods, regardless of its ground-breaking success in computer vision and segmentation tasks in the medical domain. To alleviate these problems, image contrast enhancement becomes inevitable to improve every minute objects' visibility in retinal fundus images, particularly the tiny vessels for accurate analysis and diagnosis. This study investigates the impact of image contrast enhancement on the performance of DCNN based method. The network is trained with the raw DRIVE dataset in RGB format and the enhanced version of DRIVE dataset using the same configuration. The take-in of the enhanced DRIVE dataset in the segmentation task achieves a remarkably improved performance hit of 2.60% sensitivity, besides other slight improvements in accuracy, specificity and AUC, when validated on the contrast-enhanced DRIVE dataset.

Keywords: Retinal blood vessels · Deep Convolutional Neural Networks · Segmentation · Image contrast enhancement

1 Introduction

Diabetic Retinopathy (DR) is an eye disease that progresses from a mild stage to a severe stage where early treatment is lacking. It is a complication of diabetes mellitus and can cause permanent and irreversible blindness. It is common among the working-age group and worst enough; little children are also affected. DR is a global affliction to the health sector, society, and the world at large. Its consequences on the economy, quality of life, productivity, and global Gross Domestic Products is frightening. Variation in the retinal vascular structures is a clear pointer to the presence of pathologies foreshadowing DR eye disease in the retinal. The resultant effect of DR progression to the severe stage is total vision

© Springer Nature Singapore Pte Ltd. 2021
T.-P. Hong et al. (Eds.): ACIIDS 2021, CCIS 1371, pp. 278–290, 2021.
https://doi.org/10.1007/978-981-16-1685-3_23

loss [1, 2]. However, early detection of DR eye disease can arrest severe complications and total blindness. Currently, manual screening is the major approach to detecting DR eye disease. Unfortunately, the Ophthalmologists that handle the screening are few in the field, and the screening process is labour intensive, time-consuming, error-prone, and not easily accessible in terms of cost and availability in the rural areas. Therefore, Computer-aided algorithms for the detection of DR at an early stage remains the best option.

Segmentation of blood vessels from the retinal fundus image is an active area of research that has attracted many interests. These interests have led to the development of many automatic algorithms with brilliant innovations and ideas. These algorithms are to aid early detection, analysis, diagnosis, and timely treatment of DR. Nevertheless, the task remains challenging due to low contrast between the foreground and the background of retinal fundus image, non-uniform illumination and noise challenge. Degradation in image quality often leads to loss of information that might be useful for medical analysis and diagnosis. Hence, optimal segmentation performance becomes illusive since every minute detail needed for accurate prediction is missing or not visible. Given this, this study proposes Contrast Enhancement in Deep Convolutional Neural Networks (DCCN) for Segmentation of Retinal Blood Vessels from Fundus Images. DCNN based models have proven to be very robust in speech recognition, character recognition, detection, classification, and segmentation tasks. Some of the DCNN based architectures include AlexNet [3], VGGNet [4], ResNet [5], U-Net [6], FCN [7].

1.1   Related Work

Many novel and groundbreaking segmentation tasks in medical imaging are DCNN based. Despite the great achievement of DCNN based approaches, its performance accuracy can be affected by low contrast, noise, uneven illumination, and other artifacts that degrade the quality of an image. Soomro et al. [9] showed that proper noise removal and contrast enhancement techniques could significantly improve the performance of a CNNs based model. The choice and adequate application of these techniques can go a long way to restore a degraded image quality back or close to its original condition. Sahu et al. [10] presented a denoising and contrast enhancement techniques for retinal fundus images.

A useful contrast enhancement technique will preserve vital information in an image, restore the missing information due to low contrast, and enhance the image for better analysis. Many of the existing CNNs based algorithm for segmentation of blood vessels from retinal fundus image applied image enhancement techniques either at the pre-processing or post-processing stage to enhance the CNNs segmentation predictive accuracy.

Aforetime, many authors have employed some of the image processing and enhancement techniques to improve CNNs segmentation performance. These include; Lisowski et al. [11] acknowledged that CNNs based models can conveniently learn from raw input images but that they learn better from properly enhanced or pre-processed images. The Authors, however, applied Global Contrast Normalization, Zero-phase Compound Analysis (ZCA) pre-processing techniques before passing the images to CNNs for training. Soomro et al. [12] presented the sensitivity boosting effect in combining robust pre and post-processing techniques with CNNs based architecture. The pre and post-processing techniques were able to address noise and uneven illumination problems, which boosted the sensitivity with a slight performance hit of 0.007. The model can eventually detect tiny vessels. Similarly, Guo et al. [13] introduced a blurring and denoising algorithm using short connections, and this brilliant innovation yielded an outstanding segmentation performance of extracting blood vessels from the retinal fundus image. The model can detect tiny vessels, solve pathological interference, remove noise, and segment disc boundary.

Dharmawan et al. [14] explored Contrast Limited Adaptive Histogram Equalization (CLAHE), matched filter for contrast enhancement on a CNNs based model, and achieved a very robust sensitivity of 70.71. Jebaseeli et al. [15] preprocessed the images using CLAHE, segmented with the Tandem Pulse Coupled Neural Net- work (TPCNN) model, and assigned values to all the training parameters using Particle Swarm Organisation (PSO). The model achieved an average sensitivity of 70.23% on Digital Retinal Images for Vessels Extraction (DRIVE) datasets [16], amongst other datasets. Sule et al. [17] applied some enhancement techniques on RGB retinal fundus image to enhance CNNs performance for the segmentation of blood vessels from the retinal fundus image. Fraz et al. [18] proposed an ensemble model consisting of bagged and boosting decision trees to segment blood vessels from the retinal. Marin et al. [19] applied both pre and post-processing techniques to enhance the retinal images before using Neural Network for the pixel classification.

The main contribution of this study is to show the impact of contrast-enhanced retinal fundus images on the predictive accuracy of a DCNN based method for the extraction of retinal blood vessels from fundus images.

2   Method

The major steps considered in this proposed study are data preparation, pre-processing, and the segmentation processing stages. The flow diagram and the proposed framework are shown in Fig. 1.

2.1   Data Preparation

The DRIVE database has 40 fundus images, splits into the training set of 20 images, and testing sets of 20 images as well. There are two ground truths (manually annotated images), but this study used the first ground truth. Each data point in the dataset is 565×584 pixels in size. Data Augmentation is considered necessary because the DRIVE dataset contains a training set of only 20 images, which contradicts the large data requirement for CNNs based models for proper training and learning. To avoid over-fitting and inadequate learning during training, we used augmentation to increase the data point size for adequate learning and training. The augmentation is implemented online using the following transformation operations: random crop, random rotation, random shear, random vertical flip, random shift, random horizontal flip, random zoom, and random translation.

2.2 Pre-processing

To achieve the goal of this study, this section focuses on the enhancement of the input datasets for better image quality before the DCNN segmentation process. Some contrast enhancement techniques applied to improve the quality of retinal fundus images are CLAHE for contrast enhancement, and Colour space for colour constancy. Each of these enhancement techniques is detailed below:

Data Cropping: Cropping is applied to enable the divisibility of the size of the input image by 16 (2^4)and also to alleviate the noise in the background.

Data Normalization: This is to make sure that all the image pixels have the same data distribution and are in the 0 to 1 range, also to avoid instability during network training [20], [21]. The mean value and standard deviation calculations for the normalization are represented in Eq. (1) and Eq. (2) respectively.

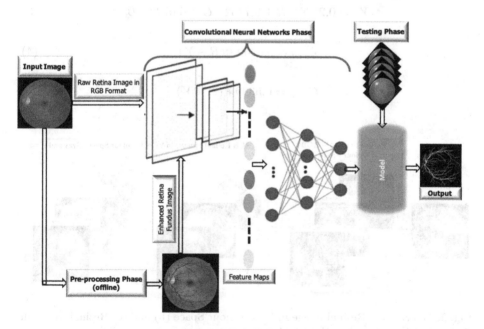

Fig. 1. The overview framework for the proposed method.

$$\bar{A} = a_i - \bar{A} \tag{1}$$

$$\bar{a}_i = \frac{a_i - \bar{A}}{\sigma_a} \tag{2}$$

Where $\bar{a}_i$ and a represents pixels old and new values, $\bar{A}$ represents mean pixel value and σ_a signifies standard deviation for pixel value.

Colour Space. Colour spaces play a vital role in the recognition both at human and machine levels. In computer vision, they are effective in addressing the problem of varying illumination in an image. RGB is the most popular colour space, but is limited in maintaining colour constancy under the direct application of some image contrast enhancement techniques. YCrCB colour model is an RGB encoder, calculated from RGB through a linear transformation. It has three components, Y-luminance, which contains the grayscale information, while the remaining two chrominance(Cr and Cb) contain the colour information. The YCrCb colour components are separable in terms of intensity, which makes it a perfect colour space that can maintain colour constancy when faced with image processing applications. The transformation of RGB to YCrCb colour space is in Eqs. (3) to Eq. (5), and well detailed in Fig. 1. The outcome of the transformation is in Fig. 2.

$$Y_L = 0.2126 * R + 0.7152 * G + 0.0722 * B \tag{3}$$

$$C_{red} = 0.6350 * (R - Y) \tag{4}$$

$$C_{blue} = 0.5389 * (B - Y) \tag{5}$$

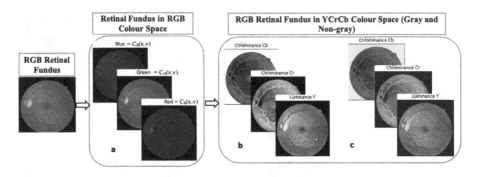

Fig. 2. Showing (a) Retinal image in RGB Colour Space (b) and (c) Retinal Image in (Gray) and (Non-Gray) in YCrCb Colour space.

CLAHE: This study applied a contrast enhancement algorithm called Contrast Limited Adaptive Histogram Equalization (CLAHE). It is often used for contrast enhancement in the medical domain [22] because of its ability to handle the limitation of contrast limiting in AHE [23]. The introduction of the clip limit in the CLAHE algorithm prevents the over-amplification of noise [24]. The quality of enhanced image using CLAHE depends majorly on three variables; the tile size, clip limit, and the distribution function. In this study tile size of 8×8 is used with the clip limit of 2.0. The outputs of all the enhancement techniques are shown in Fig. 3.

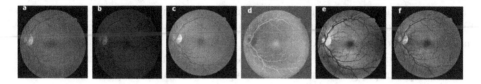

Fig. 3. Pre-processing results: (a) Original input image. (b) Normalized Mean Value Image. (c) Normalized Standard Image. (d) The normalized RBG Image in YCrCb Colour Space. (e) CLAHED image. (d) RGB of the Enhanced image.

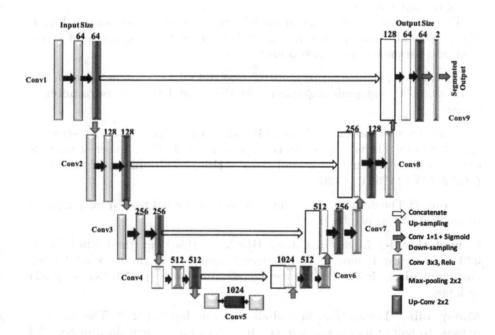

Fig. 4. The description of the U-Net architecture for the proposed method.

2.3 Architecture

The U-Net architecture, also known as encoder-decoder, is an efficient DCNN based method for segmentation tasks. It contains two paths called encoding and decoding paths. The contracting path contains five convolution blocks labeled conv1, conv2, conv3, conv4, and conv5 in Fig. 4. Each of the convolution blocks has 2 consecutive 3×3 convolution layers, the activation layer, max-pooling layers, and a dropout layer in between the 2 successive convolution layers. The convolution layer receives the input image and convolves the image, while the ReLU activation layer introduces a non-linear property to the network, the 2×2 Max-pooling layer downsamples the size of the input image while, the dropout drops the redundant neurons to ameliorate over-fitting. At each convolution block in the contracting path, the feature maps double while the size of the input image reduces by half.

The decoding path also contains five convolution blocks. Each block has a 2×2 Up-sampling layers, 3×3 transpose convolution layers, and a layer for the concatenation of output tensors from each of the convolution blocks on the contracting paths. The Up-sampling layer up-samples to gradually increase the reduced high resolution of the input image back to its original size. The final predicted output from the segmentation process is obtained from a 1×1 convolution with two classes of pixels belonging to blood vessels and pixels belonging to non-blood vessels. The size of the output image is the same as the size of the input image.

The summary of the proposed model's architecture is presented in Fig. 4. The number of convolution kernel in each layer, feature maps, size of kernel used, and parameters presented below:

Input Layer: The input is of shape **(None, 64,64,1)**, a **kernel-size** of **2*2** with **stride 1**, and **padding=same**. At the input layer, the **parameter** is 0(zero).

Convolutional-1 Layer in Conv1 Block: Kernel-size is **3*3** with **stride 1** and **32 filters**. The number of **features learned** is **1**. The number of **parameters** is calculated using **(((kernel-size)*stride)+1)*number of filters) = (((3*3)*1)+1)*32) = 320**.

Dropout-1: The dropout drops the neurons that are not needed for activation to the next at this point.

Convolutional-2 Layer in Conv1 Block: At this layer, the **kernel-size** is **3*3** with **stride 1**, number of **features learned** is **32** (from conv-1 Layer), number of **filters** is **32**. The number of **parameters = (((3*3)*32)+1)*32) = 9248**.

Max-pooling Layer: The **kernel-size** at this layer is **2*2**. The image is reduced to half of its original size i.e., from **64** to **32** as explained in Sect. 2.3 and shown in Fig. 6.

Convolutional-3 Layer in Conv2 Block: At this layer, the **kernel-size** is **3*3**, number of **features learned** is **32** (from conv-2 Layer), the number of **filter** is **64**. The number of **parameters** is $(((3*3)*32)+1)*64 = \mathbf{18496}$.

The same process of increase in the depth and decrease in the size of the input image continues through all the layers in all the convolutional blocks in the encoding path. While, the down-sampled image size is up-sampled in the decoding path.

3   Experiments

3.1   Experimental Setup and Training

The experiments are carried out on the Ubuntu Operating System using Keras platform as the front end and Tensor-Flow as the back end platform. The entire software training environment is on the desktop with the hardware configuration of Intel®Core ^{TM} i7- 7700 CPU @ 3.60GHz processor and 16GB of RAM. The network training is validated on the raw DRIVE dataset and enhanced DRIVE dataset. The two training are carried out under the same experimental setup, parameters, and hyper-parameters. The input data are cropped to 64×64 by size to ensure that the retinal image is divisible by (2^4) so that the max-pooling operation can take a random crop of 64 * 64 from each of the retinal images. We used a batch size of 16 with a training cycle of 50 epochs for the network training. ReLU activation function in Eq. (6) is used because of its low computational cost and its ability to prevent varnishing gradient challenge. Adam optimizer with the learning rate of le-3 is used to optimize Binary Cross-Entropy loss function presented in Eq. (7) for binary classification. A dropout of 0.05 is applied between the 2×2 consecutive convolution layers to avoid over-fitting. The first training was implemented using the raw DRIVE dataset while, the second used the enhanced DRIVE dataset.

$$f(y) = max(b, 0) \tag{6}$$

where b and y represent input and output respectively.

$$Loss_{BCE(gt,pre_i)} = \sum gt_i log pre_i + (1 - gt_i) log(1 - pre_i) \tag{7}$$

where gt and pre represent the groundtruth and the predicted output respectively.

3.2   Evaluation Metrics

The evaluation metrics presented in Eq. (9) to Eq. (11) are used to evaluate the performance of the two models:

$$\textbf{Accuracy} \;\; (Acc) = \frac{TP + TN}{TP + FP + FN + TN} \tag{8}$$

$$\text{Sensitivity} \quad (Sen) = \frac{TP}{TP + FN} \tag{9}$$

$$\text{Specificity} \quad (Spe) = \frac{TN}{TN + FP} \tag{10}$$

$$\textbf{Area under the ROC curve} \; (AUC) = \int_0^1 SendSpe \tag{11}$$

4   Results and Discussion

Contrast enhancement techniques are used to enhance the contrast of the images in the DRIVE dataset. Some tiny vessels and fine details are more visible and pronounced in the enhanced image, whereas the reverse is the case in the original image due to low contrast, uneven illumination, and noise.

Presented in Fig. 5 is the histogram shape plot of pixel values against their corresponding intensities for both the raw retinal fundus images in DRIVE dataset and the contrast-enhanced version. The histogram for the raw image shows that the intensity distribution is confined to a region, which is an indication of low contrast. While the histogram shape for the contrast-enhanced image stretches out showing better distribution of intensity and good contrast-enhanced image.

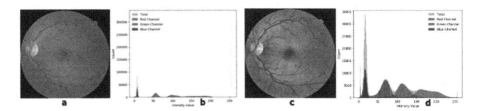

Fig. 5. Showing (a) The Original RGB Retinal Image, (b) The Intensity Histogram of the Original Image, (c) The Contrast-Enhanced Image and (d) The Intensity Histogram of the Contrast-Enhanced Image.

This study performed two experiments, one with the raw DRIVE dataset and the other trained with an enhanced DRIVE dataset using the same configuration so that we can observe, monitor, measure, and evaluate the impact of the two datasets on the performance of the proposed DCNN. We carried out the training one after the other and measured the accuracy, sensitivity, specificity, loss, and AUC of the two training. The performance of the proposed DCNN on the two datasets are presented in Table 1. The evaluation of the results in Table 1 reveals that the performance of the proposed model trained on the enhanced DRIVE dataset achieved all-round slight improvement responsiveness over the one trained with the raw DRIVE dataset.

Table 1. Performance of the Model Trained with Original DRIVE dataset and the Model Trained with the Enhanced DRIVE dataset.

Performance Metrics	Accuracy (Acc)	Loss	Sensitivity (Sn)	Specificity (Sp)	AUC
Without Contrast Enhancement	0.9445	0.1299	0.6878	0.9724	0.9724
With Contrast Enhancement	0.9460	0.1206	0.7138	0.9826	0.9730

These slight improvements are evident in the increase in accuracy of 94.45% to 94.60% (0.15%), the sensitivity of 68.78% to 71.38% (2.60%), the specificity of 98.17% to 98.26% (0.09%), AUC of 97.24% to 97.30% (0.06%) and reduction in a loss that is 12.99% to 12.06% (0.93%). More interesting is the significant performance hit of 2.60% sensitivity in the model trained on enhanced DRIVE dataset. This is a remarkable improvement indeed because sensitivity means the true positive rate, and medically it recognizes those having abnormalities as truly having defects (correct positive rates). This shows that properly enhanced images have a substantial impact on the performance accuracy of DCNN based method for the segmentation of retinal blood vessels from the retinal fundus images.

Table 2. Comparison of the Proposed Methods with other Existing Segmentation Methods on DRIVE Datasets.

Methods	Accuracy	Sensitivity	Specificity	AUC
Marin et al. [18]	0.9452	0.7067	NA	NA
Jebaseeli et al. [14]	**0.9883**	0.7023	**0.9931**	NA
Sule et al. [16]	0.9447	0.7092	0.9820	0.9721
Fraz et al. [17]	0.9430	0.7152	0.9768	NA
Soomro et al. [11]	0.9480	**0.0.7390**	0.9560	0.8440
Proposed Method Without Contrast Enhancement	0.9445	0.6878	0.9724	0.9724
Proposed Method With Contrast Enhancement	0.9460	0.7138	0.9826	**0.9730**

The comparison of the results for the two methods with the ground-truths is in Fig. 6, while the comparison of the segmented outputs for the two methods is in Fig. 7. The close comparison of the segmented outputs from the two models reveals that the model trained with the enhanced DRIVE dataset can detect few tiny vessels that are dropped out in the model trained with the raw

| 01 to 04_Test.tiff (DRIVE) | 01 to 04 Manual1.gif (DRIVE) | Proposed Output (Without Enhancement) | Proposed Output (With Enhancement) |

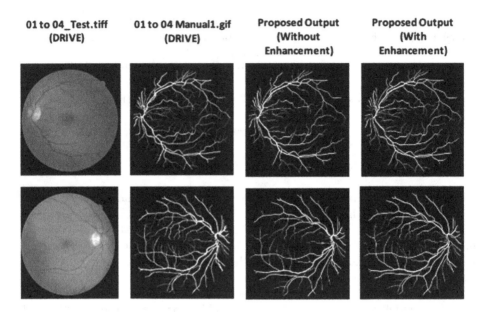

Fig. 6. Shows Original input images in 1st Column, Ground truth in the 2nd column, Segmented output from Without Contrast-Enhanced Model in 3rd column, and Segmented output from With Contrast-Enhanced Model in the 4th column.

| 01_Manual1.gif (DRIVE, GT) | Proposed Output (Without Enhancement) | Proposed Output (With Enhancements) |

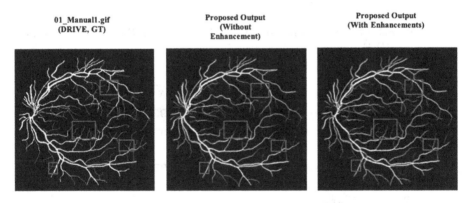

Fig. 7. Comparison of the Segmented Vessels from Proposed Methods in Column 2 (Without Contrast Enhancements) and Column 3 (With Contrast Enhancement) with the Ground truth in Column 1.

DRIVE dataset. Red boxes in Fig. 7 highlights the detection comparisons. The improvement in the DCNN performance when trained with enhanced DRIVE dataset proves the necessity of adequate contrast enhancements on medical imaging before the segmentation task in DCNN based method.

The proposed method is compared with some existing methods as shown in Table 2. Our proposed method demonstrated a promising performance of 97.30% AUC as the highest score.

5   Conclusion

This study used contrast enhancement techniques to enhance RGB retinal fundus images to boost the processing output for better diagnosis and analysis. This becomes necessary because the visual complexities of retinal fundus images can hinder the optimal performance of DCNN segmentation accuracy. We proposed training a DCNN with the raw and the enhanced DRIVE dataset using the same configuration. The proposed method is to investigate the influence of image contrast enhancement on the segmentation performance of the DCNN based method. The incorporation of the enhanced DRIVE dataset in the segmentation task remarkably improved the performance of the DCNN with a performance hit of 2.60% in sensitivity. However, the slight improvements in accuracy, specificity, and AUC seem small but, significant. However, better enhancement techniques may yield a better considerable performance outcome. More advanced and robust enhancement techniques will be considered in future research to make the DCNN predictive accuracy more dependable for medical diagnosis and analysis.

References

1. Salazar-Gonzalez, A., Kaba, D., Li, Y., Liu, X.: Segmentation of the blood vessels and optic disk in retinal images. IEEE J. Biomed. Health Inform. **18**(6), 1874–1886 (2014)
2. Franklin, S.W., Rajan, S.E.: Computerized screening of diabetic retinopathy employing blood vessel segmentation in retinal images. Biocybern. Biomed. Eng. **34**(2), 117–124 (2014)
3. Krizhevsky, A., Sutskever, I., Hinton, G.E.: ImageNet classification with deep convolutional neural networks. In: Advances in Neural Information Processing Systems, pp. 1097–1105 (2012)
4. Simonyan, K., Zisserman, A.: Very deep convolutional networks for large-scale image recognition. arXiv preprint arXiv:1409.1556 (2014)
5. Zhang, K., Zuo, W., Chen, Y., Meng, D., Zhang, L.: Beyond a Gaussian denoiser: residual learning of deep CNN for image denoising. IEEE Trans. Image Process. **26**(7), 3142–3155 (2017)
6. Ronneberger, O., Fischer, P., Brox, T.: U-Net: convolutional networks for biomedical image segmentation. In: Navab, N., Hornegger, J., Wells, W.M., Frangi, A.F. (eds.) MICCAI 2015. LNCS, vol. 9351, pp. 234–241. Springer, Cham (2015). https://doi.org/10.1007/978-3-319-24574-4_28
7. Long, J., Shelhamer, E., Darrell, T.: Fully convolutional networks for semantic segmentation. In: Proceedings of the IEEE Conference on Computer Vision and Pattern Recognition, pp. 3431–3440 (2015)
8. Budai, A., Hornegger, J., Michelson, G.: Multiscale approach for blood vessel segmentation on retinal fundus images. Invest. Ophthalmol. Vis. Sci. **50**(13), 325 (2009)
9. Soomro, T.A., et al.: Impact of image enhancement technique on CNN model for retinal blood vessels segmentation. IEEE Access **7**, 158183–158197 (2019)
10. Sahu, S., Singh, A.K., Ghrera, S.P., Elhoseny, M.: An approach for de-noising and contrast enhancement of retinal fundus image using CLAHE. Opt. Laser Technol. **110**, 87–98. 9 (2019)

11. Liskowski, P., Krawiec, K.: Segmenting retinal blood vessels with deep neural networks. IEEE Trans. Med. Imaging **35**(11), 2369–2380 (2016)
12. Soomro, T.A., et al.: Boosting sensitivity of a retinal vessel segmentation algorithm with convolutional neural network. In: 2017 International Conference on Digital Image Computing: Techniques and Applications (DICTA), pp. 1–8. IEEE (2017)
13. Guo, S., Wang, K., Kang, H., Zhang, Y., Gao, Y., Li, T.: BTS-DSN: deeply supervised neural network with short connections for retinal vessel segmentation. Int. J. Med. Inform. **126**, 105–113 (2019)
14. Dharmawan, D.A., Li, D., Ng, B.P., Rahardja, S.: A new hybrid algorithm for retinal vessels segmentation on fundus images. IEEE Access **7**, 41885–41896 (2019)
15. Jebaseeli, T.J., Durai, C.A.D., Peter, J.D.: Retinal blood vessel segmentation from depigmented diabetic retinopathy images. IETE J. Res., 1–18 (2018)
16. Hoover, A.D., Kouznetsova, V., Goldbaum, M.: Locating blood vessels in retinal images by piecewise threshold probing of a matched filter response. IEEE Trans. Med. Imaging **19**(3), 203–210 (2000)
17. Sule, O., Viriri, S.: Enhanced convolutional neural networks for segmentation of retinal blood vessel image. In: 2020 Conference on Information Communications Technology and Society (ICTAS), pp. 1–6. IEEE (2020)
18. Fraz, M.M., et al.: An ensemble classification-based approach applied to retinal blood vessel segmentation. IEEE Trans. Biomed. Eng. **59**(9), 2538–2548 (2012)
19. Marín, D., Aquino, A., Gegúndez-Arias, M.E., Bravo, J.M.: A new supervised method for blood vessel segmentation in retinal images by using gray-level and moment invariants-based features. IEEE Trans. Med. Imaging **30**(1), 146–158 (2010)
20. Han, J., Pei, J., Kamber, M.: Data Mining: Concepts and Techniques. Elsevier, Amsterdam (2011)
21. Liu, Z.: Retinal vessel segmentation based on fully convolutional networks. arXiv preprint arXiv:1911.09915 (2019)
22. Karel, Z.: Contrast Limited Adaptive Histogram Equalization, pp. 474–485. Press Professional Inc., Randallstown (1994)
23. Pizer, S.M., et al.: Adaptive histogram equalization and its variations. Comput. Vis. Graph. Image Process. **39**(3), 355–368 (1987)
24. Reza, A.M.: Realization of the Contrast Limited Adaptive Histogram Equalization (CLAHE) for real-time image enhancement. J. VLSI Signal Process. Syst. Signal Image Video Technol. **38**(1), 35–44 (2004). https://doi.org/10.1023/B:VLSI.0000028532.53893.82

Real-Time Social Distancing Alert System Using Pose Estimation on Smart Edge Devices

Hai-Thien To[1], Khac-Hoai Nam Bui[2], Van-Duc Le[1], Tien-Cuong Bui[1], Wen-Syan Li[1], and Sang Kyun Cha[1(✉)]

[1] Seoul National University, Seoul, Korea
{haithienld,levanduc,cuongbt91,wensyanli,chask}@snu.ac.kr
[2] Korea Institute of Science and Technology Information, Daejeon, Korea

Abstract. This paper focuses on developing a social distance alert system using pose estimation for smart edge devices. Recently, with the rapid development of the Deep Learning model for computer vision, a vision-based automatic real-time warning system for social distance becomes an emergent issue. In this study, different from previous works, we propose a new framework for distance measurement using pose estimation. Moreover, the system is developed on smart edge devices, which is able to deal with moving cameras instead of fixed cameras of surveillance systems. Specifically, our method includes three main processes, which are video pre-processing, pose estimation, and object distance estimation. The experiment on coral board, an AI accelerator device, provides promising results of our proposed method in which the accuracies are able to achieve more than 85% from different datasets.

Keywords: Real-time · Social distance alert system · Deep Learning · Computer Vision · Pose estimation · Smart Edge Devices

1 Introduction

Recently, with the rapid development of the Internet of Things (IoT), embedding Artificial Intelligence (AI), in particular, Machine Learning (ML) techniques into connected devices becomes an emergent research issue. Specifically, the applications of on-device AI are capturing both consumer and industrial attention, which making connected devices (e.g., automobiles, cameras, smartphones, industrial sensors) are smarter and faster [13]. Recently, Deep Learning (DL)-based models have been successfully developed for various applications [3,23]. However, implementing DL models at the edge devices, where data is generated, is still a real challenge. Specifically, executing algorithms of AI on devices with limited resources and allowing local data processing is an important step [9].

In this study, we take the implementation of AI in Smart Edge Devices (SED) for a real-solving problem into account. Particularly, we develop a pose

© Springer Nature Singapore Pte Ltd. 2021
T.-P. Hong et al. (Eds.): ACIIDS 2021, CCIS 1371, pp. 291–300, 2021.
https://doi.org/10.1007/978-981-16-1685-3_24

estimation-based social distance alert system using on-moving cameras. Specifically, Fig. 1 depicts the pipeline of the general system for the proposed system.

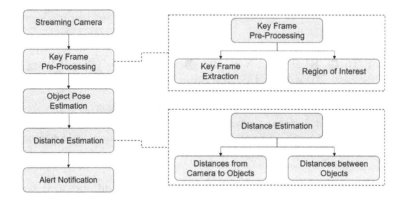

Fig. 1. The general framework of the proposed system.

Accordingly, there are three main processes, which are key frame pre-processing, pose estimation, and distance calculation. Moreover, executing the proposed system for smart devices is an emergent trend and still a challenge in this research field. Specifically, SEDs use lightweight models that cause low performance in terms of accuracy and inference time (FPS). To the best of our knowledge, our paper is the first study which focuses on using pose estimation for the social distancing alert systems. Consequently, the main contribution of this paper is threefold as follows:

- Developing an end-to-end social distancing framework for real-time processing on SEDs.
- Proposing a new method for social distancing measurement in which we synchronize the position of objects in cameras to 2D map.
- The experiment on coral board device indicates the promising results of the proposed method with different datasets.

The rest of this paper is organized as follows: We take a brief review of the trend for embedding AI into edge devices in Sect. 2. Moreover, recent research on pose estimation using DL models are also reviewed in this section. The methodology for implementing the social distance alert system using pose estimation on SEDs is presented in Sect. 3. Section 4 is a demonstration of our proposed system. Discussion and future work of this study is concluded in Sect. 5.

2   Literature Review

2.1   Deep Learning for Human Pose Estimation

Pose estimation is defined as the problem of localization of object key-points in images/videos. The classical approach for pose estimation based on mixtures-of-parts models which match the global template and part template in an image to detect an object [22]. With the rapid development of DL, the current approaches are able to deal with the cost of limited expressiveness and the global context of the classical approaches [7]. Specifically, recent pose estimation systems adopt Convolutional Neural Network (CNN) as their main building block for replacing hand-crafted features and graphical models [14, 17–20]. Consequently, inspired by the idea of transfer learning where pre-trained CNN-based methods are used for a limited amount of training data, Kendall et al. [8] have proposed PoseNet, a modified truncated GoogLetNet architecture [15] for Real-time human pose estimation. Specifically, the difference with GoogLeNet, the softmax layer (for classification) is replaced with deep fully-connected (FC) layers to regress the pose of an image from cameras. In this regard, for executing human pose estimation on SEDs in the alert system for social distance measurement, we adopt this model to enable real-time monitoring.

2.2 Social Distance Alert System Using Computer Vision

Social distancing is an effective measure for alert systems [6]. Obviously, an automatic warning system is necessary to develop methods of observation without the physical presence of humans. With the rapid development of DL for Computer Vision (CV) such as object detection and tracking [4], recent studies explore how AI enables automatic alert systems for social distancing measurements from video surveillance using CV-based methods [16]. Authors in [10] propose a DL-based framework for automating monitoring social distance using surveillance video by using YoLo and Deep Sort for object detection and tracking, respectively. Yang et al. [21] use a pre-trained CNN to detect real-time social distancing and develop a warning system. In this paper, we propose a new approach of social distance measurement which is different compared with previous systems such as i) We focus on real-time social distance measurement for SEDs, which is the trend research in terms of embedding AI to the edge devices [2]; ii) We use pose estimation technique for object detection instead of using bounding box technique, which is able to improve the accuracy of measurement social distancing between objects. More details of the proposed system are presented in the following section.

3 Methodologies

As we mention above, we focus on developing the real-time alert system for SEDs using pose estimation. Consequently, three main processes for executing the proposed system are sequentially described as follows:

3.1 Key Frame Pre-processing

We adopt a keyframe extraction method for reducing blurry frames in order to deal with real-time processing and improving the performance of the detection. Specifically, instead of using models (e.g., DL) for extracting keyframes that are not able to execute the system in real-time processing, we implement a simple process by using the variance of the Laplacian method for detecting non-blurry frames [11].

Furthermore, SEDs are fundamentally moving cameras in which objects (i.e., human) in the lower and upper part of the frame/video are near and far from the cameras, respectively. Therefore, we only try to detect objects which are near the camera by using region of interest (ROI) in each video, as shown in Fig. 2 [5].

Fig. 2. Illustration of an input video using ROI (yellow line). (Color figure online)

3.2 Pose Estimation for Edge Devices Using PoseNet Model

Human pose estimation is an important issue in the CV. It is a critical step to understand the people in images and videos. Recently, DL-based human pose estimation has been developed for emergent applications in CV, such as action recognition, animation, and video understanding. Figure 3 depicts the result of the human pose estimation process with 17 key points.   Consequently, for the proposed system, we adopt pose estimation since this approach is able to calculate the distance between objects better than using other object detection methods (e.g., bounding boxes).

However, DL models are usually heavy, ranging from a few hundred MB to several GB, which are often suitable for running on computers with high hardware configurations. Therefore, this paper focuses on developing recent approaches for optimizing DL models to fit on edge devices. Specifically, we adopt PoseNet [12], a quantized model that has a size is around 1.7–2.3 MB for detecting the position and orientation of objects (i.e., human). Technically, there are several advantages of PoseNet, such as short interference times and

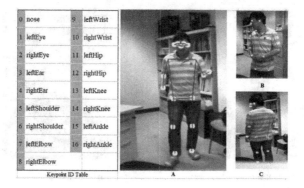

Fig. 3. The keypoints for the human pose estimation

low memory footprint. Moreover, by using transfer learning, the network is able to learn from limited-sized datasets. Consequently, the pipeline of this process is described in Fig. 4.

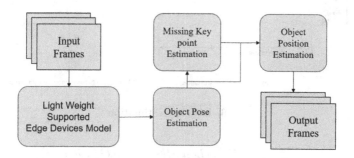

Fig. 4. Pose estimation process using light weight model.

Specifically, for calculating the distances between objects, we determine the position of objects based on all the key points. Therefore, the missing key point process is adopted by in case of there are several missing key points after object pose estimation as illustrated in Fig. 5.

3.3   Object Distance Estimation

This process calculates the distance between objects based on their position estimation. Specifically, since we focus on moving cameras of SEDs, we first calculate the distance from the camera to the object. Then, the distances between the objects are measured. For more detail, Fig. 6 depicts the distance measurement process. In particular, given the focal length value of the camera (FocalLength), the distance between the moving camera (C) to an object (O) is calculated based on the focal length as follows:

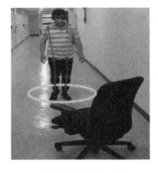

a) Non-missing keypoints   b) Missing keypoints

Fig. 5. Illustration for position estimation with missing key points.

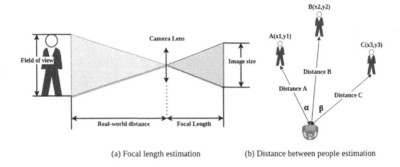

(a) Focal length estimation   (b) Distance between people estimation

Fig. 6. Distance estimation

$$d(C,O) = \frac{FocalLength * FrameOfView}{FieldOfView} \tag{1}$$

where $FrameOfView$ and $FieldOfView$ represent the height of the object in the real-world and the frame, respectively. Specifically, by applying a light-weight pose estimation model on the object near the camera, we can compute the person's height correctly in each frame. As shown in Fig 6, since we are able to estimate the position of the parts of the skeleton, the person's height is determined in the video frame. Consequently, the distances between objects are calculated based on their distance from the camera. For instance, as shown in the Fig. 6b, given the distances and position of objects to the camera, respectively, the distance among objects are determined by using the distance equation as follows:

$$d(A,B) = \sqrt{(x2 - x1)^2 + (y2 - y1)^2)} \tag{2}$$

4   Experiment

For the experiment, we test our proposed system on TownCentre [1], a well-known dataset for human detection and tracking. Moreover, since our method focuses on moving cameras of SEDs, a self-taken video using Coral Board is also taken into account to evaluate our proposed system's performance. Specifically, Coral Board[1] is a single-board computer that contains an edge TPU co-processor. Particularly, comparing with other edge devices such as Jetson Nano or Raspberry Pi, the Coral board is supported by a USB accelerator, which is able to enable real-time processing for DL models. Table 1 shows the characteristics of the device that we use for the experiment.

Table 1. Characteristics of the edge device.

Features	Google Coral Board
DL Accelerator	Google EdgeTPU (4 TOPs, Int8 only)
Memory	8MB SRAM + 1 GB RAM
DL Frameworks	Tensorflow Lite
Power	5 W

a) Pose Estimation   b) Distance Estimation   c) Rea-time Processing

Fig. 7. Real-time social distance alert system for the moving camera

Figure 7 demonstrates the interface of our system for the self-taken video. Specifically, with the value of the focal length is 660 cm, our system is able to enable real-time processing with the frame rate is around 55–85 FPS. In order to evaluate the accuracy, we measure the total number of human pose estimation (PE) with the ground truth (GT). Moreover, each video is split into k segments, and we consider the cumulative human counts from the start of the video to the end of each segment to reduce the labeling discrepancies problem. For more detail, Table 2 shows the results of our method on two considered datasets. As shown in the results, the accuracy achieves more than 85% of the

---

[1] Coral. Available online: https://www.coral.ai/ (accessed on 22 November 2020).

Table 2. Pose estimation results compared with the ground truth with test case videos.

No	Video	k values	GT	PE	Alert (True/k)	Accuracy (%)
Vdo1	TownCentre	42	205	210	37/42	0.881
Vdo2	Self-taken	42	106	112	36/42	0.854

estimation, which is a promising result since we use lightweight models for real-time processing on SEDs. Specifically, Fig. 8 depicts several cases that the system fails to locate the person accurately in the frame. Therefore, a better DL model

a) Vdo1: Wrong human pose estimation   b) Vdo2: Missing human pose estimation

Fig. 8. Failure cases of the pose estimation process

for improving the accuracy of human pose estimation is required in this research field.

5   Conclusion and Future Work

In the next few years, there will be billions of smart devices that will be connected. The research on implementing ML models for SEDs becomes emergent research nowadays. In this paper, we take the development of a social distance alert system using SEDs into account. Specifically, we present an end-to-end social distance framework for real-time processing on SEDs using a lightweight human pose estimation model. The experiment shows promising results of our method that it is worth to keep working on.

From our point of view, there were several issues that could improve the proposed system as follows: i) improving the accuracy by fine-tuning some of the hidden parameters (e.g., the maximum number of poses to detect, pose confidence score threshold, and the number of pixels); ii) extending the system by adopting multi-camera tracking using the proposed framework, which is able to enable the large-scale social distance alert system. There are several interesting issues that we take into account regarding the future work of this study.

Acknowledgment. This work was supported by the New Industry Promotion Program (1415166913, Development of Front/Side Camera Sensor for Autonomous Vehicle) funded by the Ministry of Trade, Industry Energy (MOTIE, Korea).

References

1. Benfold, B., Reid, I.D.: Stable multi-target tracking in real-time surveillance video. In: Proceedings of the 24th IEEE Conference on Computer Vision and Pattern Recognition (CVPR 2011), pp. 3457–3464. IEEE Computer Society (2011)
2. Bui, K.H.N., Jung, J.J.: Computational negotiation-based edge analytics for smart objects. Inf. Sci. **480**, 222–236 (2019)
3. Bui, K.N., Oh, H., Yi, H.: Traffic density classification using sound datasets: an empirical study on traffic flow at asymmetric roads. IEEE Access **8**, 125671–125679 (2020)
4. Bui, K.H.N., Yi, H., Cho, J.: A multi-class multi-movement vehicle counting framework for traffic analysis in complex areas using CCTV systems. Energies **13**, 2036 (2020)
5. Bui, K.N., Yi, H., Cho, J.: A vehicle counts by class framework using distinguished regions tracking at multiple intersections. In: Proceedings of the 33rd IEEE Conference on Computer Vision and Pattern Recognition Workshop (CVPRW 2020), pp. 2466–2474. IEEE Computer Society (2020)
6. Charles, C., Joseph, G., Anh, L., Joshua, P., Aaron, Y.: Strong social distancing measures in the united states reduced the COVID-19 growth rate: study evaluates the impact of social distancing measures on the growth rate of confirmed COVID-19 cases across the United States. Health Aff. 10–1377 (2020)
7. Chen, Y., Tian, Y., He, M.: Monocular human pose estimation: a survey of deep learning-based methods. Comput. Vis. Image Underst. **192**, 102897 (2020)
8. Kendall, A., Grimes, M., Cipolla, R.: PoseNet: a convolutional network for real-time 6-DOF camera relocalization. In: Proceedings of the 27th IEEE Conference on Computer Vision and Pattern Recognition (ICCV 2015), pp. 2938–2946. IEEE Computer Society (2015)
9. Merenda, M., Porcaro, C., Iero, D.: Edge machine learning for AI-enabled IoT devices: a review. Sensors **20**(9), 2533 (2020)
10. Punn, N.S., Sonbhadra, S.K., Agarwal, S.: Monitoring COVID-19 social distancing with person detection and tracking via fine-tuned YOLO v3 and deepsort techniques. CoRR abs/2005.01385 (2020). https://arxiv.org/abs/2005.01385
11. Raghav, B., Gaurav, R., Tanupriya, C.: Blur image detection using Laplacian operator and open-cv. In: Proceedings of the 5th Conference on System Modeling & Advancement in Research Trends (SMART 2016), pp. 63–67. IEEE (2016)
12. Shavit, Y., Ferens, R.: Introduction to camera pose estimation with deep learning. CoRR abs/1907.05272 (2019). http://arxiv.org/abs/1907.05272
13. Sodhro, A.H., Pirbhulal, S., de Albuquerque, V.H.C.: Artificial intelligence-driven mechanism for edge computing-based industrial applications. IEEE Trans. Ind. Inform. **15**(7), 4235–4243 (2019)
14. Sun, K., Xiao, B., Liu, D., Wang, J.: Deep high-resolution representation learning for human pose estimation. In: Proceedings of the 32nd IEEE Conference on Computer Vision and Pattern Recognition (CVPR 2019), pp. 5693–5703. IEEE Computer Society (2019)

15. Szegedy, C., et al.: Going deeper with convolutions. In: Proceedings of the 28th IEEE Conference on Computer Vision and Pattern Recognition (CVPR 2015), pp. 1–9. IEEE Computer Society (2015)

16. Nguyen, C.T., et al.: A comprehensive survey of enabling and emerging technologies for social distancing-part i: Fundamentals and enabling technologies. IEEE Access **8**, 153479–153507 (2020)

17. Tompson, J., Goroshin, R., Jain, A., LeCun, Y., Bregler, C.: Efficient object localization using convolutional networks. In: Proceedings of the 28th IEEE Conference on Computer Vision and Pattern Recognition (CVPR 2015), pp. 648–656. IEEE Computer Society (2015)

18. Toshev, A., Szegedy, C.: DeepPose: human pose estimation via deep neural networks. In: Proceedings of the 27th IEEE Conference on Computer Vision and Pattern Recognition (CVPR 2014), pp. 1653–1660. IEEE Computer Society (2014)

19. Wei, S., Ramakrishna, V., Kanade, T., Sheikh, Y.: Convolutional pose machines. In: Proceedings of the 29th IEEE Conference on Computer Vision and Pattern Recognition (CVPR 2016), pp. 4724–4732. IEEE Computer Society (2016)

20. Xiao, B., Wu, H., Wei, Y.: Simple baselines for human pose estimation and tracking. In: Ferrari, V., Hebert, M., Sminchisescu, C., Weiss, Y. (eds.) ECCV 2018. LNCS, vol. 11210, pp. 472–487. Springer, Cham (2018). https://doi.org/10.1007/978-3-030-01231-1_29

21. Yang, D., Yurtsever, E., Renganathan, V., Redmill, K.A., Ozguner, U.: A vision-based social distancing and critical density detection system for COVID-19. CoRR abs/2007.03578 (2020). https://arxiv.org/abs/2007.03578

22. Yang, Y., Ramanan, D.: Articulated human detection with flexible mixtures of parts. IEEE Trans. Pattern Anal. Mach. Intell. **35**(12), 2878–2890 (2013)

23. Yi, H., Bui, K.N.: An automated hyperparameter search-based deep learning model for highway traffic prediction. IEEE Trans. Intell. Transp. Syst. (2020, Early Access)

Automatic Video Editor for Reportages Assisted by Unsupervised Machine Learning

Leo Vitasovic, Zdravko Kunic, and Leo Mrsic$^{(\boxtimes)}$ (iD)

Algebra University College, Ilica 242, 10000 Zagreb, Croatia
{zdravko.kunic,leo.mrsic}@algebra.hr

Abstract. In this paper we explain the methodology for making a computer program that will automatically edit videos with the only input being files containing raw video interviews. Program produces two files as its outputs – an MP4 video file and an XML file compatible with the most popular video editing programs. MP4 video file will be the edited video program made and the XML file will contain the timeline information which the program puts together with clip durations, in and out points on the timeline, and information about the timeline itself. Such a program is useful for two general types of users. The first group would be video editors for whom it would save time - upon generating the XML they could continue improving the cut, without wasting time manually cutting and sorting clips. The second type of users that could benefit using this program are people who are not professional video editors and would use this program as a simple tool that enables them to quickly edit the interviews they shot and publish them on the web. In processing the inputted video files, among other technologies, program uses unsupervised machine learning. The program is written in Python and uses FFmpeg for converting audio and video files. The program has been tested on a limited sample of video interviews carried out in English .

Keywords: Automatic video editing · Topic modeling · Unsupervised machine learning · Video reportages · Documentary

1 Introduction

In the last couple of years, the development of Internet and web technologies has contributed to the usage of video content as the favored source of information. [1] A program like the one described in this paper could speed up the process of creating such content. This would be achieved by doing the technical part of setting up a video sequence, content and context analysis and making a rough cut instead of the editor – leaving them more time for the creative work. Figure 1 shows a basic flowchart of such program, describing the goals of each separate function. Every square represents a module with the description of specific functions the module should do.

After a user has chosen the video files containing footage of raw interviews, the program will first create their audio versions. These audio versions will be of the same content as original files, but without video. Then program analyzes the newly created

© Springer Nature Singapore Pte Ltd. 2021
T.-P. Hong et al. (Eds.): ACIIDS 2021, CCIS 1371, pp. 301–309, 2021.
https://doi.org/10.1007/978-981-16-1685-3_25

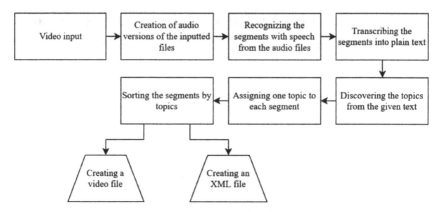

Fig. 1. Program's flowchart

audio files with the goal of detecting parts where any type of human speech is present. These segments need to be cut – for each of them a new, separate clips need to be created. For each of these clips the program will then get their transcriptions. The transcripts we got will be used as a *corpus* (structured set of documents) in determining the topics from the *documents* (smallest part of the list of input texts). These topics will be determined using the statistical method Topic Modeling by using Latent Dirichlet Allocation. After the topics have been determined, the program allocates one topic to each document. Sorting the documents by topics produces a sequence of clips which represent the rough cut of the reportage. Apart from making a video file out of this sequence, program will create an XML file, compatible with popular nonlinear editing systems such as Adobe Premiere Pro [2–4].

For it to work the program uses metadata (data about data, in this case number and text values gotten by analyzing the input files), which is stored in a single file which, a database of sorts. This approach makes it easier to observe the functionality of specific functions contained within the program – since there is language analysis heavily involved throughout the project, we can suspect the results to sometimes be only partially precise [5]. Therefore, it is important to interpret the results of each program's module individually. The metadata is saved tabularly, in a CSV (Comma Separated Values) file in which every line holds information about one clip. This file contains all information that various program modules need and the file itself is human-readable and easy to edit [34].

2   Background: Existent Voice Transcription Technologies

Today there are various applications and technologies for voice transcription. Companies like Google, Microsoft, IBM (etc.) all have their own systems that use machine learning and support multiple languages. For the purposes of this paper we have chosen to use Google's service because of the fact that it is free (for basic usage) and it supports multiple languages.

2.1 Trends in Natural Language Processing

Natural language processing is a field that deals with the relation of computers and human (natural) languages. It is a broad field that encompasses various methods which don't necessarily use machine learning. Challenges in natural language processing include distinguishing human speech from other noise, understanding and generating sentences [35–38]. Today there are many systems which deal with these challenges to some degree [6]. Topic modeling is a subset of natural language processing. It is a branch that deals with discovering the topics from a given set of documents. The core assumption is that the words which represent a topic will appear more frequently in sentences that talk about these topics. As an example, words "justice" and "verdict" will appear more often in a text that deals with judicial systems, while words "dog" and "cat" are more probable appear in texts about animals [7]. LDA is a generative statistical model meaning that the model produces new instances of the inputted data. LDA makes it possible to analyze big sets of data, by generating topics which describe the given sets. LDA assumes that every document (in our case every sentence) is a probability distribution between topics.

Every topic is defined as a mix of keywords – words that appear most often in a document about that topic. Every topic is a probability distribution between the keywords. Every keyword has a defined weight, which represents the relevance of this keyword in the topic. The goal of a trained LDA model is to discover the topics in the input set of documents [8]. Extraction of topics with LDA consists of three basic steps: (i) preparing the data, (ii) creating a model and (iii) connecting the discovered topics with documents (sentences). Preparing the data consists of tokenization, sentence cleaning, creating models for bigrams and trigrams and lemmatization. Tokenization includes transforming every single word in every document into an independent entity within the original sentence. By tokenizing the sentence: "The dog is black." we will get a list of tokens: ["the", "dog", "is", "black"]. Cleaning the sentences, we exclude the unwanted sequences of characters such as e-mail addresses and newline characters. Now it is time to create models for bigrams and trigrams (*Bigrams* and *trigrams* are sets of elements (words) in a sequence; *Bigram* is an *n-gram* for which $n = 2$, and for *trigram* $n = 3$). These sets are made of n number of words which frequently appear next to one another. N-grams are important because they let us get a better understanding of specific keywords. In example, words "French" has a different meaning than an n-gram "French revolution". Given the model works properly, it will detect "French revolution" as a bigram and not as two separate tokens ("French" and "revolution"). Trigrams follow the same rules with having n equal to three.

By lemmatizing we get canonical word forms (from linguistics, describes the fundamental form of the given word). This is important so that we can properly count how many times has a certain word appeared. In creating an LDA model we define how many topics will the model support and the number of iterations we will go over our corpus. At the start of the process, each word is allocated to represent a random topic. One iteration includes traversing through every keyword in each document and comparing the locations of that in the document. The algorithm then looks at the frequency of the topic in a document and the frequency of the keyword inside the topic. If they are not matching this means that the keyword is in a wrong topic and it gets assigned to another one. After multiple iterations, the topics will be more precisely formed and observing the keywords in each topic we can assume what is the subject matter of each topic [9, 10].

3   Methodology

ControllerFile.py calls upon all function which will then do specific steps from the course of program execution, as shown of Fig. 2.

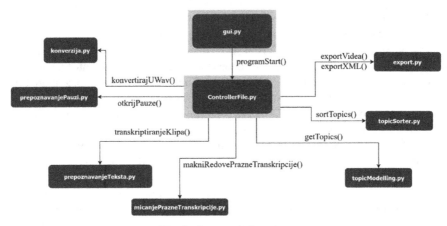

Fig. 2. Program's flowchart

3.1   Processing the Input Files and Preparing the File with Results

As aforementioned we will create a project.csv file in which we will be writing data about each clip. The columns of this file will be: Original Filename, Clip Name, Location, In point, Out point, Duration and \n (Universal newline symbol). We load a list of names of the files that the user has selected (final version of the program includes a graphical user interface through which users can select the files). The program loop is started, and it goes through every element of the said list. The program copies all the raw video files in its root directory. Then they get converted to WAV. To do so, the program will use FFmpeg (complete, cross-platform solution to record, convert and stream audio and video (*FFmpeg* web).

3.2   Processing the Input Files and Preparing the File with Results

We analyze each newly created WAV file – we need to determine the segments in which there is speech and out of those segments create new audio and video clips. For this we use the silence_removal() function from the Audio Segmentation library which uses pseudo supervised machine learning (type of machine learning in which a small part of the input data is annotated, and the larger part is unannotated). This function first splits the files into short parts ("windows") and analyzes their amplitude. It traverses the files in lengths defined as steps, then takes the 10% of the loudest and 10% of the quietest signal and uses that to train an SVM. After incorporating the whole signal on an SVM probability graph ("Support vector machine (SVM) is a computer algorithm that learns by example to assign labels to objects" (Noble, 2006)), every part of the signal that surpasses the set threshold value will be deemed a talking segment. As can be seen from Fig. 3, it took some calibrating the functions' parameters in order to achieve optimum results [11–13].

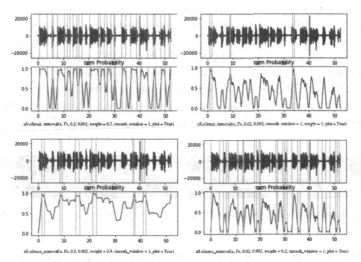

Fig. 3. Results of running the segment analysis with various parameters (upper graphs on each picture represent the sound signal and the lower ones are SVM probability graphs)

After we have a list of segments, we use Ffmpeg to create new clips. FFmpeg can split existing video files into smaller ones when given certain arguments, including the original clip, point from which we want to cut and the duration of the new clip. We create audio and video clips of the for each detected segment. Audio clips will be used for getting transcriptions and video clips will be used in the process of exporting the final video [14]. Apart from this, we write the results (Original filename, Clip name, Location on the disk, In point, Out point and Clip duration) into project.csv.

3.3 Transcribing the Audio Clips and Discovering Topics

In the next step we need to transcribe the clips and write the results into project.csv. We achieve this by using Google Cloud Speech API. The audio signal of each clip gets loaded into an empty object and then gets sent to Google's service. We do this by calling the recognize_google_cloud() function with the needed parameters, including the object containing our audio signal. The function returns a text string – the transcript of our audio clip. After this has been done for all audio clips, we write the results inside the project.csv, next to the corresponding clips [15, 16].

Before we run the function for discovering the topics, we delete all rows from project.csv which have an empty transcription field – this means that the Google's service hasn't been able to transcribe them and we will not be able to work with them anymore [17]. This usually means that the clip was wrongly identified as a talking segment in the first place.

3.4 Fetching and Preparing the Sentences and Creating an LDA Model

For LDA to produce precise results it is first necessary to prepare the input text. We do this as described in Chapter 2.2. This means we go through the process of tokenization, sentence cleaning, creating n-gram models and lemmatize the sentences. Now everything is ready for creating the LDA model. LDA model is, from the program perspective, a class which as an attribute has topics of the inputted sentences. Every topic is comprised of ten keywords which are assumed to be semantically connected and which represent the given topic. We use the class LDAModel [18, 19, 21]. After running the model on our text, we get topics and the weights of each specific keyword. Example of this result is shown on Picture 3.9. The first keyword is the one with the biggest weight, so it can be seen as the representative for the given topic. Therefore, we extract the first keyword for each topic and write that in project.csv under the column "Topic" (Fig. 4).

```
[(0,
  '0.467*"color" + 0.133*"red" + 0.133*"orange" + 0.133*"purple" + '
  '0.133*"green"')]
```

Fig. 4. A topic with the keywords it is made of with their weights, respectively

Now we sort the project.csv by the column "Topic". The goal here is to get a report composed of a few groups – where every part is comprised only of one topic. After sorting the clips, they get shuffled inside their respective groups. We do this to prevent the clips of the same person to be one after another in the final edit.

3.5 Exporting Video and XML

ControllerFile.py calls functions that export the final video and make a CSV file. For creating the video, we use FFmpeg which will in this case only concatenate the video clips – since they are all already in the correct video format, FFmpeg can just concatenate the text strings of each clip and produce the final video that way. In order to check if we have succeeded, we used VLC Player to open the resulting video file.

XML files enable the interchangeable transfer of information about the cut between various nonlinear video editing programs. XMLs use the original files as source for all the media – meaning the audio and video quality will not be altered. This enables the clips on the timeline to be trimmed and expanded as the editors want. XML file is in its structure a long textual sequence with specific hierarchical structure and standards designed specifically for nonlinear video editing systems. This means that in order to create an XML we first prepare individual strings which represent clips which we then concatenate together into a single file [22–24].

3.6 User Interface

In order to make the program easier to test and use we have implemented a graphical user interface which enables the users to select files, check if they want to export XML, video or both and run the program once they are ready. For this we have the tkinter pack, which is an existing system for creating GUIs for Python-written applications. The code for the GUI is contained in a gui.py file and the "Export" button, when clicked, starts the ControllerFile.py, thus starting the process.

4 Results

Testing this program comprised of inputting various video material, observing and analyzing the results it produced. Results were evaluated based on the criteriums of: (i) technical correctness of the CSV and XML file, (ii) semantic cohesiveness of the video and (iii) fluency of the cuts between clips. Technical correctness of the video refers to being able to run the video file within popular video players such as Windows Media Player, VLC Player, QuickTime etc. Every test video has passed this check. XML correctness is expressed in the output XML file being compatible with Adobe Premiere Pro – which is a program we chose to use as a test program. This has worked on every program's trial. Semantic cohesiveness is reflected in how much the video makes sense from the content point of view and is mostly subjective.

The aspect of this that is quantifiable the most is the quality of clips' grouping – if the statements said in each clip tell a cohesive story and make sense in a sequence in which they were stacked then this criterium can be seen as met. By fluency of the cuts we refer to the individual clips not being cut in the middle of the subject's sentence. In order to meet this criterium an amount of time has been put into calibrating the parameters for recognizing the voice segments, as can be seen on Fig. 3.

4.1 Comparing the Results in Relation to the Complexity of the Input Material

Testing the program has been carried out over three series of interviews: (i) simple test about favorite colors, (ii) attendees of the Algebra Summer and Winter school and (iii) speeches of American politicians. Simple test about favorite colors contained two people saying what their favorite colors were. This footage was shot specifically for the purposes of testing this program. The accumulated amount of time it took the program to process the videos was around a minute and the results were completely correct – the program was able to use only the parts of the clips that contained talking and it sorted them in a logical manner. Footage featuring the attendees of the summer and winter schools had been much more complex [25, 26]. This is because here every subject had been giving answers to different questions. Interviews from the summer school turned out to be good – program had managed to discover the topics rather precisely and structured the sentences in a way that makes sense.

There were errors concerning the students answering the same question multiple times – for example when they wanted to do a better take than the one, they already did. Other parts contained the journalist's voice recorded too loud – so the program thought those segments were sentences it should use as well. The winter school footage contained of a previously edited video (by a human video editor) and the program's job was to split the video into clips and then reedit them back together. The results once again were satisfactory, however on some parts, program struggled to get "the flow" of the interview so it seemingly cut random from one topic into the next one. American politician's speeches were technically the best shot content – especially when it comes to the audio signal, since the only sound present is the sound of the person talking [27, 29, 40]. All of them were native English speakers which made it easier for the program to transcribe their words. Thanks to these factors, the cuts were almost flawless. However, determining the topics was quite problematic as each politician used very elaborate words to describe their views and the speeches were not mutually connected, apart from the fact that they all talked about the Black Lives Matter movement in USA. All in all, the results were passable since it was clear which kind of position did each politician take on the subject, however the cut lacked finesse [28, 30, 39].

4.2 Potential for Further Improvement

This program could be further improved by adding specific options for specific target audiences. In example there could be an option of automatically adding music in the background. A common error that occurs when we hear the interviewer's voice could be solved by comparing the subjects' facial movement with sound – if there is a correlation (if they open their mouth while the speech sound can be heard) we can assume that the clip in question is valid. Automatically adding b-roll (b-roll is all visual material used to complement other footage) could also potentially benefit the user. The program could, without the b-roll ever being selected by the user, automatically crawl the web for videos based on the keywords it got from LDA and then edit those videos in during clips with the topics that include those keywords [31, 33].

5 Conclusion

In this paper we have shown the functioning of a program for automatic video editing of reports. The results vary depending on an array of aspects of the input material. Nonetheless we conclude that the program creates plausible and watchable output. We deem this field being potentially prolific for many innovations yet to come, as it can benefit from the rapidly developing artificial intelligence pool of knowledge. We doubt that programs like these will ever be able to completely replace human video editors, however they could change the way they approach their work.

References

1. Juang, B.H., Rabiner, L.R.: Automatic speech recognition – a brief history of the technology development. Bell Syst. Tech. J. (2005)
2. Boden, M.: Mind as Machine. Clarendon Press, Oxford (2008)
3. Papadimitriou, C., Tamaki, H., Raghavan, P., Vempala, S.: Latent semantic indexing. In: Proceedings of the Seventeenth ACM SIGACT-SIGMOD-SIGART Symposium on Principles of Database Systems - PODS 1998 (1998)
4. Blei, D.M., Ng, A.Y., Jordan, M.I.: Latent Dirichlet allocation. J. Mach. Learn. Res. 3, 993–1022 (2003)

5. Collobert, R., Weston, J., Bottou, L., Karlen, M., Kavukcuoglu, K., Kuksa, P.: Natural language processing (almost) from scratch. Comput. Res. Repos. - CORR. **12**, 2493–2537 (2011)
6. Newman, D., Chemudugunta, C., Smyth, P., Steyvers, M.: Analyzing entities and topics in news articles using statistical topic models. In: Mehrotra, S., Zeng, D.D., Chen, H., Thuraisingham, B., Wang, F.-Y. (eds.) ISI 2006. LNCS, vol. 3975, pp. 93–104. Springer, Heidelberg (2006). https://doi.org/10.1007/11760146_9
7. Stockhammer, T., Hannuksela, M.M.: H.264/AVC video for wireless transmission. IEEE Wirel. Commun. **12**(4), 6–13 (2005)
8. Rabai, L.B.A., Cohen, B., Mili, A.: Programming language use in US academia and industry. Inform. Educ. **14**(2), 143–160 (2015)
9. Darwish, A., Lakhtaria, K.: The impact of the new web 2.0 technologies in communication, development, and revolutions of societies. J. Adv. Inf. Technol. **2**(4), 204–216 (2011)
10. Tuđman, M.: Modeli znanja i obrada prirodnog jezika. Zavod za informacijske znanosti, Zagreb (2003)
11. Krivec, R.: Video postproduction. Algebra, Zagreb (2019)
12. Noble, W.S.: What is a support vector machine? Nat. Biotechnol. **24**(12), 1565–1567 (2006)
13. https://www.machinelearningplus.com/nlp/topic-modeling-visualization-how-to-present-res ults-LDA-models. Accessed 20 June 2020
14. https://www.machinelearningplus.com/nlp/lemmatization-examples-python/. Accessed 20 June 2020
15. https://www.machinelearningplus.com/nlp/gensim-tutorial/. Accessed 20 June 2020
16. https://en.wikipedia.org/wiki/Latent_Dirichlet_allocation. Accessed 20 June 2020
17. https://en.wikipedia.org/wiki/Natural_language_processing. Accessed 20 June 2020
18. https://en.wikipedia.org/wiki/Semi-supervised_learning. Accessed 20 June 2020
19. https://en.wikipedia.org/wiki/Speech_recognition. Accessed 20 June 2020
20. https://en.wikipedia.org/wiki/Rough_cut. Accessed 20 June 2020
21. https://en.wikipedia.org/wiki/Application_programming_interface. Accessed 20 June 2020
22. https://spacy.io/usage. Accessed 20 June 2020
23. https://radimrehurek.com/gensim/models/LDAmodel.html. Accessed 20 June 2020
24. https://www.geeksforgeeks.org/removing-stop-words-nltk-python/. Accessed 20 June 2020
25. https://www.nltk.org/data.html. Accessed 20 June 2020
26. https://pypi.org/project/SpeechRecognition/. Accessed 20 June 2020
27. https://realpython.com/python-speech-recognition/. Accessed 20 June 2020
28. https://FFmpeg.org/FFmpeg.html. Accessed 20 June 2020
29. https://cloud.google.com/speech-to-text. Accessed 20 June 2020
30. https://docs.python.org/3/library/tkinter.html. Accessed 20 June 2020
31. https://realpython.com/python-gui-tkinter/. Accessed 20 June 2020
32. https://www.youtube.com/watch?v=D8-snVfekto. Accessed 20 June 2020
33. https://www.tutorialspoint.com/python/python_gui_programming.htm. Accessed 20 June 2020
34. https://github.com/tyiannak/pyAudioAnalysis/wiki/5.-Segmentation. Accessed 20 June 2020
35. https://youtu.be/NYkbqzTlW3w. Accessed 20 June 2020
36. https://en.wikipedia.org/wiki/Digital_image_processing. Accessed 20 June 2020
37. https://www.ietf.org/rfc/rfc4180.txt. Accessed 20 June 2020
38. https://www.microsoft.com/hr-HR/microsoft-365/p/excel/cfq7ttc0k7dx. Accessed 20 June 2020
39. https://cloud.google.com/speech-to-text/docs/data-logging. Accessed 20 June 2020
40. https://radimrehurek.com/gensim/about.html. Accessed 20 June 2020

Residual Attention Network vs Real Attention on Aesthetic Assessment

Ranju Mandal[1], Susanne Becken[3], Rod M. Connolly[2], and Bela Stantic[1(✉)]

[1] School of Information and Communication Technology, Griffith University, Brisbane, Australia
[2] Australian Rivers Institute, Griffith University, Brisbane, Australia
[3] Griffith Institute for Tourism, Brisbane, Australia
{r.mandal,s.becken,r.connolly,b.stantic}@griffith.edu.au

Abstract. Photo aesthetics assessment is a challenging problem. Deep Convolutional Neural Network (CNN)-based algorithms have achieved promising results for aesthetics assessment in recent times. Lately, few efficient and effective attention-based CNN architectures are proposed that improve learning efficiency by adaptively adjusts the weight of each patch during the training process. In this paper, we investigate how real human attention affects instead of CNN-based synthetic attention network architecture in image aesthetic assessment. A dataset consists of a large number of images along with eye-tracking information has been developed using an eye-tracking device (https://www.tobii.com/group/about/this-is-eye-tracking/) power by sensor technology for our research, and it will be the first study of its kind in image aesthetic assessment. We adopted a Residual Attention Network and ResNet architectures which achieve state-of-the-art performance image recognition tasks on benchmark datasets. We report our findings on photo aesthetics assessment with two sets of datasets consist of original images and images with masked attention patches, which demonstrates higher accuracy when compared to the state-of-the-art methods.

Keywords: Photo aesthetic assessment · Image aesthetic evaluation · Great Barrier Reef · Aesthetic scoring · Deep learning

1 Introduction

Image quality assessment and predict photo aesthetic values have been a challenging problem in image processing and computer vision, as aesthetic assessment is subjective (i.e. influenced by individual's feelings, tastes, or opinions) in nature. A significant number of existing photo aesthetics assessment methods are available ([7, 13, 16, 20, 21, 30, 33]) in the literature using extraction of visual features and then employ various machine learning algorithms to predict photo aesthetic values. Aesthetic assessment techniques aim to quantify semantic level characteristics associated with emotions and beauty in images, whereas technical quality assessment deals with measuring low-level degradations such as noise, blur, compression artifacts, etc.

© Springer Nature Singapore Pte Ltd. 2021
T.-P. Hong et al. (Eds.): ACIIDS 2021, CCIS 1371, pp. 310–320, 2021.
https://doi.org/10.1007/978-981-16-1685-3_26

Recently, deep learning methods have shown great success in various computer vision tasks, such as object recognition [17,27,28], object detection [9,26], and image classification [32]. Deep learning methods, such as deep convolutional neural network and deep belief network, have also been applied to photo quality/aesthetics assessment and have shown good results. As most deep neural network architectures require fixed-size inputs, recent methods transform input images via cropping, scaling, and padding, and design dedicated deep network architectures, such as double-column or multi-column networks, to simultaneously take multiple transformed versions as input.

Based on the available image assessment techniques in the literature, full-reference and no-reference approaches are the two main categories of image quality assessment. While the availability of a reference image is assumed in the former (metrics such as PSNR, SSIM [35], etc.), typically blind (no-reference) approaches rely on a statistical model of distortions to predict image quality. A quality score is to predict that relates well with human perception is the main goal in both cases.

Broadly, the task involved to distinguish computationally the aesthetic attributes of an image [10] for a related assumption. The literature proposes several methods to solve such challenging classification and scoring problems. The earlier approaches can be categorised into two groups, based on visual feature types (hand-crafted features and deep features based on Convolutional Neural Network), and evaluation criteria, dataset characteristics and evaluation metrics (examples include: Precision-recall, Euclidean distance, ROC curve, and mean Average Precision). More specifically, the term "hand-crafted" features refer to properties derived employing various algorithms using the information present in an image. As an example, edges and corners are two simple features that can be extracted from images. A basic edge detector algorithm works by finding areas where the image intensity "suddenly" changes. For example, the shell of a turtle can be identified as an edge. Likewise, the so-called Histogram of Gradients (HoG) [5] is another type of handcrafted feature that can be applied in many different ways.

Earlier proposed techniques designed hand-crafted aesthetics features according to aesthetics perception of people and photography rules [2,6,7] and obtained encouraging results while handcrafted feature design for aesthetic assessment is a very challenging task. More robust feature extraction techniques were proposed later on to leverage more generic image features (e.g. Fisher Vector [18,23,24] and the bag of visual words [30]) for photo aesthetics evaluation. Generic feature-based representation of images is not ideal for image aesthetic assessment as those features are designed to represent natural images in general, and not specifically for aesthetics assessment (Fig. 1).

In contrast, Convolutional Neural Network (CNN)-based features are learning from the training samples, and they do this by using dimensionality reduction and convolutional filters. Recent approaches to image aesthetic assessment mostly apply more complex and robust deep Convolutional Neural Networks (CNN) architectures ([3,4,14,37]). Availability of large-scaled labeled and scored

(a)   (b)   (c)

(d)   (e)   (f)

Fig. 1. Some example images from GBR dataset with score $\mu(\pm\sigma)$, where μ and σ represent Mean and Standard Deviation (SD) of score, respectively. (a) high aesthetics and low SD ($\mu = 9.583$, $SD = 0.64$), (b) high aesthetics and low SD ($\mu = 9.462$, $SD = 0.746$), (c) high aesthetics and high SD ($\mu = 5.916$, $SD = 3.773$), (d) low aesthetics and low SD ($\mu = 2.273$, $SD = 1.42$), (e) low aesthetics and low SD ($\mu = 3.0$, $SD = 1.0$), (f) low aesthetics and high SD ($\mu = 3.454$, $SD = 3.23$)

images from online repositories have enabled CNN-based methods to perform better than previously proposed non-CNN approaches [19,25]. Moreover, having access to pre-trained models (e.g. ImageNet [27]) for network training initialization and fine-tuned the network on subject data of image aesthetic assessment have been proven more effective technique for typical deep CNN approach [34] (Fig. 2).

2   Methodology

We propose two novel Deep CNN architecture for image aesthetics assessment adapted from the recently published state-of-the-art image classification model [11,34]. Both the models used in our experiments have been well tested as image classifier for a large number of classes. In experiments, our aim for predictions with higher correlation with human ratings, instead of classifying images to low/high score or mean score regression, the distribution of ratings are predicted as a histogram [8]. The squared EMD (Earth Mover's Distance) loss-based assessment was proposed by Talebi et al. [8], which shows a performance boost inaccurate prediction of the mean score. All network models for aesthetic assessment

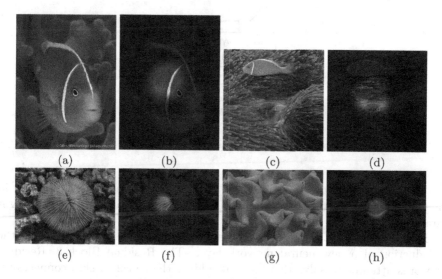

Fig. 2. Four example images from GBR dataset along with same processed images after studying eye movements. (a), (c), (e), (g) are original images. (b), (d), (f), (h) are masked images of (a), (c), (e), (g) respectively, after masked using attention patch information obtained from the Tobii eye-tracking device.

are based on image classifier architectures. Two different architectures (with and without attention mechanism) state-of-the-art networks are explored for the proposed applications. Networks used in our experiments were first trained and then fine-tuned using the large-scale aesthetics assessment AVA dataset [19]. The AVA dataset has 250,000 images, which is very useful for training such a large deep neural network model. The complete architecture of this project consists of different sub-modules, and each of these sub-modules consists of building blocks, such as pooling, filters, activation functions, and so forth. The following sections provide more information on the sub-modules. A more detailed description of the architecture and different modules is provided below.

ResNet: Theoretically, neural networks should get better results as added more layers. A deeper network can learn anything a shallower version of itself can, plus possibly some more parameters. The intuition behind adding more layers to a deep neural network was that more layers progressively learn more complex features. The first, second, third, layers learn features such as edges, shapes, objects, respectively, and so on. He et al. [11] empirically presented that there is a maximum threshold for depth with the traditional CNN model. As more layers are added, the network gets better results until at some point; then as continue add extra layers, the accuracy starts to drop. The reason behind failures of the very deep CNN was mostly related to optimization function, network weights initialization, or the well-known vanishing/exploding gradient problem. Vanishing gradients are especially blamed, however, He et al. [11] argue that the use of Batch Normalization ensures that the gradients have healthy norms. In

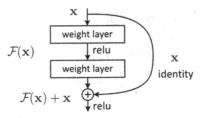

Fig. 3. Residual learning: a building block.

contrast, deeper networks are harder to optimize due to add more difficulty in the process of training; it becomes harder for the optimization to find the right parameters.

The problem of training very deep networks has been attenuated with the introduction of a new neural network layer -The Residual Block 3. Residual Networks attempt to solve this issue by adding the so-called skip connections. A skip connection is depicted in Fig. 3. If, for a given dataset, there are no more things a network can learn by adding more layers to it, then it can just learn the identity mapping for those additional layers. In this way, it preserves the information in the previous layers and can not do worse than shallower ones. So, the most important contribution to ResNet architecture is the 'Skip Connection' identity mapping.

This identity mapping does not have any parameters and is just there to add the output from the previous layer to the layer ahead. However, sometimes x and F(x) will not have the same dimension. Recall that a convolution operation typically shrinks the spatial resolution of an image, e.g. a 3×3 convolution on a 32×32 image results in a 30×30 image. The identity mapping is multiplied by a linear projection W to expand the channels of shortcuts to match the residual. This allows for the input x and F(x) to be combined as input to the next layer. A block with a skip connection as in the image above is called a residual block, and a Residual Neural Network (ResNet) is just a concatenation of such blocks.

$$y = F(x, W_i) + W_s x$$

The above equation shows when F(x) and x have a different dimensionality such as 32×32 and 30×30. This Ws term can be implemented with 1×1 convolutions, this introduces additional parameters to the model. The Skip Connections between layers add the outputs from previous layers to the outputs of stacked layers. This results in the ability to train much deeper networks than what was previously possible. He et al. [11] proposed their network with 100 and 1,000 layers and tested on bench mark datasets such as CIFAR-10, ImageNet dataset with 152 layers and achieved state-of-the-art performance.

Residual Attention Network: Residual Attention Network (Fig. 4), a convolutional neural network that incorporates both attention mechanism and residual units which can incorporate with state-of-art feed forward network architecture in an end-to-end training fashion. Residual Attention Network is constructed by

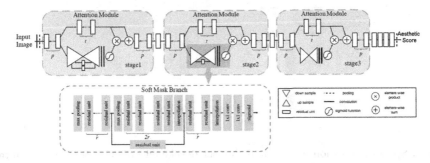

Fig. 4. An architecture the Residual Attention Network Architecture [34]. The output layer is modified to produce image aesthetic score instead class label.

stacking multiple Attention Modules which generate attention-aware features. Residual unit is a basic component that utilizes skip-connections to jump over few layers with nonlinearities and batch normalizations which is the prominent feature is the attention module.

Each Attention Module (see Fig. 4) is divided into two branches: mask branch and trunk branch. The trunk branch performs feature processing with Residual Units and can be adapted to any state-of-the-art network structures. Mask Branch uses bottom-up top-down structure softly weight output features with the goal of improving trunk branch features. The Bottom-Up step collects global information of the whole image by downsampling (i.e. max pooling) the image. The Top-Down step combines global information with original feature maps by upsampling (i.e. interpolation) to keep the output size the same as the input feature map. Inside each Attention Module, bottom-up top-down feedforward structure is used to unfold the feedforward and feedback attention process into a single feedforward process. The attention-aware features from different modules change adaptively as layers going deeper. Importantly, an attention residual learning to train very deep Residual Attention Networks which can be easily scaled up to hundreds of layers. The experiment also demonstrates that Residual Attention Network is robust against noisy labels.

In Residual Attention Network, a pre-activation Residual Unit [12], ResNeXt [36] and Inception [31] are used as basic unit to construct Attention Module. Given trunk branch output T(x) with input x, the mask branch uses bottom-up top-down structure [1,15,22] to learn same size mask M(x) that softly weight output features T(x). The bottom-up top-down structure mimics the fast feedforward and feedback attention process. The output mask is used as control gates for neurons of trunk branch similar to Highway Network [29]. The output of Attention Module H is:

$$H_i, c(x) = M_i, c(x) \times T_i, c(x)$$

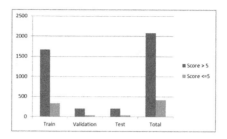

Fig. 5. Score distribution of the GBR dataset developed in-house, containing a total of 5417 images.

3   Results and Discussions

Dataset: For experimental purposes, both publicly available datasets and dataset developed in-house were used. For the dataset specific to the Great Barrier Reef (i.e. the GBR dataset), we used 5,417 underwater GBR images, which were downloaded from the Flickr social media platform. These images were sorted based on the content and then rated by participants in an online survey for their aesthetic beauty. At least 10 survey participants provided an aesthetic score for each image and the mean score was calculated. Most of the images (i.e. 80%) served as training material to enable the proposed Neural Network model to learn key feature parameters. The remaining 20% of the data were used during the test and validation phases. The validation dataset helps to understand the system performance in terms of accuracy during the training phase, whereas the test set is normally used once the training phase is completed and ready for deployment. To better understand the distribution of 'beautiful' and 'ugly' pictures in the dataset, Fig. 5 presents the number of images with scores above or equal/below 5. More than 2,080 images were scored as highly aesthetic ($score > 5$) and only 420 ($score <= 5$) images were scored as having low aesthetics. Figure 5 also shows how many images of high and low scores were used in each experimental stage.

The GBR dataset of 5,471 images is comparatively small for training a multi-layered deep Convolutional Neural Network. It was, therefore, necessary to complement the GBR data with a large-size, publicly available dataset (AVA, see Murry et al., 2012). This helped to train the system and allowed us to use the in-house GBR dataset for fine-tuning the algorithm. The detailed dataset description of the AVA [19] is described below.

- AVA1: We adopted the score of 0.5 (mean aesthetic score ranges between 0 and 1) as the threshold value to divide the dataset into high aesthetic value and low aesthetic value classes. By doing this, we obtained 74,056 images in the low aesthetic value class and 181,447 images in the high aesthetic value class. 229,954 and 25,549 (approximately 10% images) were used for testing system performance.

– AVA2: In a different experimental setup, and to increase the gap between images with high aesthetic and low aesthetic value, all images were sorted based on their mean scores. Then, the top 10% of images were considered as highly aesthetic and the bottom 10% images were classed as low aesthetic. Thus, 51,100 images (approximately, 20% of the full dataset) then formed the AVA dataset that was used for training (Fig. 6 and Table 1).

Fig. 6. Ranking some sampled images from GBR dataset with high and low aesthetic values using our proposed aesthetic assessment model using NIMA on Xception. Predicted (and ground truth) scores are shown together for every images (a) 9.583 (9.43), (b) 9.462 (9.22), (c) 5.916 (5.09), (d) 5.916 (5.60), (e) 2.273 (2.55), (f) 3.0 (3.5), (g) 3.454 (3.23), (h) 4.1 (4.22)

Table 1. Performance of the proposed method and few recently published architectures in predicting images from GBR dataset. Accuracy values are based on classification of photos to two classes high and low aesthetics(column 2). LCC (linear correlation coefficient) and are computed between predicted and ground truth mean scores (column 3) and standard deviation of scores (column 5). EMD measures closeness of the predicted and ground truth rating distributions.

Model	Accuracy	LCC-Mean	LCC-std.dev	EMD
NIMA(MobileNet)	80.36%	0.518	0.152	0.081
NIMA(VGG16)	80.60%	0.610	0.205	0.052
NIMA (Inception-Resnet)	81.51%	0.636	0.233	0.050
Residual Attention Network	81.74%	0.641	0.211	0.045

4   Conclusion

The objective of the work was to study the effectiveness of real human attention obtained using an eye-tracking device on deep Convolutional Neural Network architecture for automatic image aesthetic assessment and predicting an aesthetic score for images. The significance of aesthetic evaluation system was studied in details from the literature. A state-of-the-art deep Residual Attention Network architecture and ResNet were adapted to our need for the modeling of our GBR aesthetic assessment task. A wide range of experiments was conducted using a comprehensive analysis of performance is presented. Findings on photo aesthetics assessment of original images and images with masked attention patches demonstrates the higher accuracy of proposed method when compared to the state-of-the-art methods.

Acknowledgement. This research was partly funded by the National Environment Science Program (NESP) Tropical Water Quality Hub Project No 5.5.

References

1. Badrinarayanan, V., Handa, A., Cipolla, R.: SegNet: a deep convolutional encoder-decoder architecture for robust semantic pixel-wise labelling. CoRR abs/1505.07293 (2015). http://arxiv.org/abs/1505.07293
2. Bhattacharya, S., Sukthankar, R., Shah, M.: A framework for photo-quality assessment and enhancement based on visual aesthetics. In: Proceedings of the the ACM International Conference on Multimedia, pp. 271–280 (2010)
3. Bianco, S., Celona, L., Napoletano, P., Schettini, R.: On the use of deep learning for blind image quality assessment. arXiv preprint arXiv:1602.05531 (2016)
4. Bosse, S., Maniry, D., Wiegand, T., Samek, W.: A deep neural network for image quality assessment. In: Proceedings of the Computer Vision and Pattern Recognition, pp. 3773–3777 (2016)
5. Dalal, N., Triggs, B.: Histograms of oriented gradients for human detection. In: Proceedings of the Computer Vision and Pattern Recognition (CVPR 2005), vol. 1, pp. 886–893 (2005)
6. Datta, R., Joshi, D., Li, J., Wang, J.Z.: Studying aesthetics in photographic images using a computational approach. In: Leonardis, A., Bischof, H., Pinz, A. (eds.) ECCV 2006. LNCS, vol. 3953, pp. 288–301. Springer, Heidelberg (2006). https://doi.org/10.1007/11744078_23
7. Dhar, S., Ordonez, V., Berg, T.: High level describable attributes for predicting aesthetics and interestingness. In: Proceedings of the Computer Vision and Pattern Recognition, pp. 1657–1664 (2011)
8. Esfandarani, H.T., Milanfar, P.: NIMA: neural image assessment. CoRR abs/1709.05424 (2017). http://arxiv.org/abs/1709.05424
9. Girshick, R.B.: Fast R-CNN. In: ICCV (2015)
10. Haas, A., et al.: Can we measure beauty? Computational evaluation of coral reef aesthetics. PeerJ **3**, e1390 (2015). https://doi.org/10.7717/peerj.1390
11. He, K., Zhang, X., Ren, S., Sun, J.: Deep residual learning for image recognition. In: Proceedings of the IEEE Conference on Computer Vision and Pattern Recognition (CVPR), June 2016

12. He, K., Zhang, X., Ren, S., Sun, J.: Identity mappings in deep residual networks. CoRR abs/1603.05027 (2016). http://arxiv.org/abs/1603.05027

13. Jiang, W., Loui, A., Cerosaletti, C.: Automatic aesthetic value assessment in photographic images. In: Proceedings of the International Conference on Multimedia and Expo (ICME), pp. 920–925 (2010)

14. Kang, L., Ye, P., Li, Y., Doermann, D.: Convolutional neural networks for no-reference image quality assessment. In: Proceedings of the Computer Vision and Pattern Recognition, pp. 1733–1740 (2014)

15. Long, J., Shelhamer, E., Darrell, T.: Fully convolutional networks for semantic segmentation. CoRR abs/1411.4038 (2014). http://arxiv.org/abs/1411.4038

16. Lu, X., Lin, Z., Jin, H., Yang, J., Wang, J.Z.: Rapid: rating pictorial aesthetics using deep learning. In: Proceedings of the ACM International Conference on Multimedia, pp. 457–466 (2014)

17. Mandal, R., Connolly, R.M., Schlacher, T.A., Stantic, B.: Assessing fish abundance from underwater video using deep neural networks. In: 2018 International Joint Conference on Neural Networks, IJCNN 2018, Rio de Janeiro, Brazil, 8–13 July 2018, pp. 1–6. IEEE (2018). https://doi.org/10.1109/IJCNN.2018.8489482

18. Marchesotti, L., Perronnin, F., Larlus, D., Csurka, G.: Assessing the aesthetic quality of photographs using generic image descriptors. In: Proceedings of the International Conference on Computer Vision, pp. 1784–1791 (2011)

19. Murray, N., Marchesotti, L., Perronnin, F.: Ava: a large-scale database for aesthetic visual analysis. In: Proceedings of the Computer Vision and Pattern Recognition, pp. 2408–2415 (2012)

20. Nishiyama, M., Okabe, T., Sato, I., Sato, Y.: Aesthetic quality classification of photographs based on color harmony. In: Proceedings of the Computer Vision and Pattern Recognition, pp. 33–40 (2011)

21. Niu, Y., Liu, F.: What makes a professional video? A computational aesthetics approach. IEEE Trans. Circuits Syst. Video Technol. **22**(7), 1037–1049 (2012)

22. Noh, H., Hong, S., Han, B.: Learning deconvolution network for semantic segmentation. CoRR abs/1505.04366 (2015). http://arxiv.org/abs/1505.04366

23. Perronnin, F., Dance, C.: Fisher kernels on visual vocabularies for image categorization. In: Proceedings of the Computer Vision and Pattern Recognition, pp. 1–8 (2007)

24. Perronnin, F., Sánchez, J., Mensink, T.: Improving the fisher kernel for large-scale image classification. In: Daniilidis, K., Maragos, P., Paragios, N. (eds.) ECCV 2010. LNCS, vol. 6314, pp. 143–156. Springer, Heidelberg (2010). https://doi.org/10.1007/978-3-642-15561-1_11

25. Ponomarenko, N., et al.: A new color image database TID2013: innovations and results. In: Blanc-Talon, J., Kasinski, A., Philips, W., Popescu, D., Scheunders, P. (eds.) ACIVS 2013. LNCS, vol. 8192, pp. 402–413. Springer, Cham (2013). https://doi.org/10.1007/978-3-319-02895-8_36

26. Ren, S., He, K., Girshick, R.B., Sun, J.: Faster R-CNN: towards real-time object detection with region proposal networks. In: NIPS (2015)

27. Russakovsky, O., et al.: ImageNet large scale visual recognition challenge. IJCV **115**(3), 211–252 (2015). https://doi.org/10.1007/s11263-015-0816-y

28. Simonyan, K., Zisserman, A.: Very deep convolutional networks for large-scale image recognition. CoRR abs/1409.1556 (2014)

29. Srivastava, R.K., Greff, K., Schmidhuber, J.: Training very deep networks. CoRR abs/1507.06228 (2015). http://arxiv.org/abs/1507.06228

30. Su, H.H., Chen, T.W., Kao, C.C., Hsu, W.H., Chien, S.Y.: Scenic photo quality assessment with bag of aesthetics-preserving features. In: Proceedings of the ACM International Conference on Multimedia, pp. 1213–1216 (2011)
31. Szegedy, C., Ioffe, S., Vanhoucke, V.: Inception-v4, inception-resnet and the impact of residual connections on learning. CoRR abs/1602.07261 (2016). http://arxiv.org/abs/1602.07261
32. Szegedy, C., et al.: Going deeper with convolutions. In: Computer Vision and Pattern Recognition (CVPR) (2015). http://arxiv.org/abs/1409.4842
33. Tang, X., Luo, W., Wang, X.: Content-based photo quality assessment. IEEE Trans. Multimedia **15**(8), 1930–1943 (2013)
34. Wang, F., et al.: Residual attention network for image classification. In: Proceedings of the Computer Vision and Pattern Recognition (CVPR), pp. 6450–6458 (2017)
35. Wang, Z., Bovik, A.C., Sheikh, H.R., Simoncelli, E.P.: Image quality assessment: from error visibility to structural similarity. IEEE Trans. Multimedia **13**(4), 600–612 (2004)
36. Xie, S., Girshick, R., Dollár, P., Tu, Z., He, K.: Aggregated residual transformations for deep neural networks. In: Proceedings of the Computer Vision and Pattern Recognition (CVPR), pp. 5987–5995 (2017)
37. Xue, W., Zhang, L., Mou, X.: Learning without human scores for blind image quality assessment. In: Proceedings of the Computer Vision and Pattern Recognition, pp. 995–1002 (2013)

ICDWiz: Visualizing ICD-11 Using 3D Force-Directed Graph

Jarernsri Mitrpanont, Wudhichart Sawangphol$^{(\boxtimes)}$, Wichayapat Thongrattana, Suthivich Suthinuntasook, Supakorn Silapadapong, and Kanrawi Kitkhachonkunlaphat

Faculty of Information and Communication Technology, Mahidol University, Nakhon Pathom, Thailand
{jarernsri.mit,wudhichart.saw}@mahidol.ac.th,
{wichayapat.tho,suthivich.sut,
supakorn.sil,kanrawi.kit}@student.mahidol.ac.th

Abstract. WHO released ICD-11 in 2018 and will be used in 2022. ICD-11 has been changed drastically in its complex structure, enormous size, multi-source of classifications and terminologies, and code of diseases multiplication. The transition from ICD-10 to ICD-11 will be extremely intricate and long process, especially for the international members. In this paper, we present the 3D visualization part of the ICDWiz system to uncover the ICD-11 using our modified 3D Force Directed Graph to visual the information and relationship of the multi-knowledge sources biomedical Concept-terms-strings-atoms or phrases. The Visualization construction and functions such as Initial graph, Collapsed Function, Ring Notation and Constructed Text Label are described. The testing is elaborated using the ICDWiz database which is developed based on UMLS-Metathesaurus structure. The result reveals the complex visual of the ICD-11 medical information such as diseases, symptoms and relationship integrated from multi-medical classifications such as ICD-11, ICD-10, ICD-10 TM (Thai Modification), MeSH, and SNOMED-CT. The mapping between ICD-10 and 11 is also visualized as well.

Keywords: Medical classification · ICD · ICD-11 · ICD-10 · ICD10-TM · MeSH · SNOMED-CT · UMLS · Hierarchy structure · Visualization · 3D force-directed graph

1 Introduction

The medical classifications such as the International Code of Disease (ICD), Medical Subject Headings (MeSH), and Systematized Nomenclature of Medicine - Clinical Terms (SNOMED-CT) play a significant role in collecting the statistics of diseases and health problems worldwide. Since the first introduction of the "International Classification of Causes of Death" in 1893 [24], many health organizations such as World Health Organization (WHO), National Library

© Springer Nature Singapore Pte Ltd. 2021
T.-P. Hong et al. (Eds.): ACIIDS 2021, CCIS 1371, pp. 321–334, 2021.
https://doi.org/10.1007/978-981-16-1685-3_27

of Medicine (NLM), and International Health Terminology Standards Development Organization (IHTSDO), started to develop tools for classifying the medical information individually. The problem fell into the health organizations to decide which medical classification system is appropriate for them. Many problems emerge when health organizations use different medical classification, need an interchange of medical classification, or encounter with a revision of the medical classification, or even worst if the whole structure has been changed.

An example for this is the revision of ICD. The 10th Revision was adopted in 1990 while the 11th Revision was released in 2018 and will be enforced in 2020. Modifications of ICD encounters problems of limit data comparability, linkage to multiple knowledge bases, and lack of uniformity in translated terms. The transitions of ICD-10 to ICD-11 need a paramount of works with the changes in structures, descriptions, codes and links with other classifications and terminologies such as SNOMED-CT. ICD-11 is larger and more complex than the previous one resulted in confusion and stress for clinicians everywhere. The addition of over 40,000 new codes means a more complex system to learn and implement. Nonetheless, most of the country will take longer to transit from ICD-10 to ICD-11 due to the problem in understanding the new structure of the revision. For example in Thailand, it took more than a decade to develop the Thai Modification of ICD-10 (ICD-10 TM) [1].

Therefore, due to the above issues, we propose a system called *ICDWiz* [18] to leverage the process of understanding the structure of ICD-11 with integrated medical multi-classifications and terminologies in the UMLS-Metathesaurus structure using visualization techniques to support semantic interoperability for Health Personnel, to facilitate the collaboration between organizations, to understand the revision and interchange of medical classification. This paper describes the 3D visualization part of the ICDWiz.

2   Backgrounds and Related Works

2.1   Medical Classification

Medical classification is a standardized statistical code to classify and describe medical terms including diagnosis, procedure of treatment, cost reimbursement, and related disease or drug [12]. These codes are used by public healthcare. The examples of Medical classification related to our work are briefly described as follows:

ICD-10 [1,15]. The 10th revision of International Statistical Classification of Diseases and Related Health Problems (ICD) endorsed in 1990 to classify morbidity and mortality. It was cited in more than 20,000 scientific articles and used by more than 100 countries including Thailand. There are 22 chapters containing more than 14,000 codes. Its code structure has at most 7 characters. The first three characters indicate category of the diagnosis while extra character for additional details. WHO uses Classification Mark-up Language (ClaML), an XML based format, to share ICD-10. ClaML format captures the information and hierarchy such as parent-child relations, a level of granularity, and cross references [22]. In addition, ICD-10-TM was developed by the Thai Health Coding Center (THCC) with support from other organizations to define and add the code of disease found in Thailand but not in ICD-10. The development started in 2007 and published in 2016. Hence, it took decade to develop ICD-10-TM since its release [1].

ICD-11 [23]. The 11th revision of ICD was released in June 2018 for members to prepare for implementation such as translating ICD into their national languages before the adoption for reporting in January 2022. ICD-11 draws extensively on the method of combining several medical codes to describe a clinical condition to more level of details. Its electronic architecture allows assignment of unique identifiers to any condition listed. There are 28 chapters containing more than 55,000 codes for diseases, disorders, injuries, external causes, signs and symptoms [23,24]. ICD-11 codes are alphanumerically ranging from 1A00.00 to ZZ9Z.ZZ instead characters in ICD-10. The first four characters indicate the category. The first position of the codes in the same chapter is the same character. The second position of the codes uses a letter to differentiate from ICD-10 codes. To describe a causal relationship, it uses the term "due to". On the other hand, it uses the term "associated with" for describing the concurrence.

Medical Subject Headings (MeSH) [20]. It is an online controlled vocabulary that lists words, groups of synonyms and related medical concepts provided by the National Library of Medicine (ULM) of the United States. MESH is used for implementing index, catalog, and search for biomedical and health-related information [3,20].

Systematized Nomenclature of Medicine - Clinical Terms (SNOMED-CT) [17]. It is one of the international medical classifications which includes medical codes, terms, synonyms, and definitions for documentation and reporting. It is the most comprehensive multilingual clinical healthcare terminology in the world maintained and distributed by the International Health Terminology Standards Development Organization (IHTSDO). Its content is represented using three types of components which are concepts, descriptions, and relationships. Moreover reference sets are supplemented for additional flexible features [3,19].

2.2   Unified Medical Language System (UMLS) [4]

UMLS is an integrated and distributed terminology platform which indicates the meaning of the word, maintains and controls the coding standard, associated system, and relationship of medical science terms. It provides an effective biomedical information system which allocates multi-source biomedical data in a system and links the relationship between those terms. Metathesaurus is one of the main components in UMLS which contain codes and terms integrated from multi-sources such as ICD-10, MeSH, and SNOMED-CT but not yet including ICD-11.

ICTWiz [18] is a new software system as a result from the VISMAP-ICD: Ontology-Mapping Visualization and Recommendation for International Statistical Classification of Diseases and Related Health Problems (ICD-10 and ICD-11) research project. VisMap-ICD project aims to develop a new software tool to alleviate the complexity problem of ICD-11. It integrates multi-classifications and terminologies from muli-sources such as MeSH, SNOMED-CT, ICD-10 and ICD-10TM using the Unified Medical Language System (UMLS) knowledge sources platform. VisMap-ICD Database is created to integrate these medical multiple knowledge sources, corpuses and platforms. It also utilizes the database, keywords and relationship of these medical classifications from [5] in which the researchers develop the Automated Document Structure Extraction and Verification System for ICD-10 using WHO ClaML data format and integrating the ICD-10-TM Thai version. Their approach can generate medical corpus including important information such as Keyword, Keyphrase, and Relationship from classification entity of ICD-10, ICD-10-TM, MeSH, and SNOMED- CT. However, the ICD-11 has not yet been considered. The 3D Force-Directed Graph is used to visualize multi-relationship of the Concept-terms-strings-atoms or phrases in the ICD-11.

2.3 Visualization Technique and Related Works

In this section, three visualization techniques which are important and useful to explore the complex content of several medical classification are described.

Hierarchical Relationship [9]. It is the most effective way to show the relationship with difference degree or levels. The super-ordinate (or parent) can represent a class or container that contains the subordinate (or child) or a member of the class. Most of medical classifications use hierarchical relationships as a basis to classify the medical information into a group. For example, ICD-11 uses hierarchical relationships based to classify the large amount of disease into different 28 chapters, and each chapter also has sub-classes to increase the specificity of the group of disease which are block, category, and sub-category.

Forced Directed Graph [2,11]. It is a web component to represent a graph data structure in a 3-dimensional space using a force-directed iterative layout. It uses ThreeJS/WebGL for 3D rendering and either d3-force-3d or graph for the underlying physics engine. Force-directed graph drawing algorithm is a class of algorithms for drawing graphs automatically. The purpose is to allocate the nodes position of a graph in 3 dimensional space so that all the edges will be located properly. This algorithm aims to reduce the crossing edges as much as possible, by assigning forces among the set of edges and the set of nodes, based on their relative positions. Then these forces are used either to simulate the motion of the edges and nodes or to minimize their energy. The algorithm will assign forces among the set of edges and the set of nodes of a graph drawing. Typically, spring-like attractive forces based on Hooke's law [16] are attached in the endpoints of the graph's edges towards each other. The algorithm also simulates push-forces like those of electrically charged particles based on Coulomb's law [21] which are used to separate all pairs of nodes.

Collapsible Tree Diagram [6]. A technique to reduce the complexity of the graph by using interactive functions and graph reconstruction functions. Collapsible tree diagram contains interactive nodes which can toggle a graph function to hide and show the child nodes of the interacted node. This technique is commonly used to handle a huge tree diagram since it can reduce the amount of displayed nodes which significantly increases the focus of the user. Normally, the structure of the medical classification in UMLS always has the number of child nodes much greater than the number of parent nodes. Needless to say, the collapsible tree diagram becomes one of the techniques to be used.

Moreover, there are several researches explored to visualize some medical classification. However, each research points to different purposes of visualization. For example, in [7], they try to find the method to indicate the error in SNOMED-CT. This study leverages the abstraction networks which are integrated from a large terminology. However, the abstraction network contains only a main feature of the concept which is impossible to indicate error inside the medical classification. Their application is called BLUSNO. The study shows that the user can rely on BLUSNO system to manage the possible reduction factors for the users. In [8], the study explores the method to visualize the pathway and ancestor of a concept in SNOMED-CT by using illustration technique. The application is called TermViz (see Fig. 1) using hierarchical relationship and tree. From their experiment, the graph needs further tuning and development.

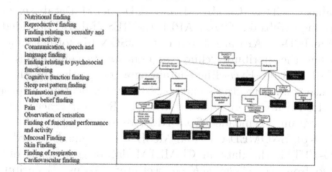

Fig. 1. TermViz Application: List of concepts (Left) and an illustration of graphical view of the same concepts (Right).

3 Proposed Approach

In this section, we emphasize on the construction of the ICD-11 visualization as a part of the ICDWiz system, in particular, it utilized the medical information from ICDWiz Integrated Database. It is important to briefly describe about the Data Preparation process. Then, the Visualization Construction will be explained.

3.1   Data Preparation for the ICDWiz Database

From Fig. 2, ICDWiz Database integrate multiple sources of medical classification including ICD-10, ICD-10 TM, ICD-11, MesH and SNOMED-CT. To uncover ICD-11 complexity linking with multiple medical sources, the UMLS-Metathesaurus structure is adopted. The preparation process consists of two main parts: Data Extraction and Data Integration.

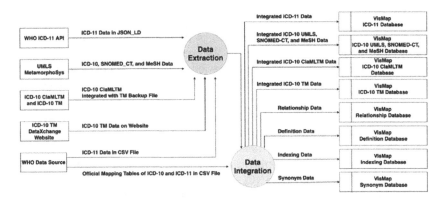

Fig. 2. ICDWiz Database Data Flow Diagram

Data Extraction extracts data from each source including ICD-11 API, UMLS Database, ICD-10 ClaMLTM and ICD-10-TM Database, and ICD-10 TM DataXchange Website. ICD-11 API is the RESTful API provided by WHO. The data from ICD-11 API are stored in the JSON_LD format as a tree structure, which has the attribute "child" to indicate its child nodes. WHO also provides ICD-11 data in Excel File containing a code, title, and class of records in which we use it for correctness and consistency verification purpose. Another source is UMLS Database provided by NLH in RRF format accessing via *Metamorphosys* program downloaded from NLH website[1]. Finally, the UMLS data including ICD-10, SNOMED-CT, and MeSH are obtained. In order to gather data of ICD-10-TM, the data in ClaMLTM database in [5] is obtained and also extracted from *DataXchangeKrabi* website[2]. We use Beautiful soup API library [14] in Python to perform data extraction from the website. Since the extracted data are in various format, thus, data integration is compulsory.

Data Integration migrates three main databases: the ICD-10 ClaMLTM, ICD-11, and UMLS databases (ICD-10, MeSH, and SNOMED-CT) to the UMLS Platform. The idea of integration is that we adopt data structure from UMLS system and map the data from other sources to the same concept as the UMLS database structure. UMLS groups the medical classification records that have similar meaning/ synonym/ title together to be one concept. Its structure starts

---

[1] https://www.nlm.nih.gov/research/umls/index.html.
[2] http://xchangekbo.moph.go.th/hdc_report/frontend/web/index.php.

from AUI which is the unique id of each medical classification record or Atom. AUIs with similar title are grouped to be SUI (String). Next, SUIs with similar title or synonym are grouped to be LUI (Term/Lexicon), and finally select one of these LUI to be CUI (Concept). This main structure are used to integrate data from ICD-10 ClaML, ICD-10 TM and ICD-11. For the ICD revision mapping, WHO provides the ICD-10 to ICD-11 official mapping table. Thus, this allows one record of ICD-11 to be mapped into multiple records of ICD-10 leading to one record of ICD-11 belongs in multiple concepts. There are two main functions in Python that are frequently used in the integration part which are Jaccard Similarity [13] Calculation and Spacy Tokenizer [10]. Finally, all of the integration is loaded into the ICDWiz database which contains the medical concepts that group the similar/related medical classifications together including ICD-10, ICD-10 TM, ICD-10 ClaMLTM, ICD-11, MeSH, and SNOMED-CT.

3.2   Visualization Construction

To visualize part of the comprehensive integrated biomedical multi-classifications, concept and terms, this section describes the proposed implementation visualization techniques. This visualization unfolds the ICD-11 integrated data linking to other medical classification in UMLS structure by allocating the biomedical concepts, terms, strings, and relationships into nodes and links in the 3-dimensional space. Using 3D-forced directed graph, however, it contains only basic graph components, node and link [2], which is insufficient to reveal all of the detail of the data from medical classifications. Therefore, there is a need to add four more dimensions to elaborate the extra component such as creating *ring notation* to represent different knowledge sources, providing *node and link color* for different node and link relationship types, and adding *text label* for a second language (Thai in this case). More details are described below.

Initial Graph. One important characteristic of the biomedical classification information is that one term has one or more relationships with the other terms or concepts which means that this characteristic fit in a network graph. Our network graph consists of two main components which are *nodes* that indicate the concepts in medical classification and *links* that indicate the relationship among nodes. 3D-forced directed graph become the most suitable approach to construct the visualization of these medical classification, which satisfies all of the requirements from above. Figure 3 displays the flow of the initial graph process. It receives the data in JSON format, then creates the index of each node from JSON data. Once the index is created, it is sent to the pruned function to add the visible node to the visible set, which is the set of all nodes. After that, the visible set will be used in `construction` function of the 3D forced directed library to construct the visualization. In addition, node colors are used to differentiate the node type as presented in Table 1.

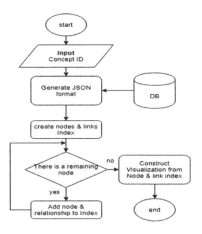

Table 1. Node Notation

Node color	Type
Yellow	Concept
Red	Term
Green	String
Blue	Medical Classification

Fig. 3. Flowchart - Initial Graph

Constructed Label Text. As the graph is constructed, the visualization still lacks the clarity to indicate the identity of information because all of the nodes and links are the same in term of color. This will be difficult for the user to distinguish each node and link. Moreover, the graph rotation in 3D space can easily cause the lost tracking for the user due to the spring-force that pushes each other nodes. Adding text label to each node in the graph is challenging since it is constructed based on 3D forced directed graph which need to rendering animated 3D computer graphics on a website, the library called *Spritetext* is used for auto-generating text object on canvas and convert into a texture based on three.js. Figure 4 shows an example.

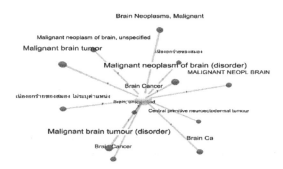

Fig. 4. 3D Forced Directed Graph with Constructed Label Text

Collapsed Function. The biomedical concept has complex semantic structure. The collapsed tree diagram technique and interactive function are applied to reduce the amount of elements visualized in the graph. Each concept consists of sub level structure which are Terms, String, and Atom. Visualizing all of the nodes and their relationships may give an overview of the structure of that concept. However, it will be unclear for the users to observe all of the elements in the graph if the concept is far too complex. Thus, each level (Term) is generated at a time to help the user to focus on each level of the whole structure. Figure 5 explains the process flow. The collapsed function will traverse through the index to check for the non-collapsed node. When the function found the non-collapsed node, it will add the child node(s) and its relationship to the visible set then send to the 3D forced directed graph construction function to generate the graph.

Ring and Color Notation. Ring and its color are used as a crucial sign to quickly notify the medical classification sources which are related to the ring's owner. The ring is implemented to be responsive generated every time when the user clicks at a node, in order to track the change of the shape in the graph. The ring is constructed during the 3D forced directed graph receiving the information data to construct each node. This function checks number of the rings to construct on the source abbreviation (SABs) arrays which contain all the child nodes then `THREE.TorusGeometry` is called from three.js library to create the ring. The color of the rings are also filled in at this step depending on the source appearing in SABs. The final step of node construction is to attach the ring to the node object in the graph. This process is iteratively done node by node. Rings have seven colors representing each medical classification sources as seen in Table 2.

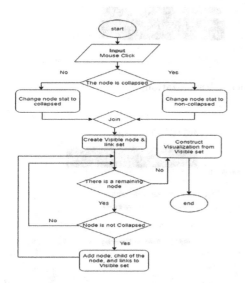

Fig. 5. Flowchart - Collapsed Function

Table 2. Ring Notation

Ring color	Knowledge sources
Blue	ICD-10
Green	ICD-10 TM
Light Blue	ICD-10 ClaMLTM
Purple	ICD-11
Yellow	MeSH
Violet	SNOMED-CT US
Pink	SNOMED-CT VET

4   Experimental Result

The medical information from ICDWiz Database, the integrated medical multi-knowledge sources and corpus based on UMLS platform are used as testing data to present the visualization and multi-relationship of the Concept-terms-strings-atoms or phrases in ICD-11. There are two main visualizations: *MED-Concept* and *ICD Mapping*. MED-Concept visuals a medical concept in ICD-11 with its related terms in other medical classifications based on UMLS ontology structure while ICD Mapping visuals the mapping of ICD-10 and ICD-11.

For the test scenario, the MED-Concept and ICD Mapping Visualization were tested using 100 medical classification samples of the ICD-10 and ICD-11. The recommended MED-Concept is used for visual then time performance is recorded for graph construction in milliseconds and the visualization result is checked. Figure 6 shows the MED-Concept Visual of "Hemorrhagic Fever, Ebola" linking to multiple terms from multi-knowledge sources indicated by the rings. Figure 7 shows the ICD-Mapping Visual of a code of ICD-10 "A98.4 Ebola virus disease" mapping to a group of 1D60.- Ebola disease code of ICD-11.

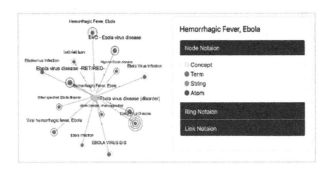

Fig. 6. MED-Concept Visualization

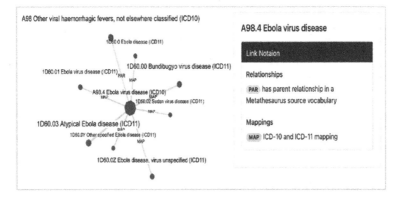

Fig. 7. ICD Mapping Visualization

Figure 8 shows the result for ICD-11 "XN83D Coronavirus" which belongs to "Coronavirus as the cause of diseases classified to other chapters". The rings around the node indicate the expandability and source of medical classification. User can view ICD-11 node with a purple ring and the related terms as a red node. The terms can be either English or Thai language. The users can view the related string in a selected term as a green node in Fig. 8.

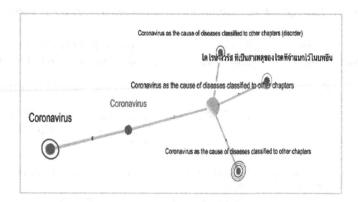

Fig. 8. Visualization - Strings in a selected medical term (Color figure online)

Figure 9 shows the result of the related medical classification in a selected string as a blue node. Each node shows a code, title, and knowledge source. For example, "XN83D" is a code for "Coronavirus" in "ICD-11". User can view the relationships to other medical classifications in green link and mapping between ICD-10 and ICD-11 in red link. For example, the "XN83D Coronavirus" in "ICD-11" has a mapping to "B97.2 Coronavirus as the cause of diseases classified to other chapters" in "ICD-10".

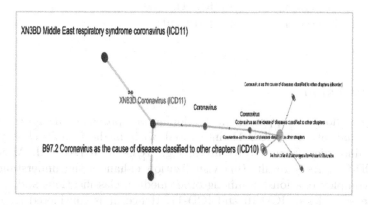

Fig. 9. Visualization - Relationships to other medical classifications and Mapping between ICD-10 and ICD-11 (Color figure online)

The test scenario for MED-Concept and ICD Mapping Visualization are performed and the run time in Millisecond (ms) for MED-Concept and ICD Mapping Visualization are approximately the same with the average of 61.418, maximum of 72.099 and minimum of 52.286 ms. The test results of the MED-Concept and the ICD-Mapping are shown in Table 3 and Table 4 accordingly.

Table 3. Example of Test Result for *Hemorrhagic Fever, Ebola* concept

Concept	Term	String	Atom (Medical Classification)
Hemorrhagic Fever, Ebola	Ebola Virus Disease	Ebola Virus Disease	Ebola Virus Disease (MeSH D019142)
			Ebola virus disease (ICD-11 1D60.01)
			...
		Ebola virus disease	Ebola virus disease (ICD-10 A98.4)
			Ebola virus disease (SNOMED-CT 123323003)
			...
	Hemorrhagic Fever, Ebola	Hemorrhagic Fever, Ebola	Hemorrhagic Fever, Ebola (MeSH D019142)
		Ebola haemorrhagic fever	Ebola haemorrhagic fever (SNOMED-CT 37109004)
		...	...

Table 4. Example of Relationship and Mapping Result for *A98.4 Ebola Virus Disease*

Type	Atom (Medical Classification)
PAR	Other viral haemorrhagic fevers, not elsewhere classified (ICD-10 A98)
MAP	Ebola disease (ICD-11 1D60.0)
	Bundibugyo virus disease (ICD-11 1D60.00)
	Ebola virus disease (ICD-11 1D60.01)
	...

5   Conclusion

This paper shows an approach in developing our proposed *3D Force Directed Graph* visualization of ICD-11 integrated with medical multi-classifications and terminologies from multi-sources, e.g., ICD-10, ICD-10TM, MeSH and SNOMED-CT. As a result, this visualization enhances the understanding of ICD-11 complex relationship among other medical classifications and the mapping disease codes of ICD-10 and ICD-11. Hence, it is considered as a useful supporting tool for the healthcare personnel in the transition of ICD-11, medical records, medical disbursements, and medical surveillance.

References

1. International Statistical Classification of Disease and Related Health Problems, 10th Revision, Thai Modification (ICD-10-TM). Ministry of Public Health, Thailand (2009)
2. Asturiano, V.: 3D Force-Directed Graph (2020). https://github.com/vasturiano/3d-force-graph
3. Baumann, N.: How to use the medical subject headings (MeSH). Int. J. Clin. Pract. (2016). https://doi.org/10.1111/ijcp.12767
4. Bodenreider, O.: The Unified Medical Language System (UMLS): Integrating biomedical terminology. Nucleic Acids Research (2004)
5. Choosang, P., Narktim, A., Nondhabongkoj, S., Noraset, T., Mitrpanont, J.: ICD-10 and ICD-10-TM Automated Document Structure Extraction and Verification System. Bachelor's thesis, Mahidol University, Thailand (2019)
6. D3.js: Collapsible tree diagram in v4 (2020). https://bl.ocks.org/d3noob/43a860bc0024792f8803bba8ca0d5ecd
7. Geller, J., Ochs, C., Perl, Y., Xu, J.: New abstraction networks and a new visualization tool in support of auditing the SNOMED CT content. In: AMIA Annual Symposium Proceedings, vol. 2012, p. 237. American Medical Informatics Association (2012)
8. Højen, A.R., Sundvall, E., Gøeg, K.R.: Methods and applications for visualization of SNOMED CT concept sets. Appl. Clin. Inform. **5**(1), 127 (2014)
9. Holten, D.: Hierarchical edge bundles: visualization of adjacency relations in hierarchical data. IEEE Trans. Vis. Comput. Graph. **12**(5), 741–748 (2006)
10. Honnibal, M., Montani, I.: Spacy 2: natural language understanding with bloom embeddings, convolutional neural networks and incremental parsing. **7**(1), (2017, to appear)
11. Hu, Y.: Efficient, high-quality force-directed graph drawing. Math. J. **10**(1), 37–71 (2005)
12. IGI Global: What is Medical Classification. https://www.igi-global.com/dictionary/optimization-medical-supervision-management-reimbursement/18206
13. Real, R., Vargas, J.M.: The probabilistic basis of Jaccard's index of similarity. Syst. Biol. **45**(3), 380–385 (1996)
14. Richardson, L.: Beautiful soup documentation (2017). Dosegljivo https://www.crummy.com/software/BeautifulSoup/bs4/doc/. Dostopano 7 July 2018
15. Rouse, M.: ICD-10 (International Classification of Diseases, Tenth Revision) (2018). https://searchhealthit.techtarget.com/definition/ICD-10
16. Rychlewski, J.: On Hooke's law. J. Appl. Math. Mech. **48**(3), 303–314 (1984)
17. Stearns, M.Q., Price, C., Spackman, K.A., Wang, A.Y.: SNOMED clinical terms: overview of the development process and project status. In: Proceedings of the AMIA Symposium, p. 662. American Medical Informatics Association (2001)
18. Suthinuntasook, S., Thongrattana, W., Sillapathadapong, S., Sawangphol, W., Mitrpanont, J.: VISMAP-ICD: Ontology-Mapping Visualization and Recommendation for International Statistical Classification of Diseases and Related Health Problems (ICD-10 and ICD-11). Bachelor's thesis, Mahidol University, Thailand (2020)
19. U.S. National Library of Medicine - National Institutes of Health: SNOMED CT FAQs (2018). https://www.nlm.nih.gov/healthit/snomedct/faq.html
20. U.S. National Library of Medicine - National Institutes of Health: Medical Subject Headings (2019). https://www.nlm.nih.gov/mesh/meshhome.html

21. Williams, E.R., Faller, J.E., Hill, H.A.: New experimental test of Coulomb's law: a laboratory upper limit on the photon rest mass. Phys. Rev. Lett. **26**(12), 721 (1971)
22. World Health Organization: ClaML format and its use in the dissemination of WHO Classifications. https://apps.who.int/classifications/apps/icd/Classification DownloadNR/ClaMLFormat.htm
23. World Health Organization: WHO releases new International Classification of Diseases (ICD 11) (2018). https://www.who.int/news-room/detail/18-06-2018-who-releases-new-international-classification-of-diseases-(icd-11)
24. World Health Organization: Classification of Diseases (ICD) (2019)

Decision Support and Control Systems

An Integrated Platform for Vehicle-Related Services and Records Management Using Blockchain Technology

Arnab Mukherjee[1] and Raju Halder[2(✉)]

[1] RCC Institute of Information Technology, Kolkata, India
cse2018053@rcciit.org.in
[2] Indian Institute of Technology Patna, Patna, India
halder@iitp.ac.in

Abstract. This paper proposes an integrated system for vehicle-related services and records management using blockchain technology, where various stakeholders (vehicle owners, regional transport offices, insurance issuers, pollution control test centers, traffic police) can avail or provide various services (vehicle's registration, driving license issuance, insurance and pollution-under-control certificates issuance, automated insurance claim, auditable trail of vehicle's documents access and verification) in a hassle-free manner without any untrusted intermediaries. While the use of blockchain technology increases accountability, transparency, and trust in the system, this allows a cohesive integration of the proposed system with other smart city digital infrastructures easily. We develop a prototype of our proposal as a proof of concept using Hyperledger Fabric, adopting Attribute-Based Access Control (ABAC) policy, and we present an experimental evaluation to demonstrate the performance of the system. To the best of our knowledge, this proposal is the first of its kind to provide a common blockchain-based platform for all vehicle-related services and records management.

Keywords: Vehicle records · Vehicle services · Blockchain technology · Smart contract

1 Introduction

Smart city technology is already starting to change lives for the better, and more change is on the way. Smart city demands an integration of multiple technological solutions securely to manage city's assets and to provide various services easily and efficiently with minimal hassles. Being transport and traffic management at the heart of smart city technology development, much attention is required to develop intelligent solutions for various services related to vehicles, transportation, and traffic management. These include the Regional Transport

© Springer Nature Singapore Pte Ltd. 2021
T.-P. Hong et al. (Eds.): ACIIDS 2021, CCIS 1371, pp. 337–351, 2021.
https://doi.org/10.1007/978-981-16-1685-3_28

Office (RTO) services, vehicle insurance, pollution-under-control certification, and most importantly, handling of traffic violations by traffic police.

As preventive measures against crimes or traffic rules violations, traffic police often ask commuters to produce their vehicle's documents which include registration certificate, pollution certificate, insurance, etc. As an alternative to carrying physical forms of vehicle documents, few cloud-based platforms are available which allow the commuters to access them in digital form during the verification process. However, such systems suffer from a number of challenges and limitations, such as lack of trust due to centralized control, susceptibility to information tampering, the existence of different service providers for different documents and lack of interoperability among them, missing payment channels, etc. As a result, commuters may need to remember different access information for different documents on different platforms or may need to pull requests by themselves in order to make the documents available on a common platform (if supported) from their corresponding service providers. Another most serious concern, which citizens often experience, is the presence of untrusted middlemen in the system, which leads to unnecessary delays and inefficiency in availing various services.

Over the past several years, blockchain [9] has emerged as a promising technology that transforms how we share information. While blockchain is most famous for its role in facilitating the rise of digital currencies, the technology is now being tested in a broad range of applications including Insurance, Supply-Chain, Health Care, E-governance, and many more [12]. Most importantly, the support of business logic in the form of smart contracts and their execution on blockchain makes this technology powerful enough to build trusted systems within an untrusted environment. The absence of centralized authority, along with the above features, makes blockchain technology a perfect choice for solutions to those systems where multiple untrusted entities are involved.

There exists a few proposals on blockchain-based vehicle insurance [6,10] and vehicle registration [11], where the processes of vehicle registration and insurance issuance are carried out on a permissioned blockchain model. CarChain [8] is another relevant proposal that allows access to a vehicle's history (repair, ownership, insurance, and taxation) using blockchain technology as a way to improve transparency in the purchase process of used-vehicles. However, all these above systems are made operational independently with limited functionalities, which leads to a lack of interoperability among the stakeholders in a broader context. Although many developed and developing countries have their online government portals [1–3] for availing related services (such as registration of a new vehicle, application for driving license, and other vehicle-related services), the biggest drawback of these systems is the lack of transparency and trust due to their centralized architecture.

In this paper, we propose an integrated solution to vehicles' records management system using blockchain technology, where various stakeholders (vehicles owners, RTOs, insurance issuers, pollution control test centers, traffic police) can avail or provide various services (vehicle's registration, driving license issuance, insurance and pollution-under-control certificates issuance, automated insurance claim, auditable trail of vehicle's documents access and verification) in a hassle-free manner without any untrusted intermediaries. While the use of blockchain technology increases accountability, transparency, and trust in the system, this allows a cohesive integration of the proposed system with other smart city digital infrastructures easily. We develop a prototype of our proposal as a proof of concept using Hyperledger Fabric, adopting the Attribute-Based Access Control (ABAC) policy. To the best of our knowledge, this proposal is the first of its kind using blockchain technology which covers a broader perspective of vehicles and records management bringing all associated stakeholders under the same platform.

To summarize, the main contributions in this paper are:

- We propose an integrated system for availing vehicle-related services and records management by leveraging the power of blockchain technology and smart contract. We consider the real-case scenario by considering all stakeholders (citizens, RTOs, traffic police, pollution control test centers, and insurance issuers) and useful services relevant to them.
- We address the security and privacy issues by enforcing appropriate access control mechanisms in the system.
- We develop a prototype as a proof of concept using the Hyperledger Fabric blockchain platform, implementing the Attribute-Based Access Control (ABAC) policy.
- Finally, we perform an experiment to demonstrate the performance of the system. Empirical evidence on response times over the number of requests issued at various time instances are quite encouraging.

The structure of the paper is organized as follows: Sect. 2 describes our proposed blockchain-based system for vehicle-related services and records management. In Sect. 3, we present the proof of concept using Hyperledger Fabric. The experimental results are discussed in Sect. 4. The related work and a comparative analysis w.r.t. the literature is discussed in Sect. 5. Finally, in Sect. 6, we conclude our work.

2   Proposed Approach

In this section, we propose a novel blockchain-based platform for availing vehicle-related services and smart management of data relating to these services in a city. We focus on transparency, increasing accountability, and improving reliability concerning the storage, automation, and sharing of vehicle details, driving license information, insurance documents of vehicles, pollution-under-control certificates, and traffic violation data among the stakeholders. The system allows various stakeholders, such as citizens of the country, regional transport authorities, vehicle insurance companies, pollution control test centers, traffic police, to participate in the system and the platform provides crucial services such as applying for a driving license, registering a vehicle, issuing pollution-under-control certificates, filing a traffic violation, etc. Overall system components comprising of all stakeholders and several smart contracts for providing relevant services are depicted in Fig. 1.

2.1   Stakeholders

Let us briefly describe the stakeholders involved in our proposed system:

- **Citizens:** Citizens of the country are at the forefront of this system and most of the services of the system are dedicated to them.

  Here citizens refer to any person in the country holding a valid driving license or owning a vehicle. Citizens can apply for a new driving license, apply for the registration of a new vehicle they purchase, initiate an insurance issue or claim for their vehicles, and check for any traffic violations involving them.

- **Regional Transport Offices (RTOs):** The RTOs have the role of evaluating applications for driving license and vehicle registration, issuing driving licenses, and providing vehicle registration numbers. Also, RTOs have the role of transferring ownership of a vehicle from one owner to another if required.

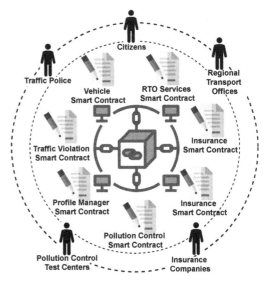

Fig. 1. Overall system components

- **Pollution Control Testing Centers (PCTCs):** The responsibility of PCTC is to test whether the vehicle complies with the prescribed emission norms and issue a pollution-under-control certificate (PUCC) for the vehicle.
- **Insurance Companies:** The insurance companies play the role to issue insurance policies for vehicles and to process insurance claims to compensate for damages that happened due to accidents. On receiving requests (either insurance issue or insurance claim) from vehicle-owners, insurance companies evaluate and approve them by following appropriate procedures.
- **Traffic Police:** The traffic policies play an important role in filing traffic violations made by citizens while driving on road. Most importantly, the system allows them to access all vehicle-related documents (ownership details, driving license information, registration and insurance certificates, PUCC, past history of the vehicle) on the proposed platform to carry out the verification process.

2.2   Citizen Services

As already stated, the proposed system provides various services to the citizens, including the application for a new driving license or for the registration of a new vehicle they purchase. The platform also facilitates citizens to issue and claim vehicle-insurances, to issue PUCC, and to view vehicle's past history on any traffic-related offenses, penalties and payment of fines. Moreover, an ownership transfer of vehicles can also be initiated by citizens with ease. The services to be availed by the citizens are depicted in Fig. 2. Let us now describe each of them.

Application for Driving License. This functionality allows any citizen to apply for a new driving license with the convenience of not visiting the RTO. All they need to do is to provide the necessary details required in the application. The supporting documents may be shared with the RTO through a peer to peer file sharing system, e.g. IPFS [5]. Once they apply for a new driving license, the designated RTO will evaluate it, and if the citizen is eligible, the RTO will issue the driving license for them. Moreover, citizens can also check for any updates regarding their application.

Application for Vehicle Registration. Citizens can apply for the registration of new vehicles that they purchase from motor-vehicle dealers. They need to provide required details along with the applications which the RTO evaluates and accordingly provides valid registration papers for the vehicles. The citizens can also check for any updates on the submitted applications.

Purchase Insurance Policy. The platform supports the issuance of vehicle-insurance. Citizens are free to choose a policy from any of the insurance companies registered on the platform. After request submission, citizens also will be able to view the status of their applications.

Initiate an Insurance Claim. The citizens may require to initiate an insurance claim to compensate for the vehicle damages if it is covered by a motor vehicle insurance policy. The system allows the citizens to upload proof images of the damages as supportive documents while applying for the claim. The peer-to-peer file-sharing system (e.g., IPFS [5]) facilitates to store the documents robustly. The citizens can also check the progress of their claims.

Apply for Pollution Testing. The system permits the citizens to apply for PUCC by choosing a nearby PCTC with a suitable visiting schedule for performing a test of their vehicle's pollution compliance. On visiting the PCTC on the scheduled date/time, the test is performed and a valid PUCC is issued against the vehicle tested.

Accessing Vehicle Profile. The citizens can check the profile of the vehicle they own. In a vehicle profile, information such as vehicle registration details, pollution certificate records, insurance policies, traffic violations, and ownership history is available. Legitimate stakeholders, e.g. traffic police, vehicle-owner, etc., will have access to this profile information for their perusal.

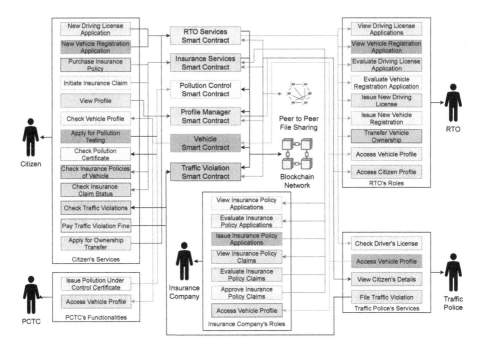

Fig. 2. Smart contracts' functionalities

Pay Traffic Violation Fine. The citizens can pay for any fines that they might have incurred by violating traffic rules. Once they pay the fines, the corresponding violations are marked as 'PAID' along with the associated payment information.

Apply for Vehicle Ownership Transfer. Citizens can apply for vehicle ownership transfer if they are selling the vehicle to someone else. Once the citizen applies for the vehicle ownership transfer, the concerned RTO will start evaluating the application to process it.

Check Profile. The citizens can view their profile (Unique Identification number (UID), address, contact details, etc.), including the list of vehicles they own and their driving licenses.

2.3   Regional Transport Offices (RTOs) Activities

RTOs are the most important stakeholders who are at the core of the system to provide crucial vehicle-related services, including document issuance, storage, safeguarding, and sharing among the stakeholders. The functionalities are depicted in Fig. 2. Let us describe them below.

Viewing and Evaluating Driving License Applications. One of the major roles of the RTO is to view and evaluate pending driving license applications submitted through the platform. It is the responsibility of the concerned RTO to properly evaluate the applications and to issue valid driving licenses if the applicants are found eligible. Note that the evaluation may require fulfilling different test criteria depending on the country's own rules and regulations. Once licenses are issued, they will be added to the corresponding citizen's profile.

Viewing and Evaluating Vehicle Registration Applications. Once a citizen applies for the registration of a vehicle, the RTO needs to evaluate the citizen- and vehicle-details. Accordingly, RTO issues a valid registration certificate to the concerned vehicle and updates the status of the application. The process may go through multiple phases in case any discrepancies are found or any corrections are required from the citizen.

Accessing Vehicle Profile. The RTOs are entitled to access records of any vehicle profile whenever required for official business. As already mentioned above, the vehicle profile includes information such as registration details, pollution certificate records, insurance policies, traffic violations, and ownership history.

Evaluate and Process Transfer Ownership. Apart from issuing the driving license and registering vehicles, the RTO also has the role of transferring ownership of any vehicle from one owner to another. Once the parties agree on the legal terms and payments, the current vehicle-owner will apply for the ownership transfer. Once the application is submitted, the RTO will evaluate the application and process the ownership change.

2.4   Roles of Insurance Companies

In this section, we describe in detail the role of insurance companies on the platform. Figure 2 depicts it as a block diagram.

View, Evaluate and Issue Insurance Policy Purchase Applications. On receiving a policy purchase request from a vehicle-owner, the corresponding insurance company accesses the vehicle's details and its history to evaluate the request and determines the policy-amount accordingly. Once the payment is done, the insurance company issues the requested policy and adds it to the vehicle's profile for which the policy is purchased.

View, Evaluate and Approve Insurance Claims. The insurance company has the role of evaluating any new insurance claims filed by citizens. On receiving such claim-requests along with the proofs uploaded by the claimants, the designated authorities of the insurance company perform physical verification and release the insurance money based on the evaluation. Observe that, once they start evaluating any claim, they update the status of the progress to make the claimants aware of this.

Access Vehicle Profile. The insurance companies should have access to the vehicle's profile when they receive requests for insurance issues or claim. This requires permission from the vehicle-owner as well. As already mentioned, a vehicle profile consists of information such as vehicle details, pollution certificate records, insurance policies, traffic violations, and ownership history.

2.5   Pollution Control Testing Centers (PCTCs) Activities

Here we describe the functionalities of Pollution Control Testing Centers (PCTCs) on the platform, depicted in Fig. 2.

Issue Pollution-under-Control Certificates. The primary function of PCTCs is to issue valid PUCC to any vehicle that tests for emissions at their test center. Once they issue PUCC to the concerned vehicle, the certificate is added to the vehicle's profile so that legitimate stakeholders can check for the latest PUCC of the vehicle.

Access Vehicle Profile. Like an insurance company, PCTCs also have access to vehicle profiles if the vehicle-owner gives permission.

2.6   Activities of Traffic Police

In this section, we describe in detail the activities of the traffic police. The list of supported functionalities is depicted in Fig. 2.

Verification of Citizen Profiles. On suspicion, the traffic police might need to verify the profiles of citizens, where the citizen is a vehicle driver or a vehicle owner. The traffic police have full access to the citizens' profiles.

Accessing Vehicle Profiles. Like citizen's profile verification, the traffic police are also entitled to access and verify the records of any vehicle profile that they need for official police business. This is worthwhile to note that, as the vehicle profile contains all vehicle-related information (such as registration details, pollution certificate records, insurance policies, traffic violations, and ownership history), the vehicle-owner does not need to carry physical copies of the documents or the traffic police do not need to access documents from multiple untrusted platforms.

File Traffic Violation. The primary role of the traffic police is to file any traffic violation that takes place. In order to do this, they invoke the corresponding function providing required inputs, such as vehicle registration number and driving license along with appropriate violation-proof. Once they submit the traffic violation reports, the concerned drivers and vehicle owners will be notified about this and the amount of fine issued.

2.7   Access Control

Since access control to critical information on the platform is a crucial aspect, we consider Attribute-Based Access Control (ABAC) which considers the attribute of subjects, objects, permissions, and environment [7]. Inclusion of the notion of the environment to the model allows specification of dynamic access rules, which is one of the unique characteristics of ABAC that sets it apart from traditional access control models like DAC (Discretionary Access Control), MAC (Mandatory Access Control), and RBAC (Role-Based Access Control). It regulates whether to grant access to the user by determining whether the request contains valid attributes. The policies can be updated in proportion to the circumstances by appending or removing the attributes. According to NIST specifications [7], various components of ABAC are:

U :	Set of users
O :	Set of objects
E :	Set of environmental conditions
$\overrightarrow{UA}$:	$\{a_1^u, a_2^u, a_3^u, ..., a_n^u\}$ is an ordered list of user attributes
$\overrightarrow{OA}$:	$\{a_1^o, a_2^o, a_3^o, ..., a_m^o\}$ is an ordered list of object attributes
$\overrightarrow{EA}$:	$\{a_1^e, a_2^e, a_3^e, ..., a_p^e\}$ is an ordered list of environmental attributes
V_i^x :	$\{v_{i1}^x, v_{i2}^x, v_{i3}^x, ..., v_{ik}^x\}$ is the set of values that can be taken up by the attribute a_i^x, where $x \in \{u, o, e\}$
OP :	Set of allowable operations on the object
P :	$\{r_1, r_2, r_3, ..., r_l\}$ is a set of rules that governs the access to an object depending on the values of the user- and the object-attributes and the prevalent environmental conditions

Let us provide an example of ABAC policy for our proposed system. Consider a set of users $U = \{u_1, u_2, u_3, u_4, u_5\}$, a set of objects $O = \{o_1, o_2, o_3\}$, a set of allowed operations $OP = \{create, read, write\}$ and a set of environmental conditions $E = \{e_1, e_2\}$. Consider the following attributes which are associated with users, objects and environment conditions respectively: (1) $\overrightarrow{UA} = \{user-type, department, designation\}$, (2) $\overrightarrow{OA} = \{document-type, expiry-date\}$, and (3) $\overrightarrow{EA} = \{date, state\}$. Suppose the values that can be taken by the attributes are:

Algorithm 1: Algorithm CheckPolicy

Data: ABAC Policy P, ABAC Request R
Result: Allow or Deny

1   $\langle\, \overrightarrow{\mathsf{UA}}.val,\ \overrightarrow{\mathsf{OA}}.val,\ \overrightarrow{\mathsf{EA}}.val,\ action\, \rangle \leftarrow R$
2   permission $\leftarrow true$
3   **for** $rule \in P$ **do**
4   Let $rule$ be of the form $\langle\, f_1(\overrightarrow{\mathsf{UA}},\ \overrightarrow{v}^{\,u}),\ f_2(\overrightarrow{\mathsf{OA}},\ \overrightarrow{v}^{\,o}),\ f_3(\overrightarrow{\mathsf{EA}},\ \overrightarrow{v}^{\,e}),\ op\, \rangle.$
5   **if** $(\overrightarrow{\mathsf{UA}}.val \nvdash f_1(\overrightarrow{\mathsf{UA}},\ \overrightarrow{v}^{\,u}))$ **OR** $(\overrightarrow{\mathsf{UO}}.val \nvdash f_2(\overrightarrow{\mathsf{OA}},\ \overrightarrow{v}^{\,o}))$ **OR** $(\overrightarrow{\mathsf{EA}}.val \nvdash f_3(\overrightarrow{\mathsf{EA}},\ \overrightarrow{v}^{\,e}))$
     then
6   permission $= false$;
7   break;
8   **end**
9   **for** operation $\in action$ **do**
10   **if** operation $\notin op$ **then**
11   permission $= false$;
12   break;
13   **end**
14   **end**
15 **end**
16 Return permission;

- $V_1^u = \{citizen,\ rto\text{-}personnel,\ insurance\text{-}personnel,\ traffic\text{-}police\}$, set of possible values for user attribute 'user-type'.
- $V_2^u = \{license\text{-}dept,\ testing\text{-}dept,\ insurance\text{-}sales,\ insurance\text{-}inspection,\ pollution\text{-}dept,\ law\text{-}enforcement\}$, set of possible values for user attribute 'department'.
- $V_3^u = \{traffic\text{-}sergeant,\ scrutiny\text{-}officer,\ insurance\text{-}inspector,\ pollution\text{-}examiner\}$, set of possible values for user attribute 'designation'.
- $V_1^o = \{driving\text{-}license,\ registration\text{-}cert,\ insurance,\ pollution\text{-}cert\}$, set of possible values for object attribute 'document-type'.
- $V_2^o = \{mm/yy \mid mm \in \{00,\ \dots,\ 12\},\ yy > 80\}$, set of possible values for object attribute 'expiry-date'.
- $V_1^e = \{today\}$, set of possible values for environment attribute 'date'.
- $V_1^e = \{driving,\ non\text{-}driving\}$, set of possible values for object attribute 'state'.

Few among many other access rules as a part of the policy P are:

- *rule-1*: Traffic-police with designation traffic-sergeant under law-enforcement department can read any non-expired documents for a car under driving, which is expressed as: "user-type = *traffic-police* ∧ department = *law-enforcement* ∧ designation = *traffic-sergeant* ∧ document-type = * ∧ date = *today* ∧ expiry-date> date ∧ state=*driving* ∧ OP = *read*", where * denotes any attribute value.
- *rule-2*: RTO personnel with designation scrutiny-officer under license-dept can create, read and write driving-license of any citizens anytime. This can be formally expressed as: "user-type = *rto-personnel* ∧ department = *license-dept* ∧ designation = *scrutiny-officer* ∧ document-type = *driving-license* ∧ expiry-date= * ∧ date = *today* ∧ state = * ∧ OP = *"".

Algorithm 1 depicts the overall algorithmic steps of Attribute-Based Access Control Policy adopted in our proposed system. The access request R by a user contains an ordered list of values corresponding to the attributes in $\overrightarrow{UA}$, $\overrightarrow{OA}$ and $\overrightarrow{EA}$ (denoted by $\overrightarrow{UA}.val$, $\overrightarrow{OA}.val$ and $\overrightarrow{EA}.val$ respectively) and a set of actions (denoted by $action$) which user wants to perform. Given a rule in policy P which is expressed in a logical expression AND-ed by three subexpressions $f_1(\overrightarrow{UA}, \overrightarrow{v}^u))$, $f_2(\overrightarrow{OA}, \overrightarrow{v}^o))$ and $f_3(\overrightarrow{EA}, \overrightarrow{v}^e))$ over the set of attributes and their permissible values, the steps 5–8 check whether the values $\overrightarrow{UA}.val$, $\overrightarrow{EA}.val$ and $\overrightarrow{EA}.val$ satisfy (denoted by $\models$) the rule and the permission will be decided accordingly. On success of the above verification, steps 9–14 check if all operations under the requested actions '$action$' conform with 'op' in the specified rule.

3 Proof of Concept

This section describes a prototype implementation of the proposed system. The detailed source-code of our implementation is available on GitHub at https://github.com/mukherjeearnab/vehicle-platform-blockchain. Since our system is based on a permissioned blockchain model, we use the Hyperledger Fabric [4] blockchain platform for the implementation. The overall architecture is depicted in Fig. 3. The whole system comprises of three parts: Hyperledger Fabric blockchain network (back-end), Node.JS REST API[1] (middle-ware), React JS web-based GUI[2] (front-end).

We establish a blockchain network on the Hyperledger Fabric platform, considering five organizations: *Citizens, Regional Transport Office, Insurance Companies, Pollution Control Testing Center*, and *Traffic Police*. Our implementation considers six Smart Contracts (SC), namely **RTO Services SC, Vehicle SC, Insurance SC, Pollution SC, Profile Manager SC** and **Traffic Violation SC**, which execute all the functionalities of the system. We implement ABAC (Attribute-Based Access Control) to manage the access rights to critical information on the blockchain, assuming all five organizations connected through a single channel instead of multiple channels just to minimize the complexity of the implementation without loss of generality. In this prototype, we provide every organization with one Peer node, one Fabric CA (Certificate Authority) node and one instance of CouchDB state database. The network consists of one Orderer node using solo to serve the purpose of Ordering Service. For user authentication on the platform, we opt for token-based authentication. In particular, we implement a Node.JS and MongoDB authentication server which performs the task of authenticating issuing JSON Web Token (JWT). Moreover, we use an instance of InterPlanetary File System (IPFS) [5] as our go-to peer to peer file-sharing network to store data files associated with various services, such as copies of certificates, damage-proofs during an insurance claim, etc.

---

[1] http://www.ics.uci.edu/~fielding/pubs/dissertation/rest_arch_style.htm.
[2] https://reactjs.org/.

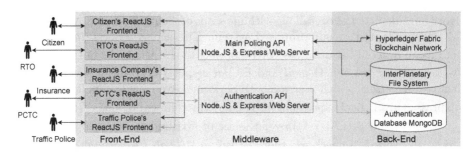

Fig. 3. Prototype architecture

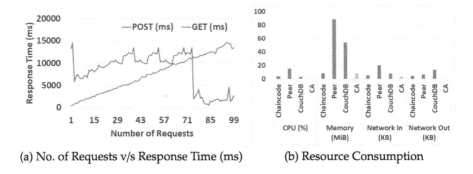

(a) No. of Requests v/s Response Time (ms)   (b) Resource Consumption

Fig. 4. Benchmark results

4   Experimental Results

In this section, we discuss the results of our experiments to demonstrate the performance of our system. Here, in particular, we assess the response times over the number of requests issued at various time instances to the blockchain network. The experiment is conducted on a machine with an AMD Phenom(TM) II X6 1090T clocked at 3.6 GHz, 8 GB 1333 MHz dual-channel memory running Canonical Ubuntu Server 20.04 LTS. We sent HTTP requests using Apache JMeter to the system prototype, where load, ramp-up period, and loop count were set to 4000 threads, 500 s, and 1 respectively. Also, we used solo as the Ordering Service where BatchTimeout was set to 2 and MaxMessageCount was set to 10. During the experiment, we noted that the CPU utilization and memory usage reached a maximum of 99% and 72% respectively.

Figure 4(a) depicts the relation between the number of requests and response time (in milliseconds) for both GET requests and POST requests on the platform. The X-axis represents the number of requests issued at an instance and Y-axis represents the response time in milliseconds. Observe that, for GET requests, the response time increases with the increase in the number of requests. However, due to the limitation of the computation power of our computer system on which the experiment is conducted, we consider only a single peer for each

Table 1. Hyperledger Caliper benchmark summary.

Name	Succ.	Fail	Send rate (TPS)	Max latency (s)	Min latency (Sec)	Avg. latency (s)	Throughput (TPS)
Read	100	0	30.6	2.75	0.83	1.91	22.9
Query	100	0	100.3	3.56	1.08	1.54	75.9
Write	67	33	25.6	15.4	3.84	9.54	5.46

organization and this leads to a failure when the number of requests crossed 74. Observe that all POST requests change the state of the blockchain ledger, hence they take more time on average than the GET requests. Notably, the experiment demonstrates that our system showed its ability to handle all requests by responding with HTTP status indicating successful request submission up to a threshold limit of 74 requests at an instance.

Apart from the stress-test using Apache JMeter, we also conducted a performance test on our blockchain network using Hyperledger Caliper. Table 1 depicts the performance summary under three different operations, indicating success rate, transaction send rate, transaction latency, and transaction throughput of the blockchain network, where TPS denotes transaction per second. The resource consumption (CPU, Memory, Network IO) by the system components are depicted in Fig. 4(b). Note that CPU utilization of the Certificate Authorities (CA) during the benchmark was zero units.

5 Related Work and Comparative Analysis

Many developed and developing countries have their online government portals [1–3] for availing vehicle-related services (such as registration of a new vehicle, application for driving license, and other services). The biggest drawback of these systems is the lack of transparency, reliability, and trust due to their centralized architecture. Moreover, these centralized systems are always prone to various attacks and information tampering. To overcome this, we leverage the power of immutable and tamper-proof distributed ledger technology which eliminates the need for untrusted centralized authority and builds a firm pillar of trust in the system.

In recent years, few proposals on blockchain-based vehicle insurance systems are proposed [6,10]. The authors in [10] proposed a framework for auto-insurance claims and adjudication using a permissioned blockchain model with two partitions to share the information on a need to know basis. The system acts as a vehicular forensic system for connected and automated vehicles that facilitates the collection of relevant evidence from potentially liable entities. In [6], the authors proposed an insurance record system that includes all aspects of insurance transactions improving the experience around providing insurance, especially as evidence in the event of a dispute. The process of vehicle registration proposed in [11] is carried out on a permissioned blockchain model. Besides vehicle registration service, the proposed system supports few other functionalities including car-ownership transfer, giving a car on a lease, car-ownership as a guarantee to a creditor, and vehicle seizure. CarChain [8], on the other hand, proposed a system of vehicle history reporting, where all the past history of a vehicle are recorded on the blockchain making their access more easy and transparent during the purchasing of used-vehicles.

Observe that all the above systems are designed with limited functionalities aiming to fulfill specific objectives only. For example, [6,10] consider only insurance-related activities, whereas [11] deals with car registration and ownership. Similarly, [8] stores the history of the car including the owners, accidents, and repairs. Unlike existing systems, our proposal is more generic in the sense that it considers all associated stakeholders relevant to both vehicle- and traffic-management and provides a complete set of functionalities including insurance issuance/claim. The support of access-control, in addition, provides selective access rights of digital contents to legitimate stakeholders only.

6   Conclusion

This paper proposes an integrated platform for availing vehicle-related services and smart management of data relating to these services in a city. Transparency, accountability, reliability, and trust concerning the storage, automation, and sharing of vehicles' profile data are established by exploiting the power of blockchain technology. We overcome several challenges and limitations of the existing centralized system by providing seamless interactions among the stakeholders. The prototype is implemented using Hyperledger Fabric and the experimental results on response times for several requests issued at various time instances are encouraging. Currently, we are in the process of extending it for adoption to a smart city solution, bringing a smart policing system and other traffic management activities also on board.

References

1. Apply for your first provisional driving licence—gov.uk. https://www.gov.uk/apply-first-provisional-driving-licence
2. Driving in Canada. https://www.canada.ca/en/immigration-refugees-citizenship/services/new-immigrants/new-life-canada/driving.html
3. Sarathi service. https://sarathi.parivahan.gov.in/sarathiservicecov5/stateSelection.do?stCd=WB
4. Androulaki, E., et al.: Hyperledger fabric: a distributed operating system for permissioned blockchains. arXiv-1801 (2018)
5. Benet, J.: Ipfs-content addressed, versioned, p2p file system. arXiv (2014)
6. Demir, M., Turetken, O., Ferworn, A.: Blockchain based transparent vehicle insurance management. In: 2019 Sixth International Conference on Software Defined Systems (SDS), pp. 213–220 (2019)
7. Hu, V.C., et al.: Guide to attribute based access control (ABAC) definition and considerations (draft), vol. 800, p. 162. NIST special publication (2013)
8. Masoud, M., Jaradat, Y., Jannoud, I., Zaidan, D.: Carchain: a novel public blockchain-based used motor vehicle history reporting system, April 2019

9. Nakamoto, S.: Bitcoin: a peer-to-peer electronic cash system (2008)
10. Oham, C., Jurdak, R., Kanhere, S.S., Dorri, A., Jha, S.: B-fica: blockchain based framework for auto-insurance claim and adjudication. In: IEEE International Conference on iThings and GreenCom and CPSCom and SmartData, pp. 1171–1180 (2018)
11. Rosado, T., Vasconcelos, A., Correia, M.: A Blockchain Use Case for Car Registration, pp. 205–234, November 2019
12. Xie, J., et al.: A survey of blockchain technology applied to smart cities: research issues and challenges. IEEE Commun. Surv. Tutor. **21**(3), 2794–2830 (2019)

Increase Driver's Comfort Using Multi-population Evolutionary Algorithm

Szymon Tengler[(✉)] [iD] and Kornel Warwas [iD]

Department of Computer Science and Engineering,
University of Bielsko-Biala, Bielsko-Biala, Poland
`{stengler,kwarwas}@ath.bielsko.pl`

Abstract. An application of own algorithm for reduction optimization calculation time is presented in the paper. The algorithm is called Distributed Multi-Population Evolutionary Algorithm and uses a genetic algorithm with real-coded genes in the chromosome. The client-server architecture was used for processing lots of populations with a simultaneous data exchange between the populations. Each generation consists of standard genetic operators: a natural selection, one-point crossing, uniform mutation and a data exchange within the computational units. This method was used for selecting values of damping coefficients of a driver's seat sub-assembly to improve driving comfort. Numerical results of a vehicle driving with different variants of velocity and obstacles are presented in the paper.

Keywords: Optimization · Active damping · Evolutionary algorithm · Distributed and parallel computing · Vehicle modelling

1 Introduction

The Multi-Population Evolutionary Algorithms are widely used in many fields of science and industry. For instance in work [7] a multi-population evolutionary algorithm was used for locating buildings playing a key role in the climate change mitigation. The similar optimization algorithm was used to solve a problem of an appropriate deployment of wireless cameras for monitoring human activity [17], including the classical capacitated vehicle routing problem (CVRP) [22]. As a further example in paper [26] the authors used the Multi-Population Genetic Algorithm for Unmanned Aerial Vehicles Path Planning. In paper [33] this method was used for optimizing the Train-Set Circulation Plan Problem, and in [8] for solving Dynamic Shortest Path Routing Problems in Mobile Ad Hoc Networks. Regardless of a field authors agree that use of many populations with emigration of the best individuals after each generation, contributes significantly to productivity growth and it enables to obtain better matching of the resultant population. These methods are less susceptible to the local minima and a selection of a start point, although their weak point is that they indicate only an approximate optimum value [20].

© Springer Nature Singapore Pte Ltd. 2021
T.-P. Hong et al. (Eds.): ACIIDS 2021, CCIS 1371, pp. 352–364, 2021.
https://doi.org/10.1007/978-981-16-1685-3_29

Driving comfort of a driver is one of the key elements in this scope. It has an influence on a level of operator tiredness and an effective time in a vehicle [24]. A lot of research focused on interaction between a human being and a seat [9, 19]. They presented works on designing the best possible shapes of a driver's seat. In this case tools like CAD CAE and FEM were used. A comfort improvement is usualy connected with modifications of the vehicle sub-assemblies [14, 23]. In this case researchers focus on changes and improvements in the vehicle elements such as a suspension rather than including the human body interaction, and those systems are commonly called an active or semi-active suspension [23, 25] and also in [2, 13, 27]. Works in this scope usually deal with reducing the vibrations present in those sub-assemblies. It can be achieved in different ways, and use of a supporting system is one of them. Those systems have to be calibrated properly to operate correctly. Constructing a test stand is usually connected with high costs and in the first stage of tests it is not justified. An application of virtual models and computer modelling allows reducing the costs and verifying basic assumptions. Optimization is particularly useful in this scope [6, 16, 31]. A virtual model and optimization results are grounds for using of the dedicated drivers of the existing systems or implementing new active elements into the standard equipment of a vehicle in a next stage. It is difficult to select an appropriate optimization method for a specific issue and it usually requires knowledge of a domain expert.

In this paper there was used a distributed variant of Multi-Population Evolutionary Algorithm (called DMPEA) where calculations are being made in a parallel way using computer units in the cluster nodes. An application of this method for solving the issue of selecting the optimum values of damping of a special vehicle seat while driving through the obstacle in a form of a speed bump on basis of proprietary computer programs in the heterogeneous environment (C++/C#), is presented. Fundamentals of the mathematical model of the object in question are presented, the optimization problem is defined and the computation results are presented for a different number of individuals in the population.

The computation results were compared with the results obtained by the classical genetic algorithm. The authors of this article described a similar subject in [32]. Nevertheless, there were other road maneuvers with the use of a non-evolutionary gradient optimization method.

2 Model of Investigations

A rescue fire-fighting vehicle was an object of the research. It is defined as a system of eleven rigid bodies: a frame, a cabin, a semitrailer, an engine, a front and rear axle, a driver's seat and wheels (Fig. 1).

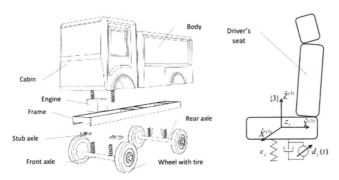

Fig. 1. The sub-assemblies of analyzed vehicle

To determine a location (position and orientation) of these bodies, joint coordinates and homogenous transformations method were used [28]. A motion of each body, is described in relation to a preceding body. The joint coordinates vector of the vehicle can be written in the following form:

$$\mathbf{q} = \begin{bmatrix} \mathbf{q}_f^T & \mathbf{q}_c^T & \mathbf{q}_s^T & \mathbf{q}_b^T & \mathbf{q}_{a,1}^T \\ \mathbf{q}_{a,2}^T & \mathbf{q}_{w,1}^T & \mathbf{q}_{w,2}^T & \mathbf{q}_{w,3}^T & \mathbf{q}_{w,4}^T \end{bmatrix}^T \tag{1}$$

where:

- $\mathbf{q}_f = \begin{bmatrix} x_f & y_f & z_f & \psi_f & \theta_f & \varphi_f \end{bmatrix}^T$ - vector of generalized coordinates of frame,
- $\mathbf{q}_c = \begin{bmatrix} \theta_c \end{bmatrix}$ - vector of generalized coordinates of cabin,
- $\mathbf{q}_s = \begin{bmatrix} z_s \end{bmatrix}$ - vector of generalized coordinates of driver's seat,
- $\mathbf{q}_b = \begin{bmatrix} \varnothing \end{bmatrix}$ - vector of generalized coordinates of vehicle body fixed to frame,
- $\mathbf{q}_{sa,i} = \begin{bmatrix} \psi_{sa,i} \end{bmatrix}$ - vector of generalized coordinates of stub axles ($i = 1, 2$),
- $\mathbf{q}_{a,i} = \begin{bmatrix} z_{a,i} & \varphi_{a,i} \end{bmatrix}^T$ - vector of generalized coordinates of front and rear axles ($i = 1, 2$),
- $\mathbf{q}_{w,i} = \begin{bmatrix} \psi_{w,i} & \theta_{w,i} \end{bmatrix}^T$ - vector of generalized coordinates of front wheels ($i = 1, 2$),
- $\mathbf{q}_{w,i} = \begin{bmatrix} \theta_{w,i} \end{bmatrix}$ - vector of generalized coordinates of front wheels ($i = 3, 4$),
- x_i, y_i, z_i - mass center coordinates of the i-th body,
- $\psi_i, \theta_i, \varphi_i$ - rotation angles of the i-th body.

The Lagrange equations of the second kind [3, 28, 29] was used for formulation of vehicle motion. It can be written in the general form:

$$\mathbf{A}^{(i)}\ddot{\mathbf{q}}^{(i)} = \mathbf{f}^{(i)} \tag{2}$$

where: $\mathbf{A}^{(i)}$ - mass matrix, $\mathbf{f}^{(i)}$ - vector of external, Coriolis, centrifugal and gravity forces.

The equations of motion of mentioned sub-assemblies can be described in fallowing form:

− sub-system 1 (frame - F, front axle - AF, stub axle - SA, front wheels - WF)

$$
\mathbf{A}^{(1)} = \begin{bmatrix} \mathbf{A}_{f,f}^{(1)} & \mathbf{A}_{f,af}^{(1)} & \mathbf{A}_{f,sa}^{(1)} & \mathbf{A}_{f,wf}^{(1)} \\ \mathbf{A}_{af,f}^{(1)} & \mathbf{A}_{af,af}^{(1)} & \mathbf{A}_{af,sa}^{(1)} & \mathbf{A}_{af,wf}^{(1)} \\ \mathbf{A}_{sa,f}^{(1)} & \mathbf{A}_{sa,af}^{(1)} & \mathbf{A}_{sa,sa}^{(1)} & \mathbf{A}_{sa,wf}^{(1)} \\ \mathbf{A}_{wf,f}^{(1)} & \mathbf{A}_{wf,af}^{(1)} & \mathbf{A}_{wf,sa}^{(1)} & \mathbf{A}_{wf,wf}^{(1)} \end{bmatrix}, \quad \mathbf{q}^{(1)} = \begin{bmatrix} \tilde{\mathbf{q}}^{(1)} \\ \tilde{\mathbf{q}}^{(4)} \\ \tilde{\mathbf{q}}_{sa}^{(1)} \\ \tilde{\mathbf{q}}_{wf}^{(1)} \end{bmatrix}, \quad \mathbf{f}^{(1)} = \begin{bmatrix} \mathbf{f}_f^{(1)} \\ \mathbf{f}_{af}^{(1)} \\ \mathbf{f}_{sa}^{(1)} \\ \mathbf{f}_{wf}^{(1)} \end{bmatrix},
$$

where: $\mathbf{q}_{sa}^{(1)} = \begin{bmatrix} \tilde{\psi}^{(6)} \\ \tilde{\psi}^{(7)} \end{bmatrix}$, $\mathbf{q}_{wf}^{(1)} = \begin{bmatrix} \tilde{\theta}^{(1)} \\ \tilde{\theta}^{(2)} \end{bmatrix}$,

− sub-system 2 (frame - F, rear axle - ar, rear wheels - wr)

$$
\mathbf{A}^{(2)} = \begin{bmatrix} \mathbf{A}_{f,f}^{(2)} & \mathbf{A}_{f,ar}^{(2)} & \mathbf{A}_{f,wr}^{(2)} \\ \mathbf{A}_{ar,f}^{(2)} & \mathbf{A}_{ar,ar}^{(2)} & \mathbf{A}_{ar,wr}^{(2)} \\ \mathbf{A}_{wr,f}^{(2)} & \mathbf{A}_{wr,ar}^{(2)} & \mathbf{A}_{wr,wr}^{(2)} \end{bmatrix}, \quad \mathbf{q}^{(2)} = \begin{bmatrix} \tilde{\mathbf{q}}^{(1)} \\ \tilde{\mathbf{q}}^{(5)} \\ \tilde{\mathbf{q}}_{wr}^{(2)} \end{bmatrix}, \quad \mathbf{f}^{(1)} = \begin{bmatrix} \mathbf{f}_f^{(2)} \\ \mathbf{f}_{ar}^{(2)} \\ \mathbf{f}_{wr}^{(2)} \end{bmatrix},
$$

where $\mathbf{q}_{wr}^{(2)} = \begin{bmatrix} \tilde{\theta}^{(3)} \\ \tilde{\theta}^{(4)} \end{bmatrix}$,

− sub-system 3 (frame - F, cabin - C, driver seat - S)

$$
\mathbf{A}^{(2)} = \begin{bmatrix} \mathbf{A}_{f,f}^{(3)} & \mathbf{A}_{f,C}^{(3)} & \mathbf{A}_{f,S}^{(3)} \\ \mathbf{A}_{C,f}^{(3)} & \mathbf{A}_{C,C}^{(3)} & \mathbf{A}_{C,S}^{(3)} \\ \mathbf{A}_{S,f}^{(3)} & \mathbf{A}_{S,C}^{(3)} & \mathbf{A}_{S,S}^{(3)} \end{bmatrix}, \quad \mathbf{q}^{(3)} = \begin{bmatrix} \tilde{\mathbf{q}}^{(1)} \\ \tilde{\mathbf{q}}^{(2)} \\ \tilde{\mathbf{q}}^{(3)} \end{bmatrix}, \quad \mathbf{f}^{(1)} = \begin{bmatrix} \mathbf{f}_F^{(3)} \\ \mathbf{f}_C^{(3)} \\ \mathbf{f}_S^{(3)} \end{bmatrix},
$$

Taking these formulations into account the dynamic equations of motion of the whole vehicle can be written as:

$$
\begin{cases} \mathbf{A}\ddot{\mathbf{q}} - \boldsymbol{\Phi}\mathbf{r} = \mathbf{f} \\ \boldsymbol{\Phi}^T \ddot{\mathbf{q}} = \mathbf{w} \end{cases} \tag{3}
$$

where:

− $\mathbf{A}$ - mass matrix,
− $\mathbf{q}$ - vector of generalized coordinates of the system,
− $\boldsymbol{\Phi} = \begin{bmatrix} 0\,0\,0\,1\,0\,0\,0\,0\,0 \\ 0\,0\,1\,0\,0\,0\,0\,0\,0 \end{bmatrix}^T$ - constraints matrix,
− $\mathbf{f}$ - vector of external, Coriolis and centrifugal forces,
− $\mathbf{r} = \begin{bmatrix} r_1 & r_2 \end{bmatrix}^T$ - vector of unknown constraint reactions corresponding to torques acting on the stub axles connected with the wheels,

$-\mathbf{w} = \begin{bmatrix} \ddot{\psi}_{w,1} & \ddot{\psi}_{w,2} \end{bmatrix}^T$ - vector of right sides of constraint equations,

A procedure for detailed formulation of Eq. (3) with a description of elements in matrix **A** and vector **f** is presented in [12, 28, 29].The road reaction forces were obtained by using Fiala tire model [10].

3   Optimization Problem

As mentioned in the previous chapter the main goal of optimization process was to select appropriate vehicle parameters to improve a driver's comfort. The vector of decisive variables contains damping coefficients of the vehicle's seat in the discrete time stamps was defined as:

$$\mathbf{d} = \begin{bmatrix} d_1 & \dots & d_i & \dots & d_{n_d} \end{bmatrix}^T, \tag{4}$$

where: d_i - value of damping coefficient of vehicle seat, n_d - number of time stamps.

In order to obtain a continuous course of decisive variables a spline functions of the first degree were used. During optimization process following inequality constraints:

$$d_{min} \leq d_i \leq d_{max}, \tag{5}$$

and also the external penalty function of following form were considered [20, 21]:

$$\zeta_i(\mathbf{d}) = \begin{cases} 0 & \text{for } g_i(\mathbf{d}) \leq 0 \\ C_{1,i} e^{C_{2,i} g_i(\mathbf{d})} & \text{for } g_i(\mathbf{d}) > 0 \end{cases} \tag{6}$$

where:real.

- $g_i(\mathbf{d})$ for $i = 1, \dots, n_g$ - the inequality constraint defined based on (5),
- n_g - inequality constraints number,
- $C_{1,i}, C_{2,i}$ - empiricall weights.

Finally, the goal function can be written:

$$\Omega(\mathbf{d}, \ddot{\mathbf{q}}) = C \cdot \text{RMQ} + \sum_{i=1}^{n_g} \zeta_i(\mathbf{d}) \rightarrow min \tag{7}$$

where:

- C - empiricall weight,
- $\text{RMQ} = \dfrac{\sqrt[4]{\int_0^{t_e} [\ddot{z}_s(t)]^4 dt}}{t_e}$, the Root Mean Quad of the driver's seat acceleration,
- t_e - duration of simulation.

4   Optimization Algorithm

The problem was solved by the Evalutionary Algorithm (EA) and Distributed Multi-population Evolutionary Algorithm (DMPEA). In both methods the real-coded genes in the chromosome was used [18]. An additional, selection one-point crossover and uniform mutation have been used as genetic operators [18, 30]. Evolution methods as opposed to classical methods are ususaly used to find the global extreme in the state space. Classical methods also depend strongly on a starting point. An important element deciding about the selection of the specific optimization method is its convergence. Since pseudo-random numbers are used the computational intelligence methods should be executed a few times to achieve the desired solution accuracy. It is connected with prolongation of computation time which in the case of the dynamic optimization where in each step of the optimization procedure the dynamic equations of motion should be integrated in the entire time interval, is usually long. An application of parallel and distributed computing may shorten computing time. Since there is an easy access to the computing units having the multi-core processors the parallel computing can be made on the common use units without using the dedicated computing centers. The discussed method of the modification of the classic genetic algorithm consists in using resources of a single unit for parallel computing in which calculations of many populations making parallel computing are made. The populations after each generation exchange the data with each other in a form of the best individuals (Fig. 2).

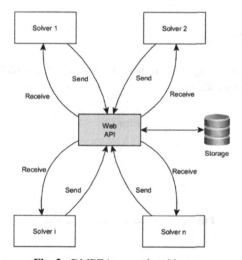

Fig. 2. DMPEA general architecture

Such an approach allows better matching of the population to be obtained in each computing node and is an additional form of entering a new genetic material into the population. From the genetic point of view the presented method can be called a new additional genetic operator being a complement of the classical operators (Fig. 3).

In the presented approach the computations can be moved to the additional computing units being the nodes of the computing cluster using possibilities of distributed

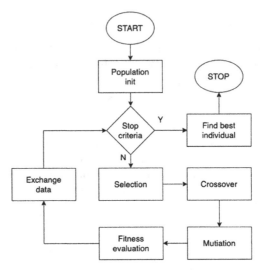

Fig. 3. DMPEA flowchart

computing. In this article a comparison of the genetic algorithm in the classical form and its variation (DMPEA) activated on one computing unit with use of lots of instances of the computing application is performed.

5   Calculation Results

A road shape as a speed bump (0.06 m height) over which a vehicle was driven with a constant velocity is presented in Fig. 4.

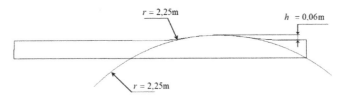

Fig. 4. The road with the bump shape

This speed bump was used for performing two different maneuvers presented in Fig. 5 and 6.

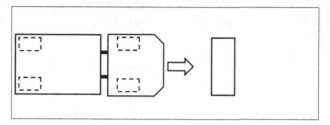

Fig. 5. Location of the velocity bumps in maneuver 1

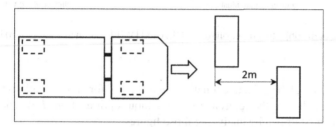

Fig. 6. Location of the velocity bumps in maneuver 2

Constant velocity was equal to 10 km/h. It was provided by PID regulator controlling the driving torque moments [28]. Two manoeuvers were taken into account – the first maneuver lasted for 3 s and second lasted for 2.6 s. The all required vehicle physical parameters to perform simulations were taken from work [28]. The driver model was used in calculations as additional weight of 80 kg, and it was joined with the driver's seat. The Runge-Kutta method of 4-th order with a constant step [21] was used to integrate equations of motion. The inequality constraints (the minimal and maximal values of the damping coefficients) were assumed as: $d_{z,min} = 10$ Ns/m and $d_{z,max} = 4000$ Ns/m, respectively. The mathematical model, the mechanism of generating dynamic equations of motion, appropriate integration methods and optimization algorithms were recorded in the proprietary computer program in C++ using the techniques of object-oriented programming. The server was also storage of the data made in technology ASP.NET Web API with use of C# programming language. The calculations were made on one computing cluster node with the use of two parallel activated instances of an application performing optimization computing and sending the results to the global store. The calculations were made for a variable number of the individuals in the population from 10 to 50 including multi start (10 times per each number of individuials and the best one was choosen). The DMPEA computing results were compared with the results obtained by the classical evolutionary algorithm [30].

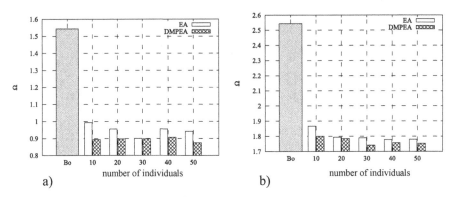

Fig. 7. Values of the objective function for both EA and DMPEA a) in maneuver 1, b) in maneuver 2

In Fig. 7 values of the objective function are presented for the calculations by methods EA, DMPEA, including the reference value without optimization (*Bo*). The following conclusions can be drawn by analyzing those figures:

- optimization with use of the DMPEA method allowed obtaining a lower value of the objective function than in the case of EA use,
- an application of the DMPEA method even on the single node of the computing cluster, due to use of the parallel computing, does not cause computing time prolongation.

Furthermore, owing to the genetic material exchange between the parallel populations the computations by the DMPEA method:

- are characterized by a great repeatability,
- are less susceptible to a number of individuals in the population,
- enable to obtain better results in the case of a single activation of calculations than an equivalent number of separate activations of the classical evolutionary algorithm.

Courses of the decisive variables for the best computing result are presented in Fig. 8a and Fig. 9a while the worst in Fig. 8b and 9b.

Big differences between the courses of the decisive variables can be noticed in the examples presented. In spite of these differences, Figs. 10 and 11 show that the subassembly of an active suspension of the driver's seat yields desirable effects. These figures present courses of the driver's seat acceleration for the best and worst result of the minimized objective function.

The considerations above confirm that the results obtained by the DMPEA method cause an increase in comfort of a driver while performing the maneuvers. In the case of the best result a significant difference can be noticed comparing to the course with use of the EA classical method.

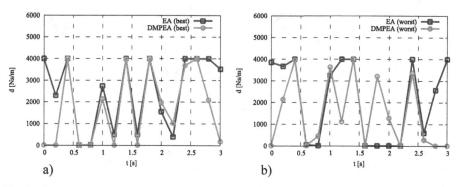

Fig. 8. Courses of the decisive variables (damping) of the driver's seat obtained in the optimization process - maneuver 1, (a) the best case, (b) the worst case

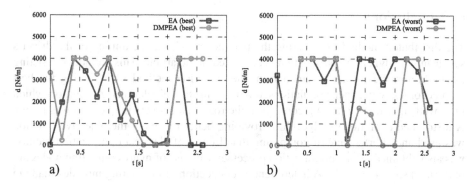

Fig. 9. Courses of the decisive variables (damping) of the driver's seat obtained in the optimization process - maneuver 2, (a) the best case, (b) the worst case

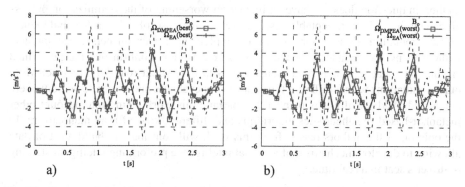

Fig. 10. Courses of the driver's seat acceleration of the optimization - maneuver 1, (a) the best case, (b) the worst case

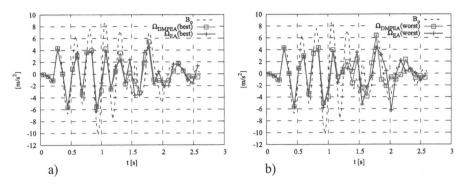

Fig. 11. Courses of the driver's seat acceleration of the optimization - maneuver 2, (a) the best case, (b) the worst case

6 Conclusions

The distributed method of selecting the parameters of active damping of the driver's seat using the dynamic optimization is presented in the article. This method, consisting in combining the results obtained from the evolution method with sharing the best chromosomes with other populations was used in the calculations. A solution which improved significantly driving comfort in the road manoeuvers in question was found. A distributed evolutionary algorithm providing better results of the objective function was used and shorter time of computing than by use of classical optimization method was also obtained. Since optimization processes with use of many computing nodes are performed simultaneously. A heterogenous connection of computing modules made in different programming languages was also shown. The computing module was made by use of language C++, whereas parallelization of the optimization process was performed with use of language C# and technology technologii REST WebAPI. The use of the method in question does not have an impact on worsening of the optimization process efficiency, it leads to acceptable results, even in the case of using a few individuals in the population. Furthermore, it can be implemented even on personal computers without using the dedicated computing clusters. Authors confirm that the performed calculations deal with two road maneuvers with a different road bump shape. More other simulations with a different types of road unevenness and a vehicle velocity were conducted. Additionally it is planned to continue this work taking into account other metaphor-based metaheuristics algorithms and different kind of road manoeuvers. In the opinion of the authors results from this work will contribute to build a set of data allowing to develop mechanism of artificial intelligence preforming intelligent steering of driver's seat in a real time.

References

1. Abbas, W., Abouelatta, O., El-Azab, M., El-Saidy, M., Megahed, A.: Genetic algorithms for the optimal vehicle's driver-seat suspension design. J. Mech. Eng. Autom. **1**, 44–52 (2011)
2. Abbas, W., Abouelatta, O., El-Azab, M., Megahed, A.: Genetic algorithms for the optimal vehicle's driver-seat suspension design. Ain Shams J. Eng. Phys. Math. **2**, 17–29 (2009)

3. Adamiec-Wójcik, I., Awrajcewicz, J., Grzegożek, G., Wojciech, S.: Dynamics of articulated vehicles by means of multibody methods. Dynamical Systems: Mathematical and Numerical Approaches, pp. 11–20 (2015)
4. Alshraideh, M., Tahat, L.: Multiple-population genetic algorithm for solving min-max optimization problems. Int. Rev. Comput. Softw. (I.RE.CO.S.) **10**, 9–19 (2015)
5. Augustynek, K., Warwas, K.: An application of PSO method in a motion optimization of a passenger car. Modelowanie Inżynierskie **27**, 5–12 (2016)
6. Augustynek, K., Warwas, K.: Real-time drive functions optimisation of the satellite by using MLP and RBF neural network. Dyn. Syst. Control Stability 53–64 (2015)
7. Bucking, S., Dermardiros, V.: Distributed evolutionary algorithm for co-optimization of building and district systems for early community energy masterplanning. Appl. Soft Comput. **63**, 14–22 (2018). https://doi.org/10.1016/j.asoc.2017.10.044
8. Cheng, H., Yang, S.: Multi-population genetic algorithms with immigrants scheme for dynamic shortest path routing problems in mobile ad hoc networks. In: Chio, C., et al. (eds.) EvoApplications 2010. LNCS, vol. 6024, pp. 562–571. Springer, Heidelberg (2010). https://doi.org/10.1007/978-3-642-12239-2_58
9. Chimote, K., Gupta, M.: Integrated approach of ergonomics and fem into truck drivers seat comfort. In: 1st International and 16th National Conference on Machines and Mechanisms, pp. 183–188 (2013)
10. Fiala, E.: Leteral forces at the rolling pneumatic tire. Technische Hochschule (2012)
11. Gong, D., Zhou, Y.: Multi-population genetic algorithms with space partition for multi-objective optimization problems. IJCSNS Int. J. Comput. Sci. Netw. Secur. **6**, 52–57 (2006)
12. Grzegożek, G., Adamiec-Wójcik, I., Wojciech, S.: Komputerowe Modelowanie Dynamiki Pojazdów Samochodowych, Wydawnictwo Politechniki Krakowskiej (2003)
13. Gupta, D., Sonawane, V., Sudhakar, D.: Optimization of vehicle suspension system using genetic algorithm. Int. J. Mech. Eng. Technol. (IJMET) **6**, 47–55 (2015)
14. Holen, P., Thorvald, B.: Possibilities and limitations with distributed damping in heavy vehicles. Veh. Syst. Dyn. Suppl. **41**, 172–181 (2004)
15. Kommenda, M., Affenzeller, M., Kronberger, G., Burlacu, B., Winkler, S.: Multi-population genetic programming with data migration for symbolic regression. In: Borowik, G., Chaczko, Z., Jacak, W., Łuba, T. (eds.) Computational Intelligence and Efficiency in Engineering Systems. SCI, vol. 595, pp. 75–87. Springer, Cham (2015). https://doi.org/10.1007/978-3-319-15720-7_6
16. Lozia, Z., Zdanowicz, P.: Optimization of damping in the passive automotive suspension system with using two quarter-car models. In: Conference on Automotive Vehicles and Combustion Engines (KONMOT 2016), vol. 148 (2016)
17. Navin, A.H., Mirnia, M.K.: Solving coverage problem in wireless camera-based sensor networks by using a distributed evolutionary algorithm. In: 5th IEEE International Conference on Software Engineering and Service Science (ICSESS), pp. 1076–1079 (2014)
18. Michalewicz, Z.: Genetic Algorithms + Data Structures = Evolution Programs. Springer, Heidelberg (2013)
19. Mircheski, I., Kandikjan, T., Sidorenko, S.: Comfort analysis of vehicle driver's seat through simulation of the sitting process. Tehnicki vjesnik **21**, 291–298 (2014)
20. Paulavičius, R., Žilinskas, J.: Simplicial Global Optimization. Springer, New York (2014)
21. Press, W., Teukolsky, W., Vetterling, S., Flannery, W.B.: Numerical Recipes 3rd edn, The Art of Scientific Computing. Cambridge University Press (2007)
22. Puljic, K., Manger, R.: A distributed evolutionary algorithm with a superlinear speedup for solving the vehicle routing problem. Comput. Inform. **31**, 675–692 (2012)

23. Dhananjay Rao, K., Kumar, S.: Modeling and simulation of quarter car semi active suspension system using LQR controller. In: Satapathy, S.C., Biswal, B.N., Udgata, S.K., Mandal, J.K. (eds.) Proceedings of the 3rd International Conference on Frontiers of Intelligent Computing: Theory and Applications (FICTA) 2014. AISC, vol. 327, pp. 441–448. Springer, Cham (2015). https://doi.org/10.1007/978-3-319-11933-5_48

24. Sammonds, G., Fray, M., Mansfield, N.: Overall car seat discomfort onset during long duration driving trials. In: Advances in Physical Ergonomics and Human Factors Part II, AHFE Conference Books, pp. 25–35 (2014)

25. Savaresi, S., Poussot-Vassal, C., Spelta, C., Sename, O., Dugard, L.: Semi-Active Suspension Control Design for Vehicles, Elsevier Ltd. (2010)

26. Silva, M., Silva, J., Motta Toledo, S., Williams, B.: A hybrid multi-population genetic algorithm for UAV path planning. In: 16 Proceedings of the Genetic and Evolutionary Computation Conference, pp. 853–860 (2016)

27. Song, C., Zhao, Y., Wang, L.: Design of active suspension based on genetic algorithm. In: Conference on Industrial Electronics and Applications, pp. 162–167 (2008)

28. Tengler, S.: Analiza dynamiki samochodow specjalnych o wysoko polozonym srodku ciezkosci, Ph.D. thesis, University of Bielsko-Biala (2012)

29. Warwas, K.: Analiza i sterowanie ruchem pojazdow wieloczlonowych z uwzglednieniem podatnosci elementow, Ph.D. thesis, University of Bielsko-Biala (2009)

30. Warwas, K., Tengler, S.: A hybrid optimization method to improve driver's comfort. In: 2017 IEEE International Conference on Innovations in Intelligent Systems and Applications, pp. 73–78 (2017)

31. Warwas, K., Tengler, S.: Dynamic optimization of multibody system using multithread calculations and a modification of variable metric method. ASME. J. Comput. Nonlinear Dyn. **12**(5) (2017)

32. Tengler, S., Warwas, K.: Driver comfort improvement by a selection of optimal springing of a seat. Tech. Trans. **10**, 217–235 (2017). https://doi.org/10.4467/2353737XCT.17.083.6440

33. Zhou, Y., Zhou, L., Wang, Y., Yang, Z., Wu, J.: Application of Multiple-Population Genetic Algorithm in Optimizing the Train-Set Circulation Plan Problem, Hindawi (2017)

The Impact of Cybersecurity on the Rescue System of Regional Governments in SmartCities

Hana Svecova$^{(\boxtimes)}$ (ID) and Pavel Blazek$^{(\boxtimes)}$ (ID)

Faculty of Informatics and Management, University of Hradec Kralove, Rokitanskeho 62,
50003 Hradec Kralove, Czech Republic
{hana.svecova,pavel.blazek}@uhk.cz

Abstract. The article's topic is the solution to the issue of cyber security of IoT in Smart Cities while ensuring the security of citizens with the cooperation of regional governments and safety officies (include police, firefighters, and paramedics).

The global COVID pandemic has highlighted the need for safety officies, and regional governments to ensure the safety of citizens using information technology and IoT in Smart Cities strategies. The pandemic revealed conceptual shortcomings of IoT in security forcesandan inadequate solution to cyber security, which manifested itself, for example, in response to cyberattacks on hospitals around the world.

The article focuses on the evaluation the concepts of coordination and the use of available technologies used by safety officies. It points to the current problems (positive and negative impacts) of IoT in connection with cyber security. It emphasizes the need to address this issue more in detail and to integrate it into Smart Cities' strategies as a separate segment. In the case of currently resolved pandemics, it is evident that the mutual coordination of regional self-government and safety officies using smart technologies and IoT leads to the saving of many human lives.

Keywords: Cyber · Security · Smart Cities · Police · Firefighters · Paramedics · Regional · Governments · Safety officies

1 Introduction

Smart Cities' strategies are very complex, inconsistent all over the world. From this non-complex solution, various documents (strategies) have been created. They aim to functions, but an essential component focused on the mutual coordination of regional self-government with state security forces in Smart Cities is neglected. State security forces are an integral part of SmartCities and should not be forgotten. During the global pandemic of COVID-19, the need for the use of smart technology, IoT, became more than evident for the components mentioned above.

In the Czech Republic, the security forces of the state, which are a part one of the parts of the IRS (Integrated Rescue System) [1], included in the competence of the Ministry of the Interior of The Czech Republic, participate in the protection of the population

© Springer Nature Singapore Pte Ltd. 2021
T.-P. Hong et al. (Eds.): ACIIDS 2021, CCIS 1371, pp. 365–375, 2021.
https://doi.org/10.1007/978-981-16-1685-3_30

safety. IRS consists of the Police, the Fire Brigade and the Medical Rescue Service of The Czech Republic. The independent armed component is the Army of The Czech Republic, and the Ministry of Defense coordinates its activities.

Many authors have addressed the issue of security forces in the context of the Smart Cities strategy. Elizabeth E. John, in her reflections, focuses on comparing the police activities with modern information technology. In her reflections, she focuses on the actions of the police as a private entity, stating that thanks to modern technologies in Smart Cities, the police will be integrated into the urban infrastructure in such a way that they will not have a sovereign position. In connection with the new sensors within the framework of intelligent traffic control, camera systems, and other sensors, the issue of restricting the freedom and movement of the population is also being considered. Residents are thus always under the surveillance of smart technologies and the police, which increases security, but at the same time, may restrict privacy [2].

To ensure cyber security in SmartCities, the National Institute of Standards and Technology has created a security framework to increase cyber security in Smart Cities [3].

In his article, Quentin Hardy wrote: "The city is developing a policy for drone-based delivery. Emergency medical goods, transported from an airport to a hospital, are likely to be the first, he said" [4].

In her article with the name: "Why Local Governments Are a Hot Target of Cyber Attacks," Cynthia Brumfield analyzed the current issue of ransomware cyber-attacks. The analysis describes the problems of local governments, cities, and municipalities in connection with cyberattacks, security, and control of the state's security forces [5].

The article aimsto acquaint readers with the current trends in the use of information technology applications, IoT and smart technologies used by the security forces of the state in ensuring the protection of the population in connection with the global pandemic COVID-19.

It also draws attention to pointing out the partial parts of Smart Cities strategies, neglected in connection with innovation strategies. In their section, the innovation strategies systematically point to procedures related to the protection of the population. The mutual coordination of regional self-government and security forces appears, and they emphasized cyber security and the use of smart technologies.

The article is divided into chapters. They are chronologically arranged: Strategies and Regional Governments in Smart Cities, Security Components and of Smart Technologies in Smart Cities, Positive and Negative Impacts on the Use of Smart Technologies in Security Forces in the Coordination of Regional Governments in Smart Cities and Conclusion.

2   Strategies and Regional Authorities in Smart Cities

The development of Smart Cities, together with the expansion of cyber security, is one of the Czech Republic's Innovation Strategy 2019–2030 under The Country For The Future [6]. The innovation strategy includes nine fundamental pillars (Fig. 1), which complement each other in digitization, security, cyber security, public administration, and smart technologies.

Fig. 1. Pillars of innovation strategy

One of the main objectives of the "Digital State, Production and Services pillar is to ensure online and shared services on an ongoing basis, including industrial enterprises and system security of complex units (cities, airports, enterprises, and power plants) using intelligent cybernetic systems and treating the most serious risks and preparing the society for trends such as IoT, AI, BigData, new types of human-machine interfaces" [7].

The Government of the Czech Republic commissioned an extensive study of the current level of the Czech Republic's involvement in the Smart Cities and Smart Region strategy in connection with new trends, including proposals for measures [8].

In the resulting study, sections focusing on more intensive cooperation between regions and cities were developed. Cities that have implemented or will gradually implement the Smart Cities strategy should follow the NRIS3 (National Regional Innovation Strategy) Strategy by following per under the regional annexes of RIS3 (Regional Innovation Strategy). Thus, issues related to protecting the security of the population, cyber security, smart technologies, IoT to improve the living conditions and security should be addressed more intensively.

The NRIS3 Strategy of the Czech Republic is a strategic document ensuring the effective targeting of European, national, regional, and private funds for activities leading to the strengthening of research innovation capacities in priority areas identified at the national and regional level. The NRIS3 strategy includes its regional annexes, which defines regional priorities and proposals for interventions to fulfil them. It specifies regional application branches (regional domains of specialization) with the the business sphere, research sphere, and non-profit entities, including cluster organizations and government officials [9].

It is highly recommended for the period 2021+ to update RIS3 strategies by supplementing the latest innovations related to services, products, tools, procedures, or applications. It is mainly about adding targets for research trends, development, and innovation (R & D & I). New trends should be primarily those in new key technologies - digital technologies and cyber technologies [7].

Recommendations for regions include the RIS3 update for the period 2021+. Part 4 of point F, which deals with societal challenges, should take into account: "employment, economic and economic growth, ecosystem degradation, population ageing, increase in diseases of civilization, social justice, urbanization and smart cities, social justice and basic human rights; migration, single digital market, personal data protection, cyber security - using advanced technologies" [9].

Regional governments were recommended to become more involved in the dissemination and use of smart technologies, IoT and at the same time, increase the support, development, and coordination of Smart Cities for the strengthening of cyber security.

3 Security Components and Smart Technologies in Smart Cities

During the COVID-19 pandemic, ensuring the safety of citizens using smart technologies was a high priority. IoT, in particular, has become known in connection with Smart Cities and cyber security.

Cyber security has become a part of Smart Cities' strategies around the world. We can characterize it as a set of technologies, processes, and procedures designed to protect networks, devices, programs, and data from attack, damage, or unauthorized access [5]. It consists of network security, application, endpoint, data, identity management, database and infrastructure, cloud service, mobile application, and data backup or recovery process.

Cyber security needs to be implemented for IoT solutions in many areas, such as cloud services, critical infrastructure, and information systems in Smart Cities or network environments. The appropriate answer always depends on the specific organizational structure of the entity on which the technologies are implemented.

All states have an organizational division defined according to their specifics - mini-trying region, regions, and microregions. These entities must ensure the protection of the population through safety, officies and the army. These components belong inseparably to territorial self-governing units, cities, or municipalities. To ensure citizens' security, they use information technology, information systems, including IoT and other smart technologies, and integrated applications. It is an important part of the entire organizational infrastructure of regions, cities, and municipalities, potentially integrating security forces into one unit. The use of smart solutions and modern information technology to protect the population should be more included in Smart Cities strategies.

The security forces are involved in the day-to-day protection of the security of the population. In the methodology for the preparation and implementation of the Smart Cities strategy at the level of cities, municipalities, and regions, the IRS components are included in the section focused on mobility (Fig. 2).

The involvement and mutual coordination of security forces and regional self-government is a neglected part of the Smart Cities strategy. This subsection is partly mentioned in the document entitled Final Report of Smart Cities and Smart Region [8].

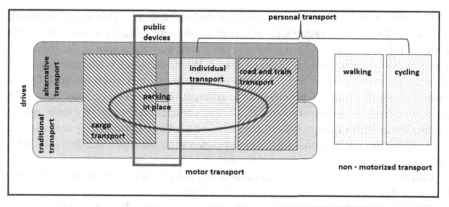

Fig. 2. Urban mobility structure

The document provides only strict information related to the involvement of IRS security forces in protecting residents, in part focused on urban mobility. However, it does not mention mutual coordination from the position of regional self-government. According to the methodology, it is possible to: "Implement clean urban mobility means addressing and promoting a balance between all these elements of urban mobility, including the requirement for effective traffic management, taking into account the needs of IRS, emergency and emergency services (network managers and energy suppliers, urban services).

Urban mobility also includes means of transport of IRS units:

- Police – mostly personal
- Emergency medical service – mostly personal (ambulances)
- Fire brigade – mostly freight (CAS)

It is possible to use the tools of modern information technologies, including the Internet of Things (IoT)" [10]. The development of intelligent technologies, including IoT, can be found in IRS components, for example, in using drones [11]. Within the EU eHealth strategy framework, it is possible to use the IRS with neighbouring states' border areas.

Research projects are being developed around the world focused on the use of IoT in the security forces of the state with simultaneous integration and at the same time applicable in Smart Cities.

The project Program for Early Response to Floods by Sea Level Rise is being implemented in the USA. The program uses 3D modelling sensors for water level monitoring and LIDAR (Light Detection and Ranging Data). It is intended to inform local governments in the riskiest areas and ensure timely warning of the population [12].

At EU level, the Reverse 112 project is being implemented to introduce a uniform standard for warning population called the Emergency Call Center (General Phone Number 112) at EU level [11]. Future visions include population warning systems using IoT through smart homes or smart appliances and links to Smart Cities [13].

However, intelligent camera systems have the largest share in protecting the security of citizens. The camera systems are operated by the city police, the state police, firefighters, and paramedics and are gradually integrated into one output to the operational centre of the security forces mentioned above. When the systems are interconnected, all components can react in time to the current emergency at a given moment. Camera systems in Smart Cities can provide image information to various systems. Apart from the supervision of IRS control rooms, the defined cameras can be part of a smart parking system. The complex camera system can be linked to other information systems, such as searching for missing persons, searching for stolen vehicles, traffic accidents, and other incidents.

Last year, a project entitled: "Accesses to the Information Systems of the Police of the Czech Republic for Municipal Police" was launched.

The essence is the interconnection of camera systems of IRS units and municipal (city) police for faster determination of the basic procedures of evaluation and immediate response to the result of information processing. Wider availability of data from the information systems of the Police of the Czech Republic and the Ministry of the Interior greatly helps to increase the security of citizens [13]. The camera system's interconnection with the city policeity police's camera system's camera system is shown below (Fig. 3).

Fig. 3. Interconnection of CCTV systems

In the future, plan and implement artificial elements throughout Artificial intelligence (AI) in camera systems for the evaluation of accidental emergencies without human intervention.

CCTV (closed-circuit television) systems are expected to be high in a few years. As part of the joint integration and coordination of cities, regions, and IRS units, camera systems should respond operatively to emergencies. Using AI, it is possible to automatically send reports or data to special information systems or notifications to individual operational centres of IRS units.

CCTV systems are considered one of the basic parts for protecting the population using IoT in Smart Cities. We can find coherence also be found in other sectors, such as education, transport, and environmental policy.

Another possibility could be creating one centrally managed comprehensive information system at the level of regional self-government for ordering and distributing protective aids to medical facilities. This system could collect selected data that could be publicly accessible to citizens, for example, on websites or mobile applications. The newly created IS could serve for more effective mutual coordination of all entities as mentioned above and at the same time could inform citizens about important facts, such as sufficient protective equipment in medical facilities. This would improve citizens' awareness, at the same time safety, and reduce the stress burden on citizens in the event of a crisis associated with, for example, pandemics or floods.

4 Positive and Negative Impacts of the Use of Smart Technologies in Security Components in the Coordination of Regional Authorities in Smart Cities

Lately, the whole world has been paralyzed by the worldwide COVID - 19 pandemics, highlighted the importance of government agencies that take care of human lives' daily protection. It is precisely the regional self-governments that participate in the coordination of security forces (safety officies) and other organizations that ensure population protection. The application is called Smart Quarantine, and other applications have been created, which enable to distribute, for example, to distribute protective equipment. Smart technologies, Smart Cities, IoT, IS, regional governments and security forces in the real world have formed a single comprehensive unit that has helped to cope with a crisis. The unit could collect selected data that could be publicly accessible to citizens, e.g., on websites or mobile applications. The newly created IS could serve for more effective mutual coordination of all entities as mentioned above and at the same time could inform citizens about important facts, such as sufficient protective equipment in medical facilities. This would improve citizens' awareness at the same time safety and reduce the stress burden on citizens in the event of a crisis associated with, for example, pandemics or floods.

4.1 Positive Effects

One of the positive impacts of Smart Technologies, IoT, and entities as mentioned above was the creation of the Smart Quarantine project. It was established under the auspices of Czech IT companies associated under the name "COVID19CZ" [14].

Smart Quarantine is based on mapping an infected person's movement over the last five days based on data from mobile operators and credit card data. The so-called "memory map" that emerged from this data helped hygienists, who then went through it step by step with the infected person during the epidemiological investigation [14]. The tool was also integrated into the mobile application Mapy.cz [15].

The above was preceded by the project and application of Zachranka, which addresses the protection of the population's health outside the pandemic period. It is used for contact with the emergency services, fire brigade and police in urgent cases when it can also pass on the coordinates of the event location to the operating workplace. With this mobile application's help, you can also contact help in selected border areas of countries neighbouring the Czech Republic. The application has an educational part, thanks to which the user can get acquainted with the procedures in various life-threatening situations [16].

Another Smart trend was the use of drones, which has been involved in protecting the population in cooperation with these entities and at the same time has been in Smart Cities. These have been used by the state security forces to monitor demarcated areas or to supply medical equipment. Data from them were processed by IRS units and further processed and published, for example, in statistical outputs within the Ministry of Health [17]. In the future, it is highly desirable to involve all entities in coordination with each other at a higher level of interconnection in Smart Cities.

China has also seen a strong development of IoT in Smart Cities. At the level of microregions, IoT is integrated into the daily lives of citizens. We can meet it in urban transport, where billing is handled based on biometric face identification. Other IoT technologies are connected, for example, to waste bins, which send a notification if filled. With the help of IoT in Smart Cities and similar technologies, continuous monitoring of various areas is performed. Based on the data thus obtained, it is possible to issue disaster warnings [18].

Other technologies used in Smart Cities and IoT include camera systems, which are highly integrated into many cities. Many camera systems have artificial intelligence implemented with the possibility of biometric face recognition technology. Even with this technology, it was possible to alleviate the global pandemics in Smart Cities, because the police, fire brigade, paramedics, and the military had the necessary overview thanks to interconnected information systems. Therefore, the overall coordination was effective, and with the help of IoT in Smart Cities, many cities/counties managed to cope with a pandemic of this magnitude. To this end, research on the impact of IoT in the context of COVID-19 has been carried out in global pandemics [19].

Within the framework of movement restrictive measures, finding other variants for mutual communication and cooperation in organizational teams was necessary. Software for video conferencing calls such as Zoom, Microsoft Teams, Skype, or Cisco Webex began to be used. Libresteam has announced an increase in the use of Onsight Connect software for remote management and various diagnostics (Fig. 4) [20].

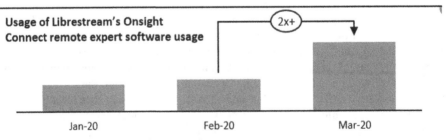

Fig. 4. Use of the Librestream's Onsight application in 1Q 2020

Furthermore, the increase in the use of health applications in IoT in connection with monitoring information systems in Smart Cities increased (Fig. 5).

Download Rank of the app "Kinsa Health" on ios in the category "Medical" in the USA

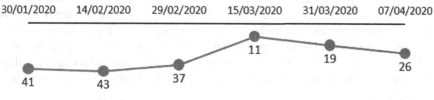

Fig. 5. Use of mobile applications in healthcare

Expert studies state that regional self-government and state security forces can respond quickly in crises.

4.2 Negative Effects

Many states were not sufficiently prepared for a pandemic in terms of the amount of protective equipment, smart technologies, and coordination of components, ensuring citizens' safety. Within the framework of Smart Cities strategies worldwide, this part has been neglected, which had negative effects on the security of the population at the beginning of the global pandemics.

The negative impacts manifested themselves in poor coordination in the distribution of protective equipment due to the lack of IS, the disconnection of IS with the security forces of the state, or the lack of crisis plans.

The problem is at the level of legislation and the ongoing decentralization of agenda management. An example is an IS operated by an emergency medical service with no connection to other health care systems, such as hospital information systems (HIS) of the catchment hospitals.

The most significant adverse effects during the first wave of coronavirus pandemics, arose at the technology level when hospitals were affected by cyber-attacks. The problem stems from insufficient security against hacker attacks, both at the level of the technologies used and staff training. It was not just the Czech Republic; hospitals worldwide faced an intense series of attacks for several months. However, the area of attacks was wider, affecting other elements of the strategic infrastructure of states. The most well-known threats to the infrastructure were Eave-dropping, Phishing, DDOS, and ransomware.

Security should be maximized in terms of the importance of the transmitted data at a high level. In some cases, it would be appropriate to choose a secure communication network at the level of security forces for crisis management" [16].

5   Conclusion

In the current, pandemics make intensive use of all information technologies to ensure population protection. Regional self-government, together with security authorities, is of utmost importance in coordinating a pandemic. These institutions are gradually implementing smart solutions (IoT, Smart technologies) to ensure the current crisis as effectively as possible and provide citizens with the best possible protection.

The development and implementation of Smart Cities strategies is a process that has already begun and will be developed within cities around the world. Along with it, the security of citizens will also increase, which is already evident in the situation of the COVID-19 pandemic. Systems that use artificial intelligence are gaining prominence. These are predominantly systems based on specific algorithms for evaluating data from IoT element networks. The new approach will have an impact on changing the concept of the work of the security forces. Complementing and interconnecting technologies will create comprehensive information systems, which will increase the importance of cyber security.

The individual parts of the previous text show the importance of solving cyber security, and the cooperation of regional governments safety officies in ensuring the safety of citizens using IoT technology in Smart Cities. It points out the neglected part of the Smart Cities strategy and the entities not included in it. Current trends in information technology and IoT are being promoted in safety officies in ensuring security. The process brings with it both positive and negative impacts associated with cyber security and IoT. The development of information technologies is permanent, so it is necessary to include all emerging technologies and strategies, such as 5G networks or artificial intelligence solutions, on the issue of citizen security and increasing cyber security using IoT in Smart Cities in cooperation with other entities.

Acknowledgement. The research has been partially supported by the Faculty of Informatics and management UHK specific research project 2107 Computer Networks for Cloud, Distributed Computing, and Internet of Things III.

References

1. Integrated Rescue System - Fire Rescue Service of the Czech Republic. https://www.hzscr.cz/hasicien/article/about-us-scope-of-activities-integrated-rescue-system.aspx. Accessed 22 June 2020
2. Joh, E.E.: Policing the smart city. Int. J. Law Context **15**(2), 177–182 (2019). https://doi.org/10.1017/S1744552319000107
3. thelma.allen@nist.gov: NIST Cybersecurity for IoT Program – Publications. NIST (2019). https://www.nist.gov/itl/applied-cybersecurity/nist-cybersecurity-iot-program/publications. Accessed 21 Feb 2021
4. Hardy, Q.: Technology is monitoring the urban landscape. The New York Times (2016). Accessed 20 July 2016
5. Smartcity: Why Smart Cities Need To Update Their IoT Cybersecurity? Smart City (2020). https://www.smartcity.press/iot-cybersecurity-threats-and-solutions/. Accessed 15 June 2020
6. Innovation Strategy of the Czech Republic 2019–2030—Research & Development in the Czech Republic. https://www.vyzkum.cz/FrontClanek.aspx?idsekce=866015. Accessed 10 Jan 2020
7. What is Cyber Security? Definition, Best Practices & More. Digital Guardian (2020). https://digitalguardian.com/blog/what-cyber-security. Accessed 15 June 2020

8. Final Report Smart City and Smart Region. https://www.vlada.cz/assets/evropske-zalezitosti/aktualne/Zaverecna-zprava_Smart_City_a_Smart_Region.pdf. Accessed 22 Feb 2020

9. Methodological Recommendations to Regions for updating Regional RIS3 Strategies in the programming period 2021+—MPO (Ministry of Industry and Trade. https://www.mpo.cz/cz/podnikani/ris3-strategie/krajska-dimenze/metodicka-doporuceni-krajum-pro-aktualizace-krajskych-ris3-strategii-v-programovem-obdobi-2021--247029. Accessed 22 Feb 2020

10. Methodology Smart Cities. https://mmr.cz/getmedia/f76636e0-88ad-40f9-8e27-cbb774ea7caf/Metodika_Smart_Cities.pdf.aspx?ext=.pdf. Accessed 14 June 2020

11. Use of Drones for the Integrated Rescue System. Smart Cities Pilsen.https://smartcity.plzen.eu/projekty-ziti/vyuziti-dronu-pro-integrovany-zachranny-system. Accessed 22 Feb 2020

12. Integrated Rescue System - Fire Rescue Service of the Czech Republic. https://www.hzscr.cz/clanek/integrovany-zachranny-system.aspx. Accessed 09 Feb 2020

13. Access to information systems of the Police of the Czech Republic for municipal police - Ministry of the Interior of the Czech Republic. https://www.mvcr.cz/clanek/pristupy-do-informacnich-systemu-policie-ceske-republiky-pro-obecni-policie.aspx. Accessed 23 Feb 2020

14. Coronavirus: Fear vs freedom. investigace.cz (2020).https://www.investigace.cz/koronavirus-strach-versus-svoboda/. Accessed 13 June 2020

15. Mapy.cz: Mapy.cz. https://mapy.cz/. Accessed 13 June 2020

16. Záchranka. https://www.zachrankaapp.cz/. Accessed 13 June 2020

17. EECC_briefing_FINAL(1).pdf. https://www.ctif.org/sites/default/files/news_files/2018-11/EECC_briefing_FINAL(1).pdf. Accessed 10 Jan 2021

18. Kummitha, R.K.R.: Smart technologies for fighting pandemics: the techno- and human-driven approaches in controlling the virus transmission. Gov. Inf. Q. **37**(3) (2020). https://doi.org/10.1016/j.giq.2020.101481. ISSN 0740-624X

19. The impact of Covid-19 on the Internet of Things Part 2. https://iot-analytics.com/the-impact-of-covid-19-on-the-internet-of-things-part-2/. Accessed 14 June 2020

20. Augmented Reality Remote Expert Solution - Onsight Connect. Librestream. https://librestream.com/products/onsight-connect/. Accessed 14 June 2020

A General Modeling Framework
for Robust Optimization

Maciej Drwal[(✉)]

Department of Computer Science, Wroclaw University of Science
and Technology, Wrocław, Poland
maciej.drwal@pwr.edu.pl

Abstract. This paper introduces PYROPT, a Python programming language package for robust optimization. The package provides a convenient way for expressing robust optimization problems using simple and concise syntax, based on the extension of popular CVXPY domain-specific language. Moreover, it offers an easy access to the collection of solution algorithms for robust optimization, which utilize the state of the art reformulation and cut-generation methods, leveraging a variety of mathematical programming solvers. In particular, the package provides readily available modeling and solution framework for min-max linear programming, min-max regret linear programming with polyhedral uncertainty, and min-max regret binary programming with interval uncertainty.

Keywords: Scientific software · Robust optimization · Mixed-integer programming

1 Introduction

Optimization problems are very often defined on the basis of parameters with imprecise or uncertain values. While there exists a wide range of well-established and efficient methods for solving many important types of optimization problems, their effectiveness relies on assuming the full knowledge of the input data. However, such assumption is usually not valid with regard to all the input data parameters at the time when the problem is being solved. Typically, at least some of the input data parameters would be subject to uncertainty; for instance, they may be obtained through a noisy measurement process, or they may come from a forecast or estimation. Consequently, there is a risk of getting vastly different solutions of an optimization problem, when using potentially wrong parameters (which was well demonstrated even for fairly simple linear programming cases, see [6]). This situation is very common in various engineering applications.

The robust optimization approach aims at mitigating the influence of uncertainty on the problem's solution. It is especially useful when also no reliable

This research was supported by the National Science Center of Poland, under grant 2017/26/D/ST6/00423.

statistical information on the parameters is available. The robust approach dictates the modeling methodology (assumptions on the sets of parameters' possible values, and the forms of optimization criteria) as well as the choice of solution algorithms.

Although the robust optimization approach has gained popularity in recent years [9,11,12,14,16,17,20], it still lacks the widespread recognition and the level of acceptance by operations research practitioners, even compared to that of stochastic programming. We believe that one of the main reasons for such state of matters is the lack of comprehensive and freely available software package for robust optimization. Usually, such optimization problems with uncertain data are solved by bespoke methods, or with the use of narrowly specialized programs, implemented from scratch exclusively for the problem at hand. Very often these methods are only variations of well known algorithms, relying on standard transformations and solver backends.

There are many software solver packages for deterministic optimization problems. Modern solvers efficiently handle not only linear programming problems, but convex problems (especially the quadratic, second-order cone and semidefinite), as well as mixed-integer problems. Various high-level domain-specific modeling languages are available, and highly optimized implementations of popular solution algorithms could be used directly from within such solution frameworks.

In this paper we introduce PYROPT[1], a Python module for robust optimization modeling and solving. The module is based on the state of the art package CVXPY for convex optimization [1,8]. It extends the Python-based modeling language with expressions allowing to formulate optimization models with uncertainty in a standard way. The module bears some resemblance to the package for stochastic optimization CVXSTOC [3], which extends CVXPY's modeling language with random variables, and provides problem transformations to deterministic counterparts via Monte-Carlo sampling.

Currently, the support for robust optimization in popular mathematical modeling languages is very limited. In particular, to the best of our knowledge, no such comprehensive library is available for Python programming language, which is currently generally considered as one of the most popular tools for writing code in scientific communities. It was in fact the most popular language according to 2019 IEEE Spectrum annual ranking [19]. In recent years, its user base has outgrown those of well-established alternatives of R, Matlab, Julia and Fortran, which all have been used by scientists and engineers for many years. Undoubtedly, the concise, readable but very expressive syntax of Python allows for quick prototyping, easy experimentation and troubleshooting. As such, it offers a relatively low barrier of entry for new users, who want to create small to medium-sized scripts or programs, without the need for diving into the details of lower level components.

Basic support for robust optimization modeling is available for AIMMS [15], however, only to the extent of min-max robustness (and only as an licensed add-on, since the software is proprietary). There also exist libraries for Matlab (e.g.,

---

[1] PYROPT is available at https://github.com/maciejdrwal/pyropt.

YALMIP [18]) offering only basic min-max robustness support. We also note one attempt at providing such package for the programming language Julia: JuMPeR Robust Optimization with JuMP [13].

2   Basic Concepts

The PYROPT provides a basic class Uncertain, which derives from the CVXPY's class Parameter. An object of such class can be used in modeling expressions in the same way as variables or constants. Before being passed to the solver, the mathematical model, which contains Uncertain parameters, will undergo special transformations. These would either produce a proper deterministic counterpart of the problem, or a sequence of derived subproblems, which will be used in the iterative cut-generation procedure. The uncertain parameters can have their values changed multiple times within such a procedure, depending on the context of usage.

The following types of uncertainty are considered in PYROPT:

a) interval uncertainty, $\mathcal{U} = \{\mathbf{c} \in \mathbb{R}^n : c_i^- \leq c_i \leq c_i^+,\ i = 1, \ldots, n\}$,
b) polyhedral uncertainty, $\mathcal{U} = \{\mathbf{c} \in \mathbb{R}^n : \mathbf{G} \cdot \mathbf{c} \leq \mathbf{h},\ \mathbf{c} \geq 0\}$,
c) ellipsoidal uncertainty, $\mathcal{U} = \{\mathbf{c} \in \mathbb{R}^n : \sqrt{\sum_{i=1}^n c_i^2/\sigma_i^2} \leq \Omega\}$,
d) discrete uncertainty, $\mathcal{U} = \{\mathbf{c}_1, \mathbf{c}_2, \ldots, \mathbf{c}_k\}$.

Uncertain parameters can be provided as scalars, vectors or matrices, using Python's standard NUMPY arrays.

The interval-uncertain parameters can be defined in two ways: either by specifying lower and upper bounds, or by specifying nominal values and interval widths (see Table 1). Note that within a vector or matrix only some elements could be uncertain; in such case the certain elements may be defined with interval width equal to 0, or the upper bound equal to the lower bound.

The polyhedral-uncertain parameters can also be defined in two ways: either by specifying the $n \times k$ matrix $\mathbf{G}$ and k-vector $\mathbf{h}$, or by enumerating vertices $v_1, v_2, \ldots, v_k$, where we define $\mathcal{U} = conv\{v_1, v_2, \ldots, v_k\}$. Observe that interval uncertainty is a special case of polyhedral uncertainty.

The ellipsoidal-uncertain parameters are defined by specifying parameters σ_i, $i = 1, \ldots, n$, and parameter $\Omega > 0$.

Table 1. Definitions of uncertain data

```
b = Uncertain(lbs=[1,2,3], ubs=[4,5,6])

A = Uncertain(mid=np.array(   [[0,-1],
                               [1, 1]]),
                 width=np.array([[2, 1],
                               [0, 3]]))
```

3 Solution Methods

Robust optimization problems can be solved by exact, approximate or heuristic methods; PYROPT enables the use of all these three types of algorithms. Exact algorithms are recommended, or at least the ones with a guaranteed approximation ratio, as otherwise the solution method may fall short of the robustness concept itself. However, robust versions of many optimization problems are significantly more difficult to solve to optimality, and very often only suboptimal solutions to their large-sized instances are attainable in practice. In this section we describe the methodology used for solving the supported types of robust problems.

3.1 Robust Linear Programming

A fundamental robust optimization model is min-max linear program (LP). Let us define the form of the considered linear programs as follows:

$$\min \mathbf{c}^T \mathbf{x} + \mathbf{d} \tag{1}$$
$$\text{s.t. } \mathbf{A}\mathbf{x} \le \mathbf{b}, \tag{2}$$

where $m \times n$ matrix $\mathbf{A}$, as well as n-vectors $\mathbf{c}$ and $\mathbf{d}$, and m-vector $\mathbf{b}$, can have uncertain parameters. Note the inequality in (2), a vector $\mathbf{d}$ in the objective function, and lack of default non-negativity constraint on decision vector $\mathbf{x}$, which depart from the canonical formulation of LP. We define the set of uncertain parameters as:

$$\mathcal{U} = \left\{ \begin{bmatrix} \mathbf{c} \ \mathbf{d} \\ \mathbf{A} \ \mathbf{b} \end{bmatrix} = \mathbf{D}_0 + \sum_{k=1}^{K} \xi_k \mathbf{D}_k : \xi \in \mathcal{Z} \right\},$$

where $\mathbf{D}_0$, $\mathbf{D}_k$ are the matrices of nominal values of the data, and $\mathcal{Z}$ is the perturbation set. With this definition it is easy to further define interval and ellipsoidal uncertainty sets as perturbations of unit box and unit ball, respectively.

The robust problem is then:

$$\min_{\mathbf{x}} \left\{ \sup_{(\mathbf{c},d,A,b) \in \mathcal{U}} [\mathbf{c}^T \mathbf{x} + \mathbf{d}] : \mathbf{A}\mathbf{x} \le \mathbf{b} \ \forall(\mathbf{c}, d, A, b) \in \mathcal{U} \right\}. \tag{3}$$

Observe that a robust feasible solution $\mathbf{x}$ is the one that is feasible for all realizations of uncertain parameters. The robust optimal solution achieves the lowest value for the worst realization of data.

To solve (3) under interval or ellipsoidal uncertainty, PYROPT transforms the problem into an equivalent deterministic counterpart. First, if the objective function contains uncertain parameters, the original problem is transformed into an equivalent:

$$\min_{\mathbf{x},t} \left\{ t : \mathbf{c}^T \mathbf{x} + \mathbf{d} \le t, \ \mathbf{A}\mathbf{x} \le \mathbf{b} \right\}. \tag{4}$$

Consequently, the uncertain parameters only remain in constraints. Following that, each constraint's row that contains uncertainty is replaced by the set of new constraints:

1. In case of interval uncertainty:

$$-u_l \leq \mathbf{a}_l^T \mathbf{x} - \mathbf{b}_l \leq u_l, \quad l = 1, \ldots, L, \tag{5}$$

$$\mathbf{a}_0^T \mathbf{x} + \sum_{l=1}^{L} u_l \leq \mathbf{b}_0, \tag{6}$$

  obtaining LP with decision variables $\mathbf{x}$, t and $\mathbf{u}$.
2. In case of ellipsoidal uncertainty:

$$\mathbf{a}_0^T \mathbf{x} + \Omega \sqrt{\sum_{l=1}^{L} (\mathbf{a}_l^T \mathbf{x} - \mathbf{b}_l)^2} \leq \mathbf{b}_0, \tag{7}$$

  obtaining second-order cone problem with decision variables $\mathbf{x}$ and t.

The remaining constraints are rewritten intact. In both cases, the deterministic counterpart can be solved to optimality in polynomial time (via convex optimization).

3.2 Min-Max Regret Linear Programming

The second type of robustness supported by PYROPT is the maximum regret robustness. Unlike the case of min-max robustness, described in the previous subsection, the linear programming problem under maximum regret robustness is already NP-hard for interval uncertainty [5] (it can, however, be solved in polynomial time for discrete uncertain parameters).

We consider only uncertain parameters $\mathbf{c}$ of the objective function in (1)–(2). However, we also consider a more general polyhedral uncertainty, of which the interval representation is a special case. We assume that parameters belong to a closed polyhedron, represented by a system of linear inequalities:

$$\mathcal{U} = \{\mathbf{c} \in \mathbb{R}^n : \mathbf{G} \cdot \mathbf{c} \leq \mathbf{h}, \ \mathbf{c} \geq 0\}.$$

Given an LP in the form (1)–(2), we define the regret of solution $\mathbf{x}$ under scenario $\mathbf{c} \in \mathcal{U}$ as the difference between the cost of the solution and the optimal value in the given scenario, i.e.,

$$R(\mathbf{x}, \mathbf{c}) = \mathbf{c}^T \mathbf{x} - \min_{\mathbf{y}: \, \mathbf{Ay} \leq \mathbf{b}} \mathbf{c}^T \mathbf{y}.$$

The robust problem is to find a solution $\mathbf{x}$ that minimizes the maximum regret over all possible scenarios $\mathbf{c} \in \mathcal{U}$:

$$\min_{\mathbf{x}} \max_{\mathbf{c}} \ R(\mathbf{x}, \mathbf{c}) \tag{8}$$

$$\text{s.t. } \mathbf{Ax} \leq \mathbf{b}, \tag{9}$$

$$\mathbf{c} \in \mathcal{U}. \tag{10}$$

Let $P = \{\mathbf{y} \in \mathbb{R}^n : \ \mathbf{Ay} \le \mathbf{b}\}$ be the set of feasible points of the original LP. The robust problem can be equivalently written as:

$$\min_{\mathbf{x},t} \ t \tag{11}$$

$$\text{s.t.} \ \ \forall \mathbf{c} \in V(\mathcal{U}) \ \forall \mathbf{y} \in V(P) \ \ \mathbf{c}(\mathbf{x} - \mathbf{y}) \le t, \tag{12}$$

$$\mathbf{x} \in P. \tag{13}$$

where $V(X)$ is the set of vertices of polyhedron X. This formulation is a linear program with potentially very large number of constraints (typically exponential in n). Thus instead of solving it directly, a more efficient approach is to decompose the problem into "Master Problem" (MP) and a sequence of "Auxiliary Problems" (AP). The Master Problem is created by relaxing the constraints (12), and maintaining a (preferably small) set $\mathcal{C}$ of pairs of vectors $(\hat{\mathbf{c}}, \hat{\mathbf{y}})$. Each element of $\mathcal{C}$ corresponds to a cut:

$$\hat{\mathbf{c}}(\mathbf{x} - \hat{\mathbf{y}}) \le t.$$

The initial set of cuts can be obtained by randomly sampling a fixed number of extreme scenarios (vertices of polyhedron $\mathcal{U}$). These can be found via standard linear programming, by solving for each sampled random vector $\mathbf{v} \in \{-1, 1\}^n$ the problem:

$$\min\{\mathbf{v}^T \mathbf{c} : \ \mathbf{Gc} \le \mathbf{h}, \ \mathbf{c} \ge 0\},$$

obtaining a vertex $\hat{\mathbf{c}}$. The corresponding vector $\hat{\mathbf{y}}$ can be obtained by solving the nominal LP.

Each feasible solution $(\hat{\mathbf{x}}, \hat{t})$ to the MP is a lower bound on the original robust problem (8)–(10). The "Auxiliary Problem" is the maximum-regret problem for a fixed $\hat{\mathbf{x}}$:

$$\max_{\mathbf{c},\mathbf{y}} \ \mathbf{c}(\hat{\mathbf{x}} - \mathbf{y}), \tag{14}$$

$$\text{s.t.} \ \mathbf{Ay} \le \mathbf{b}, \tag{15}$$

$$\mathbf{c} \in \mathcal{U}. \tag{16}$$

When a solution to that problem satisfies $\hat{\mathbf{c}}(\hat{\mathbf{x}} - \hat{\mathbf{y}}) \le \hat{t}$ then the algorithm arrived at optimal solution. Otherwise set of cuts is extended with a new pair $(\hat{\mathbf{c}}, \hat{\mathbf{y}})$. Observe that here the maximum-regret problem is a non-convex quadratic integer problem, which is typically hard to solve, and may pose a bottleneck of the computations.

3.3 Min-Max Regret Integer Programming

A similar methodology can be also applied to min-max regret linear problems with integer (binary) decision variables and interval uncertainty in cost vector $\mathbf{c}$. For this type of combinatorial optimization problems the worst-case scenario for a solution $\mathbf{x}$ is easily found as [2]:

$$c_i(\mathbf{x}) = \begin{cases} c_i^+ & \text{if } x_i = 1, \\ c_i^- & \text{if } x_i = 0. \end{cases}$$

Consequently, the Master Problem is obtained by relaxing (18) in:

$$\min_{\mathbf{x},t} \left(\sum_{i=1}^{n} c_i^+ x_i - t \right) \tag{17}$$

$$\text{s.t. } \forall \hat{\mathbf{y}} \in P \ \ t \le \sum_{i=1}^{n} (c_i^- + (c_i^+ - c_i^-) x_i) \hat{y}_i \tag{18}$$

$$\mathbf{x} \in P, \tag{19}$$

where $P = \{\mathbf{y} \in \{0,1\}^n : \ \mathbf{Ay} \le \mathbf{b}\}$. The set P in (18) is replaced by the cut set $\mathcal{C}$, initialized with a small number of solutions $\mathbf{y} \in P$ (obtained by sampling the extreme scenarios and solving the corresponding nominal problem instances).

Having solved the relaxed Master Problem for a given cut set $\mathcal{C}$, the solution $(\hat{\mathbf{x}}, \hat{t})$ allows us to form and solve the Auxiliary Problem:

$$\hat{\mathbf{y}} \in \arg\min\{\mathbf{c}(\hat{\mathbf{x}}) \cdot \mathbf{y} : \ \mathbf{y} \in P\},$$

which is the nominal problem for the worst-case scenario $\mathbf{c}(\hat{\mathbf{x}})$. This allows to compute the maximum regret:

$$R = \mathbf{c}(\hat{\mathbf{x}})(\hat{\mathbf{x}} - \hat{\mathbf{y}}).$$

If $\sum_{i=1}^{n} u_i^+ \hat{x}_i - \hat{t} \ge R$, then we have found a robust optimal solution; otherwise we extend the set of cuts $\mathcal{C}$ by adding a vector $\hat{\mathbf{y}}$ to it. Note that this time the Auxiliary Problem's complexity depends on the structure of the nominal (integer) problem. In general, the Master Problem is mixed-integer, and the Auxiliary Problem is binary programming problem, and both could be solved via general branch and bound techniques.

4 Examples

To illustrate the modeling and problem solving workflow in PYROPT, we present the following two examples.

Example 1
First, we consider a robust min-max version of the following linear program from [6] (Sect. 1.1., the drug production problem):

$$\min_{\mathbf{x}} \ 100x_1 + 199.9x_2 - 5500x_3 - 6100x_4,$$

$$\text{s.t. } 0.01x_1 + 0.02x_2 - 0.5x_3 - 0.6x_4 \ge 0,$$

$$x_1 + x_2 \le 1000,$$

$$90x_3 + 100x_4 \le 2000,$$

$$40x_3 + 50x_4 \le 800,$$

$$100x_1 + 199.9x_2 + 700x_3 + 800x_4 \le 100000,$$

$$\mathbf{x} \ge 0.$$

This problem has only 2 uncertain parameters, the coefficients of decision variables x_1 and x_2 in the first constraint. The remaining numerical data parameters are certain. The input data can be provided via standard Python's numerical arrays:

```
import numpy as np
c = np.array([100.0, 199.9, -5500.0, -6100.0])
A = np.array([[-0.01, -0.02, 0.5, 0.6],
              [1, 1, 0, 0],
              [0, 0, 90, 100],
              [0, 0, 40, 50],
              [100, 199.9, 700, 800]])
b = np.array([0.0, 1000.0, 2000.0, 800.0, 100000.0])
```

The optimization model can be written using Python-embedded CVXPY domain-specific language. We define the decision variables and the objective function as:

```
import cvxpy as cp
x = cp.Variable(4, nonneg=True)
objective = c @ x
```

Next, we specify the uncertain parameters, and define the constraints of the robust problem:

```
import pyropt as ro
Au = ro.Uncertain(
      mid=A,
      width=np.array([[5e-5 * 2.0, 4e-4 * 2.0, 0, 0],
                      [0, 0, 0, 0],
                      [0, 0, 0, 0],
                      [0, 0, 0, 0],
                      [0, 0, 0, 0]]))
constraints = [Au @ x <= b]
   prob = ro.RobustLinearProblem(cp.Minimize(objective),
constraints)
```

Finally, we solve the problem calling `prob.solve()`, obtaining optimal value $v^* = -8294.567$. Internally, the model is reformulated as described in Sect. 3.1, and appropriate solver is called (e.g., CPLEX, Gurobi, SCS, GLPK, etc.).

Example 2

For the second example, consider the following min-max regret combinatorial problem, known as the Restricted Items Selection (RIS) problem [10]. Given are m sets of items $I_1, I_2, \ldots, I_m$, each with $|I_j| = r_i$ items. Each item has uncertain cost, denoted $c_{ij} \in [c_{ij}^-, c_{ij}^+]$, $i = 1, \ldots, m$, $j = 1, \ldots, r_i$. We want to select exactly p_i items from each ith set, so that the total cost is minimized. Moreover, some of the item pairs cannot be selected simultaneously, and the set of such forbidden indices is denoted by T.

The deterministic Restricted Items Selection problem can be formulated as:

$$\text{minimize} \quad \sum_{i=1}^{m} \sum_{j=1}^{r_i} x_{ij} c_{ij}, \tag{20}$$

$$\text{s.t.} \quad \forall_{i=1,\ldots,m} \quad \sum_{j=1}^{r_i} x_{ij} = p_i, \tag{21}$$

$$\forall_{(i,k,j,l)\in T} \quad x_{ik} + x_{jl} \leq 1, \tag{22}$$

$$\forall_{\substack{i=1,\ldots,m \\ j=1,\ldots,r_i}} \quad x_{ij} \in \{0,1\}. \tag{23}$$

To model this problem as uncertain min-max regret problem in PYROPT, we start by defining the intervals of parameters c_{ij}. We create an matrix-object of type Uncertain, and define lower and upper bounds of each parameter as follows:

```
C = ro.Uncertain(lbs=np.array([[67, 18, 58, 87, 48],
                               [33, 47, 26, 37, 81],
                               [50, 56, 3, 40, 48]]),
                 ubs=np.array([[93, 99, 84, 98, 97],
                               [74, 97, 84, 79, 97],
                               [69, 68, 85, 67, 85]]))
```

Next, we define a corresponding matrix decision variable X, and an objective function (20):

```
X = cp.Variable((3,5), boolean=True)
objective = cp.sum(cp.multiply(C, X))
```

Constraints (21)–(22) can be defined as:

```
constraints = [ X @ np.ones(5) == np.array([2, 2, 2]) ]
constraints += [ X[1,0] + X[2,4] <= 1 ]
constraints += [ X[1,3] + X[2,0] <= 1 ]
constraints += [ X[1,4] + X[2,3] <= 1 ]
constraints += [ X[0,0] + X[1,1] <= 1 ]
constraints += [ X[0,1] + X[2,3] <= 1 ]
```

Finally, the problem-object is created as an instance of RegretProblem class by passing the objective and constraints objects:

```
problem = ro.RegretProblem(cp.Minimize(objective), constraints)
```

Calling `problem.solve()` yields the robust optimal solution ($v^* = 190$ in this example, where matrix $\mathbf{x}^*$ indicates the items to select). Internally, the problem is transformed into min-max regret integer program, and the cut-generating algorithm is executed on it, as described in Sect. 3.3. For larger problem instances this scheme also yields an approximation algorithm, as the solutions to Master and Auxiliary problems would give increasingly tighter upper and lower bounds on the robust optimal solution value; the cut-generation can be terminated after a predefined time limit.

Note that the deterministic problem RIS is NP-hard, unless the constraint set T has the property that the forbiddance relation between the items is transitive. In such case, the problem can be reformulated as a linear program, and its robust version solved as min-max regret linear program for interval and polyhedral uncertainty.

5   Conclusions

We have presented an initial version of PYROPT, a comprehensive Python library for robust mathematical optimization, based on CVXPY modeling language. While limited support for robust min-max models is available in other modeling environments for optimization, PYROPT supports a wider range of models, also providing automatic transformations for min-max regret linear problems with continuous and binary variables. We believe that the presented library would simplify the process of building and solving robust optimization problems in new application areas.

The core of the library consists of a flexible and powerful modeling language for convex optimization problems. Future work would include the addition of the support of other concepts of robustness (budgeted uncertainty [7], 2-stage robustness [4], adjustable robustness [20], etc.), as well as the improvement of the existing solution algorithms and the implementation of the new ones (such as approximate methods and heuristics).

References

1. Agrawal, A., Verschueren, R., Diamond, S., Boyd, S.: A rewriting system for convex optimization problems. J. Control Decis. **5**(1), 42–60 (2018)
2. Aissi, H., Bazgan, C., Vanderpooten, D.: Min-max and min-max regret versions of combinatorial optimization problems: a survey. Eur. J. Oper. Res. **197**(2), 427–438 (2009)
3. Ali, A., Kolter, J.Z., Diamond, S., Boyd, S.P.: Disciplined convex stochastic programming: a new framework for stochastic optimization. In: UAI, pp. 62–71 (2015)
4. Atamtürk, A., Zhang, M.: Two-stage robust network flow and design under demand uncertainty. Oper. Res. **55**(4), 662–673 (2007)
5. Averbakh, I., Lebedev, V.: On the complexity of minmax regret linear programming. Eur. J. Oper. Res. **160**(1), 227–231 (2005)
6. Ben-Tal, A., El Ghaoui, L., Nemirovski, A.: Robust Optimization. Princeton University Press, Princeton (2009)

7. Bertsimas, D., Sim, M.: The price of robustness. Oper. Res. **52**(1), 35–53 (2004)
8. Diamond, S., Boyd, S.: CVXPY: a Python-embedded modeling language for convex optimization. J. Mach. Learn. Res. **17**(83), 1–5 (2016)
9. Drwal, M.: Robust scheduling to minimize the weighted number of late jobs with interval due-date uncertainty. Comput. Oper. Res. **91**, 13–20 (2018)
10. Drwal, M.: Robust approach to restricted items selection problem. Optim. Lett. **15**(2), 649–667 (2021)
11. Drwal, M., Józefczyk, J.: Robust min-max regret scheduling to minimize the weighted number of late jobs with interval processing times. Ann. Oper. Res. **284**(1), 263–282 (2020)
12. Drwal, M., Rischke, R.: Complexity of interval minmax regret scheduling on parallel identical machines with total completion time criterion. Oper. Res. Lett. **44**(3), 354–358 (2016)
13. Dunning, I., Huchette, J., Lubin, M.: JUMP: a modeling language for mathematical optimization. SIAM Rev. **59**(2), 295–320 (2017)
14. Goerigk, M., Schöbel, A.: Algorithm engineering in robust optimization. In: Kliemann, L., Sanders, P. (eds.) Algorithm Engineering. LNCS, vol. 9220, pp. 245–279. Springer, Cham (2016). https://doi.org/10.1007/978-3-319-49487-6_8
15. Gorissen, B.L., Yanıkoğlu, İ., den Hertog, D.: A practical guide to robust optimization. Omega **53**, 124–137 (2015)
16. Kasperski, A., Zieliński, P.: Robust discrete optimization under discrete and interval uncertainty: a survey. In: Doumpos, M., Zopounidis, C., Grigoroudis, E. (eds.) Robustness Analysis in Decision Aiding, Optimization, and Analytics. ISORMS, vol. 241, pp. 113–143. Springer, Cham (2016). https://doi.org/10.1007/978-3-319-33121-8_6
17. Leyffer, S., Menickelly, M., Munson, T., Vanaret, C., Wild, S.M.: A survey of nonlinear robust optimization. INFOR: Inf. Syst. Oper. Res. **58**, 1–32 (2020)
18. Lofberg, J.: YALMIP: a toolbox for modeling and optimization in MATLAB. In: 2004 IEEE International Conference on Robotics and Automation (IEEE Cat. No. 04CH37508), pp. 284–289. IEEE (2004)
19. Cass, S.: The top programming languages. IEEE Spectr. **13** (2019)
20. Yanıkoğlu, İ., Gorissen, B.L., den Hertog, D.: A survey of adjustable robust optimization. Eur. J. Oper. Res. **277**(3), 799–813 (2019)

Unsupervised Barter Model Based on Natural Human Interaction

Yasmany Fernández-Fernández[1,2], Leandro L. Lorente-Leyva[2(✉)],
Diego H. Peluffo-Ordóñez[2,3], Ridelio Miranda Pérez[4], and Elia N. Cabrera Álvarez[4]

[1] Universidad Politécnica Estatal del Carchi, Tulcán, Ecuador
yasmany.fernandez@upec.edu.ec
[2] SDAS Research Group, Ibarra, Ecuador
leandro.lorente@sdas-group.com
[3] Modeling, Simulation and Data Analysis (MSDA) Research Program, Mohammed VI
Polytechnic University, Ben Guerir, Morocco
peluffo.diego@um6p.ma
[4] Universidad de Cienfuegos, Cienfuegos, Cuba
{rmiranda,elita}@ucf.edu.cu

Abstract. Human interaction is a natural process in business management. In various indigenous cultures, the natives still use a barter system to reach consensus or balances that determine the essence of their economies. The present investigation consists of the presentation of an unsupervised model based on pure barter. The main contribution sought is to visualize the balance that is achieved in an unsupervised environment of two entities that are close to reaching an agreement. Both Game Theory and Walrasian Theory deal with the problem of exchange of goods. However, the current objective is to show the barter model from its simplest bases for the construction of an unsupervised automatic learning scheme where a system of pairs of agents represent a basic model for decision making when guaranteeing an agreement.

Keywords: Bartering · Agents · Goods · Natural human interaction · Unsupervised learning · Mathematical model

1 Introduction

The economic structure that we know today has its origin in antiquity. From the relationship between barter and money in the economies of Mesopotamia and Egypt [1] to the new macroeconomic consensus issued by the banks around 1970s [2], an eminent need for the exchange of goods can be visualized not only in humans but also in a dynamic system that is encompassed by nature and everything that it contains.

Trading to facilitate exchange began as humans walked on Earth. But, before the invention of modern money, barter was used to facilitate exchange of goods for other goods or services. Bartering as a reciprocal exchange takes place between two or more individuals who are willing to exchange the goods each person has without using a monetary medium [3, 9].

© Springer Nature Singapore Pte Ltd. 2021
T.-P. Hong et al. (Eds.): ACIIDS 2021, CCIS 1371, pp. 387–400, 2021.
https://doi.org/10.1007/978-981-16-1685-3_32

1.1 General Agent Model

A commodity is a good or fully specified service in terms of its characteristics, their spatial availability, and their temporal availability [4].

Before reaching the Walras equilibrium, it is necessary to discuss some general principles that govern the exchange of goods between agents. Indistinctly, the objective of this research is not to present the Walras model again, but rather a natural variant of the barter process that is the object for unsupervised analysis in data mining.

To achieve this, it is necessary to base we on the various mathematical approaches that surround the modeling of the processes of exchange of goods between agents.

Let us consider that there is a finite number k of distinguishable goods and suppose that any real number can represent a certain quantity of each good. We assume that the goods are divisible and the amounts that an agent can exchange is real number in the space R_+^k.

An allocation of goods is a vector:

$$x = (x_1, x_2, \ldots, x_k) \in R_+^k \tag{1}$$

Where the i-th component x_i represent the amount of a good that the costumer-(i) receive.

For a common modern economy, it is assumed that each commodity (i) has associated a price p_i that can be either scarce (+), harmful (-) or free (0). So, a price system can be represented as a vector:

$$p = (p_1, p_2, \ldots, p_k) \in R^k \tag{2}$$

Where the i-th component p_i of the vector p means the money value that the costumer-i give for the good.

In a barter economy, money is not necessarily used. The model itself is perfect information or forecasting and we can have a model called static in a steady state where agents choose plans for life [5–7].

In a general model, the agents constitute the decision-making entities of the same of the costumers [8]. Assuming that there is a finite number of n consumers where; the consumer's objective is to decide on consumption plans according to his choice and under survival and wealth decisions then; There will also be an election criterion whose behavior will be determined by the search for the best options among those eligible.

Another assumption is that agents are weight takers [9]. Each consumer I is defined by the set of own consumption X^i, the initial endowments W^i and their preferences over combinations of goods [10]:

$$W^i = \left(W_1^i, W_2^i, \ldots, W_k^i \right) \in R_+^k \tag{3}$$

Where an endowment W^i represents an agent's initial set of assets.

There is also a consumption set X^i of all possible consumptions for the i-th consumer such that $X^i \subseteq R_+^k$ in such a way that a consumption plan $x^i \in X^i$ of an agent (i) is given by:

$$x^i = \left(x_1^i, x_2^i, \ldots, x_k^i\right) \in R_+^k \tag{4}$$

Where X^i is convex and closed which implies a perfect divisibility principle.

Preferences play a fundamental role in the general model of agents. In this sense, for the relationship symbol $\succeq (or \succ)$ we assume that $x \succeq y$ $(x \succ y)$ means that the consumption plan x is weak (strict) preferable than the consumption plan y for a giving agent, thus for any two consumption plans x and y there are three assumptions; i) completeness $(x \succeq y \ or \ y \succeq x)$, ii) reflexivity $(x \succeq y)$ and finally iii) transitivity $(if \ x \succeq y \ and \ y \succeq z, \ then \ x \succeq z)$.

Thus, there is a binary relationship between pairs of consumption baskets depending on the following relationship in Table 1:

Table 1. Binary peer preferences.

Symbol	Description	Meaning
$\succ$	Denotes strict preference	$x \succ y$: Means that x is strictly preferred y
$\sim$	Denotes an indifference	$x \sim y$: means that x and y are equally preferred
$\succeq$	Denotes weak preference	$x \succeq y$: It means that x is at least as preferred as y

From [11] we have the follow Lemmas 1 and 2:

Lemma 1: If $\preceq$ is complete, reflexive, transitive and continuous, there is a utility function U: X → R that represents these preferences. U is continuous and fulfills that u (x) ≥ u (y) if and only if $x \preceq y$.

Thus, we obtain a utility function U that associates a real number to each combination of goods in X. There are three extra assumptions related to preferences [11], which are iv) the assumption of continuity $(\forall x, y \in X; \{x|x \succeq y\} \ and \ \{x|y \succeq x\} \ are \ close)$, v) the assumption of strong monotonicity $(if \ x \geq y, x \neq y \ then \ x \succ y)$ or the weakest assumption of local insatiability $(\forall x \in X, e > 0 \exists y \in X \ suchthat \ |x - y| < e \rightarrow y \succ x)$ and vi) the assumption of strict convexity $(given \ x \neq y \ and \ z \int X \ if \ x \succeq z \ and \ y \succeq z \rightarrow tx + (1 - t)y \succ z \forall t \int (0, 1))$ or we could have a weak convexity $(if \ x \succeq y, \ then \ tx + (1 - t)y \succeq y \forall t \in (0, 1))$.

Lemma 2: If $\prec$ satisfies the six last assumptions, then $\prec$ can be represented by a utility function U (.), which is continuous, increasing and strictly quasi-concave.

In this *way* we would have characterized a consumer i with his consumption set X^i, his preferences $\succeq_i \rightarrow u^i(.)$ and his initial endowments w^i [12].

Finally, if we define an assignment as $x = (x^1, x^2, ..., x^n)$ in a collection of n *consumption* plans, we will have the following feasible assignment:

$$\sum_{i=1}^{n} x^i = \sum_{i=1}^{n} w^i \tag{5}$$

Where we will be able to maximize the utility function, which is known as the consumer problem, which is defined as:

Let the function $f : R^n \rightarrow R$ be a continuous function and $A \subset R^n$ compact, then there is a vector x^* that solves [13]{:}

$$\underset{\{x\}}{Max f(x)}$$
$$st. \tag{6}$$
$$x \in A$$

The consumer problem would be generalized according to the following optimization models:

$$\underset{\{x\}}{Max\, u^i(x^i)}$$
$$st. \tag{7}$$
$$\sum_{l=1}^{k} p_l x_l^i \leq \sum_{l=1}^{k} w_l^i$$

Where p represents a pricing system as a vector $p \in IR^l$ such that $p \equiv (p1, p2, ..., pl)$, $p_k \geq 0, k = 1, 2, ..., l, p_l x_l^i$ is the expenditure of the consumer i to consume the vector $x_i \in X^i, x_i \equiv (x_{i1}, x_{i2}, ..., x_{il})$ and $w_i \in IR$ represents a consumer income.

In the consumer problem it is fulfilled that if $p > > 0$, the budget set is compact, if the objective function is continuous and equal to unity, the existence of at least one vector $x^{*i} = x^i(p, w^i)$ that is satisfied by Weiertrass's theorem [13] maximizes the consumer problem and is also continuous; with the particularity that if u^i is strictly quasi concave, $x^{*i} = x^i(p, w^i)$ is unique.

Finally, by [12] the associated demand function of $i(vector) \rightarrow x^{*i} = x^i(p, w^i)$ exists and is continuous at all points if $> =$ they are strictly convex, continuous, and monotonous.

1.2   The Walrasian Equilibrium as a Classical Barter Model

One of the fundamental objectives of this contribution is to show the basis of classical bartering from an ancestral and simple perspective that is also seen as an unsupervised barter learning system for decision-making.

Starting from this idea, this research analyzes the principles of Walras equilibrium [17] in contrast to the contributions that will be presented in the subsequent sections.

Let us define a vector of weights $p = (p_1, p_2, p_3 \ldots \ldots p_k)$ where each agent i responds to the following problem:

$$\underset{\{x\}}{i \; Max \, u^i\left(x^i\right)}$$

$$st.$$

$$px^i = pw^i$$

(8)

Where the optimal solution is $x^{*i} = x^i(p, w^i)$ and the aggregate demand function is given by:

$$x(p) = \sum_i x^{*i} = \sum_i x^i\left(p, w^i\right)$$

(9)

Where $\sum_i w^i$ represents the aggregate supply function.

The fact of achieving a balance in the exchange has generated interesting contributions based on the approach of Walras, some recent research [14–20] use in a broad and rigorous sense the Walras equilibrium principle oriented to various areas.

From [8] we have the follow Lemmas 3–6:

We want to analyze if there is a price vector p^* such that $\sum_i x^i(p, pw^i) = \sum_i w^i$ and with free goods $\sum_i x^i(p, pw^i) \leq \sum_i w^i$.

Lemma 3. Let $z(p) = \sum_i x^i(p, pw^i) - \sum_i w^i$ *the excess demand function of the economy and* $z_j(p) = \sum_i x_j^i(p, pw^i) - \sum_i w_j^i$ *the excess demand function for good j. A price vector* $p^* \geq 0$, *is a Walrasian equilibrium or competitive equilibrium if:*

$$z_j(p^*) = 0, \; \text{if j is a scarce good, } (p_j^* > 0)$$

$$z_j(p^*) < 0, \; \text{if j is a free good} \left(p_j^* = 0\right)$$

The Walrasian equilibrium is going to be defined as a fixed point of an application of the price set itself. From this idea an important concept called Brower's Fixed Point Theorem of existence is defined [12]:

Lemma 4: Let S be a convex and compact subset (closed and bounded) of a Euclidean space and let f be continuous function, f: S → S, then there will be at least one fixed point, that is, there will be an x in S such that f(x) = x (f applies a point to itself).

We focus on the price vectors that belong to the unit simplex of dimension $k - 1$:

$$S^{k-1} = p\left\{\in R_+^2 : \sum_{l=1}^k p_l = 1\right\}$$

(10)

This set S^{k-1} is called a normalized set of prices and it is bounded $(0 \leq p_l \leq 1 \, \forall l = 1, .., k)$, closed $(\{0, 1\} \, f \, S^{k-1})$ or convex (if $p', p'' \in S^{k-1}$, which implies that if $\sum_{l=1}^{k} p_l' = 1$, $\sum_{l=1}^{k} p_l'' = 1$, then $p = \lambda p' + (1 - \lambda)p''$ is in S^{k-1} given that $\sum_l p_l = \sum_l \lambda p_l' + (1 - \lambda) \sum_l p_l'' = 1$).

Finally, we define Walras' law or identity and equilibrium existence theorem as follows:

Lemma 5: $\forall p \in S^{k-1}, pz(p) = 0$, *where* $z(p)$ *is the excess demand function.*

Lemma 6: If $z : S^{k-1} \to R^k$ *is continuous and* $pz(p) = 0$ *(Lemma 5), then exists some* $p^* \in S^{k-1}$ *such that* $z(p^*) \le 0$.

1.3   Recent Works

The idea of studying the behavior of the operations that are carried out from the exchange of goods between several individuals is not new. However, the study of this phenomenon that has been occurring since antiquity, is the basis of a simple model that is being of great value for the market economy and decision making. The Table 2 shows some of the recent research using pure exchange systems as a basis:

Table 2. Recent trends

Research	Description	Reference
Applied Barter Theory	Based primarily on the synergy and barter process in the economy rather than the mathematical model	[21]
Modeling Languages for Multi-Agents System	Multi-Agent Systems for Multi-Criteria Decision Making	[22]
Sharing Economy	Recent research focuses on social studies of a barter economy based on the transformation of surpluses into potential as a means of helping unstable economies	[23]
Energy Bartering and Microgrids	Use an energy bartering framework for enhancing resilience response of networked microgrids (NMGs) against extreme events	[24]

Many processes are analyzed from the perspective of exchange systems. With the rise of data analysis, models that we appreciate in a simple way in real life can be magnified into powerful schemes for processing and decision making.

In the previous topics, the existence mathematical foundations in the bibliography have been detailed and that are related to two fundamental topics related to the contribution of this research, that is, the general system of exchange of goods between agents and the mathematical foundation of the Walrasian equilibrium.

From now on, the classical barter problem will be formalized in its simplest form as the basis for structuring a business network, but whose initial foundation and object of study is centered on two simple agents.

The rest of the document is structured as follows: Sect. 2 describes the mathematical model between two agents. Section 3 presents the results and discussion, with application of unsupervised barter scenarios based on natural human exchange of goods. Finally, the conclusions are shown in Sect. 4.

2   The Barter Mathematical Function Between Two Agents

Observing the process that any two agents A and B must go through when trying to carry out a negotiation regarding the products that both handle and the possibility that exists of materializing a change, we have the following:

Let A and B be any two agents in the real world who possess objects of value to be exchanged. Let's also say that these agents can be defined by the weight of each of the objects that they possess in a very simple way using matrices $A_{mx1}(Z)$ and $B_{mx1}(Z)$ such that $m \in N - \{0\}$ as follows:

$$A_{weightObjcts} = \begin{pmatrix} w_{A1} \\ w_{A2} \\ . \\ . \\ w_{Am} \end{pmatrix}_{mx1} \tag{11}$$

$$B_{weightObjcts} = \begin{pmatrix} w_{B1} \\ w_{B2} \\ . \\ . \\ w_{Bn} \end{pmatrix}_{nx1} \tag{12}$$

Where m, n are the number of objects (Classes) that will face each other in a barter process and w_{Am} and w_{Bn} are their respective weights indicating the value of each object and assuming the following limit $\rho \leq w_{An} \leq \xi$ and $\rho \leq w_{Bn} \leq \xi$ so that $[\rho, \xi]$ is the range of values for the weights of the giving objects. Understanding the existence of the previously declared matrices (Eqs. 11 and 12), a barter function b_{ij} can be defined based on the differences of these matrices:

$$b_{nm} = \sum_{i=1}^{n} \sum_{j=1}^{m} \left| w_{An_i} - \begin{pmatrix} w_{B1_1} \\ w_{B2_2} \\ . \\ . \\ w_{Bmj} \end{pmatrix} \right| \tag{13}$$

Where b_{nm} represents the attempted barter of the i-th object of Agent A with the j-th object of Agent B, considering the following restrictions based on the values of the weights from the perspective of each agent:

Assuming that $w_{An_i} \leq w_{Bm_j}$:

Agent A's resources are insufficient to achieve a balance in the barter and he will need objects with a weight b_{ij} for a change between objects to be executed.

It is a good time to declare an array $A_{mx1}(Z)$ that stores the results that meet this condition:

$$A_{weightInsufficient} = \begin{pmatrix} b_{11} \\ b_{12} \\ . \\ . \\ b_{1j} \\ b_{21} \\ b_{22} \\ . \\ . \\ b_{2j} \\ . \\ . \\ b_{mn} \end{pmatrix}_{mx1} , 1 \leq n \leq j, n, m \in Z_+^* \tag{14}$$

Assuming that $w_{An_i} > w_{Bm_j}$:

Agent B's resources are insufficient to achieve a balance in the barter and he will need objects with a weight b_{ij} for a change between objects to be executed.

In a similar way we define the array $B_{mx1}(Z)$:

$$B_{weightInsufficient} = \begin{pmatrix} b_{11} \\ b_{12} \\ . \\ . \\ b_{1j} \\ b_{21} \\ b_{22} \\ . \\ . \\ b_{2j} \\ . \\ . \\ b_{nm} \end{pmatrix}_{nx1} , 1 \leq m \leq j, n, m \in Z_+^* \tag{15}$$

Assuming that $w_{An_i} = w_{Bm_j}$ we would be an ideal equilibrium condition to carry out a barter matrix E (Z_+^*):

$$E_{barterEquilibrium} = 0.5 * \begin{pmatrix} b_{11} + w_{A1_1} + w_{B1_1} \\ b_{12} + w_{A1_1} + w_{B2_2} \\ \text{... ... ...} \\ \text{... ... ...} \\ b_{1j} + w_{A1_1} + w_{Bj_j} \\ b_{21} + w_{A2_2} + w_{B1_1} \\ b_{22} + w_{A2_2} + w_{B2_2} \\ \text{... ... ...} \\ \text{... ... ...} \\ b_{2j} + w_{A2_2} + w_{Bj_j} \\ \text{... ... ...} \\ \text{... ... ...} \\ b_{nm} + w_{Am_n} + w_{Bn_m} \end{pmatrix}_{mx1} , 1 \leq m \leq j, n, m \in Z_+^*$$

$$(16)$$

where $E_{barterEquilibrium}$ represents an array of b_{nm} trades that can be materialized.

It is valid to point out that when analyzing matrix E, the weight reached by the barter attempt between two goods that have competed is being obtained, from which the b values of good i with good j have reached the same value and their difference is zero. Matrix E only shows the equilibrium reached in the weight of its resources, hence it is intentionally duplicated and divided for two so that it remains at the ideal equilibrium value and the process is more illustrative at the time of its understanding.

3 Results and Discussion

Equation (13) shows in a simple way the natural process of barter, the subjective elements of it will not be considered in the first instance. In this way, the idea that two agents have goods is defined in a simple way and these goods have a weight in importance and compete in a dichotomous way, facing the probability that there is a possibility of agreement between them. Next, we will shown how the resources of two agents A and B compete to achieve equilibrium, who are going to compare their resources looking for possible consensus. For the following experiments, we will fix that the weight in importance of all resources are in the interval $[1, 5]$ (according to the model these values may change). The data structure is unsupervised, the matrices of agents A and B will be generated randomly assuming that each one has 5 resources, respectively. The other case to study will be that A has 5 resources and B has 3, other case consider 1000 resources for both and 500 and 100. These tests considering the variability of a case in which two agents could interact even without knowing anything about each other, only the properties of the competing goods.

3.1 Unsupervised Barter Scenarios Based on Natural Human Exchange of Goods

The cases shown below have been simulated according to the model proposed in Sect. 2 as a conceptual simplification of the model proposed by Walras in Sect. 1. Many authors [25–28] have been based on this principle to obtain computational models based on agents and Walras' theorems on the exchange equilibrium.

Table 3. Bartering between two agents A and B with low resources on equitable basis (5 resourceseach Agent).

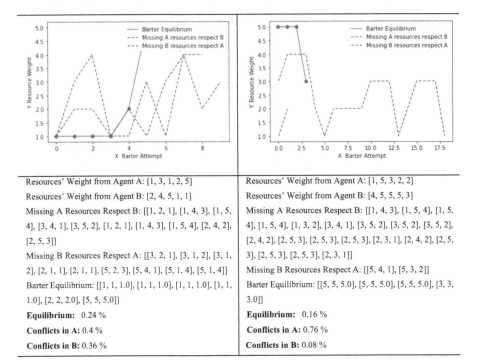

Resources' Weight from Agent A: [1, 3, 1, 2, 5]	Resources' Weight from Agent A: [1, 5, 3, 2, 2]
Resources' Weight from Agent B: [2, 4, 5, 1, 1]	Resources' Weight from Agent B: [4, 5, 5, 5, 3]
Missing A Resources Respect B: [[1, 2, 1], [1, 4, 3], [1, 5, 4], [3, 4, 1], [3, 5, 2], [1, 2, 1], [1, 4, 3], [1, 5, 4], [2, 4, 2], [2, 5, 3]]	Missing A Resources Respect B: [[1, 4, 3], [1, 5, 4], [1, 5, 4], [1, 5, 4], [1, 3, 2], [3, 4, 1], [3, 5, 2], [3, 5, 2], [3, 5, 2], [2, 4, 2], [2, 5, 3], [2, 5, 3], [2, 5, 3], [2, 3, 1], [2, 4, 2], [2, 5, 3], [2, 5, 3], [2, 5, 3], [2, 3, 1]]
Missing B Resources Respect A: [[3, 2, 1], [3, 1, 2], [3, 1, 2], [2, 1, 1], [2, 1, 1], [5, 2, 3], [5, 4, 1], [5, 1, 4], [5, 1, 4]]	Missing B Resources Respect A: [[5, 4, 1], [5, 3, 2]]
Barter Equilibrium: [[1, 1, 1.0], [1, 1, 1.0], [1, 1, 1.0], [1, 1, 1.0], [2, 2, 2.0], [5, 5, 5.0]]	Barter Equilibrium: [[5, 5, 5.0], [5, 5, 5.0], [5, 5, 5.0], [3, 3, 3.0]]
Equilibrium: 0.24 %	**Equilibrium:** 0.16 %
Conflicts in A: 0.4 %	**Conflicts in A:** 0.76 %
Conflicts in B: 0.36 %	**Conflicts in B:** 0.08 %

Tables 3, 4, 5 and 6 show, in a random and unsupervised manner, the results of applying the model proposed in Sect. 2. The results are shown clearly and briefly in the equilibrium parameters, insufficient resources of A and insufficient resources of B.

With results of this nature, events can be predicted for decision making under uncertainty. It is also possible to solve complex problems that involve a large number of data of complex structure.

Table 4. Bartering between two agents A and B with low resources in conditions of inequality (5or 3 resources each agent).

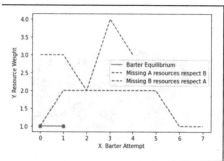	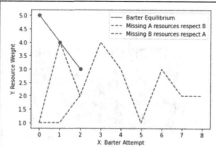
Resources' Weight from Agent A: [4, 3, 3, 1, 2]	Resources's Weight from Agent A: [5, 4, 3]
Resources' Weight from Agent B: [1, 5, 1]	Resources's Weight from Agent B: [4, 1, 3, 5, 1]
Missing A Resources Respect B: [[4, 5, 1], [3, 5, 2], [3, 5, 2], [1, 5, 4], [2, 5, 3]]	Missing A Resources Respect B: [[4, 5, 1], [3, 4, 1], [3, 5, 2]]
Missing B Resources Respect A: [[4, 1, 3], [4, 1, 3], [3, 1, 2], [3, 1, 2], [3, 1, 2], [2, 1, 1], [2, 1, 1]]	Missing B Resources Respect A: [[5, 4, 1], [5, 1, 4], [5, 3, 2], [5, 1, 4], [4, 1, 3], [4, 3, 1], [4, 1, 3], [3, 1, 2], [3, 1, 2]]
Barter Equilibrium: [[1, 1, 1.0], [1, 1, 1.0]]	Barter Equilibrium: [[5, 5, 5.0], [4, 4, 4.0], [3, 3, 3.0]]
Equilibrium: 0.14 %	**Equilibrium:** 0.2 %
Conflicts in A: 0.34 %	**Conflicts in A:** 0.2 %
Conflicts in B: 0.52 %	**Conflicts in B:** 0.6 %

Table 5. Bartering between two resourceful agents A and B on an equitable basis (1000 resources each agent).

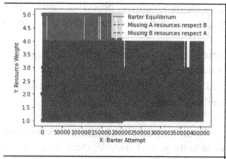	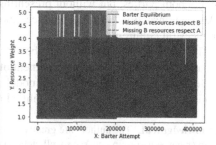
Resources' Weight from Agent A: Giant data set…	Resources' Weight from Agent A: Giant data set…
Resources' Weight from Agent B: Giant data set…	Resources' Weight from Agent B: Giant data set…
Missing A Resources Respect B: Giant data set…	Missing A Resources Respect B: Giant data set…
Missing B Resources Respect A: Giant data set…	Missing B Resources Respect A: Giant data set…
Equilibrium: 0.200066 %	**Equilibrium:** 0.199934 %
Conflicts in A: 0.406338 %	**Conflicts in A:** 0.390482 %
Conflicts in B: 0.393596 %	**Conflicts in B:** 0.409584 %

Table 6. Bartering between two resourceful agents A and B in conditions of inequality (500 or 100 each agent).

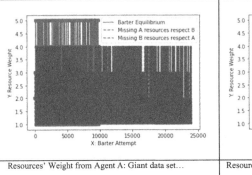 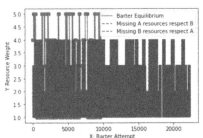

Resources' Weight from Agent A: Giant data set…	Resources' Weight from Agent A: Giant data set…
Resources' Weight from Agent B: Giant data set…	Resources' Weight from Agent B: Giant data set…
Missing A Resources Respect B: Giant data set…	Missing A Resources Respect B: Giant data set…
Missing B Resources Respect A: Giant data set…	Missing B Resources Respect A: Giant data set…
Equilibrium: 0.19634 %	**Equilibrium:** 0.19862 %
Conflicts in A: 0.48128 %	**Conflicts in A:** 0.44764 %
Conflicts in B: 0.32238 %	**Conflicts in B:** 0.35374 %

4   Conclusions

A common economic practice in many of the indigenous communities of the South American continent is associated with the exchange of goods and merchandise without the intermediation of money. These exchanges are frequently carried out, under empirical criteria and/or abstract formalities with the intention of promoting a transaction that benefits both parties, but without presenting, as far as we know, objective criteria that make the transaction fair. This practice, known as barter, has formed and is still part of our ancestral cultures.

The present investigation has been carried out to promote the mathematical formalization of barter in an environment of unsupervised data.

Both the general agent-based model and the Walras equilibrium constitute a foundation for the realization of simplified machine learning constructs.

Acknowledgment. The authors are greatly grateful by the support given by the SDAS Research Group (www.sdas-group.com). As well, acknowledge to the project "Development of new mathematical models and methods for decision making" of the Department of Mathematics and Computation of the University of Havana.

References

1. Svizzero, S., Tisdell, C.: Barter and the origin of money and some insights from the ancient palatial economies of mesopotamia and Egypt (2019)

2. Jia, D.L..: Dynamic Macroeconomic Models in Emerging Market Economies: DSGE Modelling with Financial and Housing Sectors. Springer, Heidelberg (2020). https://doi.org/10.1007/978-981-15-4588-7

3. Missos, V.: Mathematical analysis as a source of mainstream economic ideology. Econ. Thought **9**(1), 72–95 (2020)

4. Rockafellar, R.T.: Optimization and decentralization in the mathematics of economic equilibrium. other words **1**, 1–4 (2020)

5. Hua, G., et al.: The newsvendor problem with barter exchange. Omega **92**, 102149 (2020). https://doi.org/10.1016/j.omega.2019.102149

6. Dutta, C., Singh, R., Garg, K., Choudhury, T.: Modern day approach for buying and selling needs with the use of information technology: A barter system. J. Crit. Rev. **7**(12), 1451–1457 (2020)

7. Zheng, Y., Yang, T., Zhang, W., Zhao, D.: Barter Exchange via Friends' Friends. arXiv:201004933 [cs] (2020)

8. Dubovikov, M.: Mathematical model of modern economy. Handel Wewnętrzny **352**(5), 13–24 (2014)

9. Taskinsoy, J.: From Primitive Barter to Inflationary Dollar: A Warless Economic Weapon of Mass Destruction. Available at SSRN 3542145 (2020). https://doi.org/10.2139/ssrn.3542145

10. Köster, R., et al.: Model-free conventions in multi-agent reinforcement learning with heterogeneous preferences. arXiv preprint arXiv:2010.09054 (2020)

11. Espinel, R.: Las matemáticas simples del equilibrio general. Matemática **2**(2) (2020)

12. Mas-Colell, A., Whinston, M.D., Green, J.R.: Microeconomic Theory, vol. 1. Oxford University Press, New York(1995).

13. Stone, M.H.: The generalized Weierstrass approximation theorem. Math. Mag. **21**, 237–254 (1948). https://doi.org/10.2307/3029337

14. Hands, D.W.: Derivational robustness, credible substitute systems and mathematical economic models: The case of stability analysis in Walrasian general equilibrium theory. Eur. J. Philos. Sci. **6**(1), 31–53 (2015). https://doi.org/10.1007/s13194-015-0130-0

15. Zhang, W.-B.: A Simple Growth model based on neoclassical growth, monopolistic competition, and Walrasian general equilibrium theories. Int. J. Acad. Res. Bus. Soc. Sci. **9**(3) (2019). https://doi.org/10.6007/IJARBSS/v9-i3/5758

16. Deride, J., Jofré, A., Wets, R.-B.: Solving Deterministic and Stochastic Equilibrium Problems via Augmented Walrasian. Comput. Econ. **53**(1), 315–342 (2017). https://doi.org/10.1007/s10614-017-9733-1

17. Lahkar, R.: Convergence to Walrasian equilibrium with minimal information. J. Econ. Interac. Coord. **15**(3), 553–578 (2019). https://doi.org/10.1007/s11403-019-00243-8

18. Loch-Temzelides, T.: Walrasian Equilibrium Behavior in Nature (2020)

19. Rubin, G.: Oskar Lange and the Walrasian interpretation of IS-LM. J Hist Econ Thought **38**(3), 285–309 (2016). https://doi.org/10.1017/S1053837216000341

20. Zhang, B., Wei, L., Chen, M.: Implications of Barter Exchange and Decision Biases in a Pull Contract. Available at SSRN 3596186 (2020)

21. Molina-Jimenez, C., Nakib HDA, Song L, et al.: A Case for a Currencyless Economy Based on Bartering with Smart Contracts. arXiv:201007013 [cs] (2020)

22. Asici, T.Z., Tezel, B.T., Kardas, G.: On the use of the analytic hierarchy process in the evaluation of domain-specific modeling languages for multi-agent systems. Journal of Computer Languages **62**, 101020 (2021). https://doi.org/10.1016/j.cola.2020.101020

23. Drummond, T.: Exploring the Sharing Economy in Economically Distressed Environments (2020)

24. Mehri Arsoon, M., Moghaddas-Tafreshi, S.M.: Peer-to-peer energy bartering for the resilience response enhancement of networked microgrids. Appl. Energy **261**, 114413 (2020). https://doi.org/10.1016/j.apenergy.2019.114413

25. Ozturan, C.: Barter Machine: An Autonomous, Distributed Barter Exchange on the Ethereum Blockchain. Ledger, 5 (2020). https://doi.org/10.5195/ledger.2020.148

26. Bigoni, M., Camera, G., Casari, M.: Cooperation among strangers with and without a monetary system. In: Handbook of Experimental Game Theory. Edward Elgar Publishing, pp 213–240 (2020)

27. Kelejnikova, S., Shlapakova, N., Samygin, D.: Model of Decision Support on Regulation of the Vegetable Market. IOP Conference Series: Materials Science and Engineering **753**(5), 052051 (2020). https://doi.org/10.1088/1757-899X/753/5/052051

28. Sankar, C.P., Thumba, D.A., Ramamohan, T.R., Chandra, S.S.V., Satheesh Kumar, K.: Agent-based multi-edge network simulation model for knowledge diffusion through board interlocks. Expert Systems with Applications 141, 112962 (2020). https://doi.org/10.1016/j.eswa.2019.112962

Data Modelling and Processing
for Industry 4.0

Predictive Maintenance for Sensor Enhancement in Industry 4.0

Carla Silva[1](✉)(iD), Marvin F. da Silva[1](✉), Arlete Rodrigues[4], José Silva[4], Vítor Santos Costa[1,3](iD), Alípio Jorge[1,3](iD), and Inês Dutra[1,2](✉)(iD)

[1] Department of Computer Science, Faculty of Sciences, University of Porto, Porto, Portugal
carlasilva@fe.up.pt, {marvin.silva,amjorge}@fc.up.pt, {vsc,ines}@dcc.fc.up.pt
[2] CINTESIS, Porto, Portugal
[3] INESC-TEC, Porto, Portugal
[4] Bosch Security Systems, Ovar, Portugal
{Arlete.Rodrigues,Jose.Silva2}@pt.bosch.com

Abstract. This paper presents an effort to timely handle 400+ GBytes of sensor data in order to produce Predictive Maintenance (PdM) models. We follow a data-driven methodology, using state-of-the-art python libraries, such as Dask and Modin, which can handle big data. We use Dynamic Time Warping for sensors behavior description, an anomaly detection method (Matrix Profile) and forecasting methods (AutoRegressive Integrated Moving Average - ARIMA, Holt-Winters and Long Short-Term Memory - LSTM). The data was collected by various sensors in an industrial context and is composed by attributes that define their activity characterizing the environment where they are inserted, e.g. optical, temperature, pollution and working hours. We successfully managed to highlight aspects of all sensors behaviors, and produce forecast models for distinct series of sensors, despite the data dimension.

Keywords: Prognostic techniques · Predictive maintenance · Internet of Things · Industry 4.0 · Fire detection

1 Introduction

The fire detection industry is currently working on the accuracy of the early fire detection. Improvements in the schedule of the maintenance component is needed to avoid costs and losses of productivity. However, the planning of fire detection systems is hard, since it involves a large number of system variables, deals with big data and requires high computational resources. PdM, also known as online monitoring, condition-based maintenance, or risk-based maintenance, has been sought by many companies as a way of reducing costs and losses when, instead of detecting that something is wrong, it predicts that something will go wrong in order to plan ahead. In PdM, different types of approaches can be used:

© Springer Nature Singapore Pte Ltd. 2021
T.-P. Hong et al. (Eds.): ACIIDS 2021, CCIS 1371, pp. 403–415, 2021.
https://doi.org/10.1007/978-981-16-1685-3_33

(a) data-driven, (b) model-based, and (c) hybrid [13]. Most solutions currently use advanced modelling and data-driven techniques [7]. Prognostic is one of the most relevant features of PdM that attempts to predict a prospective failure of a device or component.

Nowadays, PdM unlike prognostic approaches have been presented and compared according to their requirements and performance [1]. Moreover, PdM driven approaches based in unsupervised learning techniques, such as continuous outlier monitoring, have been successfully implemented [12]. Besides, other Machine Learning (ML) techniques seem to be efficient methods, in particular Random Forests (RFs) which is one of the most used and compared ML method in PdM applications. Resorting to RFs allows for, e.g., predicting repairs for various components of commercial vehicles or performing monitoring of wind turbines using status and operational data. On the other hand, Artificial Neural Networks (ANNs) based methods (e.g. Deep Learning (DL), Convolutional Neural Network (CNN)) also have a wide range of applications, e.g., soft sensing and predictive control, mimic the operational condition of a wind turbine to monitor its conditions, predict current condition of an engine or predict faults in acoustic sensor and photovoltaic panel. In addition, Support Vector Machines (SVMs) can, e.g., identify failures in automotive transmission boxes, predict alarm faults in a bearing of a rail network, identify failures that occur on machines due to the accumulative effects of usage and stress on equipment parts or detect geometry defects in railway tracks. Furthermore, clustering methods, such as k-means, can aid in the identification of the characterization of each cluster that induces to a fault or an alert for possible maintenance actions [3]. Moreover, ML models have been used for different maintenance tasks, e.g. classification, regression, anomaly detection. For example, in classification, a model can be extended from diagnostic to prognostic to predict future times $t + 1$. Also, in regression-based anomaly detection, an alarm is raised in case a deviation occurs in consecutive intervals from a healthy state system by comparing granularity variations [17].

Currently, the industrial Internet of Things (IIoT) resort to Internet of Things (IoT) technologies in manufacturing to tackle data from multiple specific sensors and seek for algorithms to extract knowledge from the core industry business. The IoT covering the topic predictive maintenance aims to observe sensors malfunction or deterioration before it occurs. Advances in sensor technologies and predicting this type of scenarios can lead to faster and efficient productivity in industry. The capabilities of PdM have shown many advances in recent years, partly, resorting to the predictive power of Machine Learning. ML provides a crucial element in IIoT contributing for the constant improvement and enhancement of manufacturing systems [4,8,10]. However, building models based on Gbytes or even Tbytes of data is not a trivial task. In this work, we build a pipeline and methodology, resorting to state-of-the-art python libraries that handle big data, in order to build such predictive models using full data. We employ distance-based techniques [14] to describe sensors behavior, matrix profile (MP) [2] to spot abnormalities and forecasting methods [5] for sensor enhancement. By processing 400+ GBytes of data, we managed to discover critical sensor behaviour for some variables and different groups of sensors, and our models are capable of predicting future behaviour with very low error.

The remaining of the paper is organized as follows. Section 2 describes the data sources and methods used. Section 3 provides a general overview on the background. In more detail, this section describes the sliding window and distance-based methods used, as well as, the forecasting methods. Next, we present in Sect. 4 the experimental results and analysis. Finally, Sect. 5 presents the conclusion.

2 Materials and Methods

The datasets are composed by Condition Monitoring (CM) data which covers information about physical building conditions and components variables. The most important features in this type of problem are: building (temperature, EMC, chemical) and internal devices (optical, working hours, voltage) conditions. The dataset is composed by device data of many customers.

As an example, one customer has 256,417,209 entries (sensors observations) and 60 variables (sensors types of measurement e.g. optical, thermal, chemical or sensors addresses e.g. logical and physical) for the time period 2016-09-23 to 2020-03-21. For this customer, various forecasting methods were tested, and the ones that better fit the sensors behavior were selected to apply in all 54 available Remote Portal customers data. The total sample has 1,682,596,042 entries and 60 variables. The time period ranges from 2016-03-27 to 2020-05-27. Sensors of the same type can be of various production series.

2.1 Algorithm and Pipeline

A common task in machine learning, is to build algorithms that can learn from and perform predictions on data. The algorithms work by making data-driven predictions from mathematical models and input data. In order to create a predictive sensor-based system we designed an algorithm that combines forecasting methods with distance-based and sliding window methods for pattern recognition and anomaly detection in the context of PdM. We considered from a list of critical sensors D the sensor s behavior concerning an attribute a (e.g. optical, temperature, chemical) from a series S, which can be represented by x (a aggregated on S) in time t. The model is initially fit on a training dataset with time interval 0 to $t - WS$ (window size WS), and afterwards, the fitted model is used to predict the observations in a validation dataset, where the forecast is performed on the validation data according to WS. See Algorithm 1 and Fig. 1.

The algorithms used to analyse the temporal behaviour of the sensors are the classical statistical models ARIMA and Holt-Winters plus a more recent method based on matrix profiles, and an artificial neural network-based, LSTM, all described next.

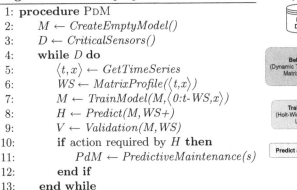

Algorithm 1. The proposed PdM
1: **procedure** PDM
2:      $M \leftarrow CreateEmptyModel()$
3:      $D \leftarrow CriticalSensors()$
4:      **while** D **do**
5:          $\langle t, x \rangle \leftarrow GetTimeSeries$
6:          $WS \leftarrow MatrixProfile(\langle t,x \rangle)$
7:          $M \leftarrow TrainModel(M, \langle 0{:}t\text{-}WS, x \rangle)$
8:          $H \leftarrow Predict(M, WS+)$
9:          $V \leftarrow Validation(M, WS)$
10:          **if** action required by H **then**
11:              $PdM \leftarrow PredictiveMaintenance(s)$
12:          **end if**
13:      **end while**
14: **end procedure**

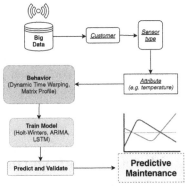

Fig. 1. PdM pipeline.

2.2 ARIMA Model

The ARIMA model can be described as the future value of the time series being a linear combination of past values and errors. The model can be shown as follows:

$$V_t = a_0 + a_1 V_{t-1} + a_2 V_{t-2} + ... + a_p V_{t-p}$$
$$+\varepsilon_t - b_1 \varepsilon_{t-1} - b_2 \varepsilon_{t-2} - ... - b_q \varepsilon_{t-q} \tag{1}$$

where, V_t is the actual measure value of the variable (e.g. optical, temperature, chemical) and ε_t is the random error at time t, a_i and b_i are coefficients, and the integers, p, an autoregressive parameter and, q, a moving average parameter.

2.3 Holt-Winters Model

Holt (1957) and Winters (1960) extended Holt's method to capture seasonality which is composed by the forecast equation and three smoothing equations that describe the models. Holt-Winters can be divided in two methods of seasonal factor modeling: additive and multiplicative.

Additive

$$Series: S_t = \alpha(V_t - c_{t-N}) + (1 - \alpha)(S_{t-1} + G_{t-1}) \tag{2}$$

$$Trend: G_t = \beta(S_t - S_{t-1}) + (1 - \beta)G_{t-1} \tag{3}$$

$$Seasonal factors: c_t = \gamma(V_t - S_t) + (1 - \gamma)c_{t-N} \tag{4}$$

$$Forecast: f_{t+m} = S_t + G_t m + c_{t-N+m} \tag{5}$$

Multiplicative

$$Series : S_t = \alpha \frac{V_t}{c_{t-N}} + (1 - \alpha)(S_{t-1} + G_{t-1}) \tag{6}$$

$$Trend : G_t = \beta(S_t - S_{t-1}) + (1 - \beta)G_{t-1} \tag{7}$$

$$Seasonal\,factors : c_t = \gamma \frac{V_t}{S_t} + (1 - \gamma)c_{t-N} \tag{8}$$

$$Forecast : f_{t+m} = (S_t + G_t m)c_{t-N+m} \tag{9}$$

where N is the length of the seasonal cycle, V_t is the actual measure value of the variable, S_t is the deseasonalized series, G_t is the trend, c_t is the seasonal factor, f_{t+m} is the forecast value for m periods ahead, and α, β and γ are smoothing constants [18].

2.4   Neural Networks

Neural networks are alternatives to statistical methods that have proven to be effective in various domains (for more details see, e.g. [6]). In this paper we use a neural network composed of two different types of layers: long short-term memory (LSTM) and fully connected (FC) layers.

Fully Connected Neural Networks. A fully connected layer is a function f from $\mathbb{R}^n$ to $\mathbb{R}^m$. If $x \in \mathbb{R}^n$ is an input to our layer, then $f(x) = \sigma(Wx + b)$; W and b are learnable parameters, σ is an activation function.

LSTM Neural Networks. LSTM layers are a type of recurrent neural network (RNN) architecture that has been shown to be effective for some time-series tasks (for more details see [6] and [9]).

2.5   Distributed Data Science and Machine Learning

Big data presents unique computational challenges. For instance, some common machine learning approaches assume that data can be held seamlessly in memory, which might not be the case for larger datasets - it certainly is not the case for our use case. Processing big datasets naïvely might also take an intolerably large amount of time. Issues of this kind require some degree of care to overcome successfully. This project utilizes several packages to handle these requirements; these packages are described below:

– Dask Data science pipelines use dataframes (two-dimensional tabular data structures with labeled axes) to represent the underlying dataset in computer memory. In Python, this functionality is provided by the popular pandas library. Pandas, however, is not able to cope either with multiprocessing or larger-than-memory datasets. Dask is used throughout this project's codebase to correct, in part, this deficiency. Using Pandas as a basic building block, Dask allows for lazy-evaluation of standard Pandas operations on multiple processors, only committing to memory a fraction of the dataset at any given time.

- RAPIDS Over the last few years, deep learning approaches have seen success on a variety of topics. Concomitant with, and essential to, this success has been the emergence of powerful dedicated hardware: GPUs. RAPIDS is a python library that allows for the execution of standard dataframe operations on GPUs - achieving tremendous (orders of magnitude) speedups in some tasks -, while at the same time integrating well with Dask.
- Modin While RAPIDS can achieve impressive speed gains for certain tasks, it is still in development (with all the attendant bugs), and it is not appropriate for other tasks. In such cases we use Modin: a python library that translates Pandas to a massively parallel setting, while still retaining easy integration with Dask for memory management. This has yielded an order of magnitude speedup on certain tasks.

Taking advantage of GPUs in the core of high-performance computing, improvements to a data science architecture, can be performed resorting to libraries for scaling data processing workloads (e.g. Dask [16] inspired by Pandas). In this scenario, the data processing task can be executed at the thread or process level which enables parallelism to occur and accelerate the data processing task. In this context, Dask can scale to multiple nodes by providing abstractions for data frames and multidimensional arrays. Furthermore, there is also Modin[1] which enables fast processing by directly replacing the Pandas data frame object and can integrate background frameworks such as Dask which can be much faster compared to pandas. In addition, along with Dask capabilities, cuML from RAPIDS[2], distribute and accelerate the machine learning process by providing GPU-accelerated ML algorithms (i.e. versions of packages in Scikit-learn), also time series algorithms (e.g. ARIMA). Besides, taking into account that Pandas data frame uses the row and column-based format, also Apache Parquet[3] which is designed for data serialization and storage on disk can accelerate the loading of the data files. This GPU-accelerated Python tools for data science improved our algorithms performance in a large scale and smoothed our work with the big data issue that we deal in the fire detection challenge. In brief, in this work the improvements at each level were provided by the following: (a) data loading (Apache Parquet), data processing (Dask and Modin) and machine learning (RAPIDS cuML) with the following steps of distributed cuML [15]:

- Training: the fit is done on all workers containing data in the training dataset.
- Model parameters after training: the trained parameters are lead to a single worker.

---

[1] https://github.com/modin-project/modin.
[2] https://rapids.ai.
[3] https://parquet.apache.org.

– Model for prediction: the trained parameters are transmitted to all workers which have partitions in the prediction dataset and predictions are done in parallel.

3   Background

This section explores the methods applied which aim to capture variable behavior and build forecasting models: DTW and Matrix Profile for capturing behavior; and ARIMA, Holt-Winters and LSTM to build forecasting models.

3.1   Sliding Window and Distance-Based Methods

Sensors behavior can be described by measuring the similarity between temporal sequences using a distance-based method such as DTW. In brief, DTW divides the series into equal points and calculates the Euclidean distance between a point in one series and another point in the other series. It keeps the minimum distance calculated, known as the time warp step. This method is useful to describe clusters of sensor behavior over critical attributes (e.g. opt1 and temp1). We focus the research in a collection of data from sensor series where the series are composed by clusters of sensors, namely, optical (O), thermal (T), optical and thermal (OT), dual optical (DO), dual optical and thermal (DOT) and chemical (C).

On the other hand, Matrix Profile (MP) allows to uncover motifs (repeated patterns) and discords (anomalies) in each cluster time series. MP uses a sliding window approach and performs calculations in two main steps: the distance profile and the profile index. In summary, the method computes the distances for the windowed sub-series and compares them with the entire series, exclude insignificant matches, updates the distance profile with the minimal values of a z-normalized Euclidean distance and sets the first nearest-neighbor index, using a sliding window approach. With a window size of m, the algorithm:

1. Computes the distances for the windowed sub-sequence against the entire time series
2. Sets an exclusion zone to ignore trivial matches
3. Updates the distance profile with the minimal values
4. Sets the first nearest-neighbor index

The distance calculations outlined above occur $n - m + 1$ times; where n is the length of the time series and m is the window size. It is relevant to notice that this method chooses the most appropriate window size that minimizes redundancy and trivial patterns.

3.2   Forecasting Methods

Forecasting methods are applicable to perform time predictions and capture future trends. In time series analysis, to model time series behavior, we usually consider: (a) a typical value (average); (b) the slope (trend); and (c) a cyclical repeating pattern (seasonality). A well-known statistical method for time series forecasting is the ARIMA model. In a time series, this method is capable of capturing a set of different temporal structures which can aid to predict future outcomes based on current and previous observations. In addition, the Holt-Winters method can be used to analyze time series where the three characteristics of the time series behavior (value, trend, and seasonality) are expressed as three types of exponential smoothing. Therefore, Holt-Winters is also known as triple exponential smoothing, and a suitable approach to model sensors behavior.

ARIMA and Holt-Winters approaches depend on parameters that can lead to a diverse number of possible input configurations, in that sense, before applying the forecasting methods over critical variables, and to find the best fit of a time series model, the Box-Jenkins method was applied. The method addresses the following: (a) study the seasonality and stationarity of the sensor temporal data; (b) perform differencing to achieve stationarity; and (c) identify the order with the aid of autocorrelation and partial autocorrelation charts. Forecasting results presented for ARIMA on the test set used a model trained with the best window size for each series. Distinct variables may have unlike best window sizes for prediction.

We also consider non-statistical methods for time series forecasting, and in particular, the use of neural networks. These types of models can learn complex non-linear patterns in the data, and present an interesting contrast to the other statistical models examined.

4   Experimental Results

4.1   Distance-Based Methods – Dynamic Time Warping

The results presented in this section are obtained using all CM customers data. However, each figure reports results for one single series. Missing values are estimated using linear interpolation. The variables studied are optical (*opt1*, units), temperature (*temp1*, °C) and working hour count (*workinghourCnt*, 1 h per digit).

Figures show the time in the x-axis (per day for the semester starting on January 1, 2020), the average of the variables absolute values in the left y-axis for all sensors measured in the respective day, and the number of sensors in the right y-axis. Blue curves and bars show all sensors of a given series, while orange curves and bars show one specific type of sensor from that series. Curves represent the average variable values, while the bars represent the number of sensors.

Figure 2a and Fig. 2b show examples of some families of sensors compared with others in the same series. For example, the family of sensors in Fig. 2a (orange line) has a high Euclidean distance (2484.4) from the whole series, with respect to optical measurements. Regarding temperature (Fig. 2b), the distance between a family of sensors to the rest of the sensors in the same series is quite small (19.2).

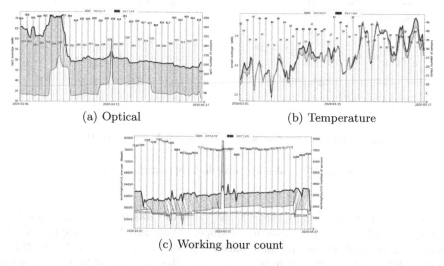

(a) Optical   (b) Temperature

(c) Working hour count

Fig. 2. DTW results for some families of sensors compared with all sensors of the same series for attributes: optical (*opt1*, units), temperature (*temp1*, °C) and working hour count (*workinghourCnt*, 1 h per digit) with x-axis: day, y-axis left: average variables values, y-axis right: number of sensors.

4.2   Sliding Window and Distance-Based Methods – Matrix Profile

The results presented in this section were obtained using all CM customers data, and uses only data corresponding to the first semester of 2020. The example variable is temperature (*temp1*). Data was aggregated on an hourly basis. For missing values, linear interpolation was used. The figures show: the distinct motifs found by the Matrix Profile method related with the behavior of a variable along the time; the timestamps associated with the occurrence of each motif in the series; the discords for a window size; and the discords for the entire series (6-month period). The method finds the top K number of motifs (patterns) given a matrix profile. By default, the algorithm finds up to 3 motifs (K) and up to 10 of their neighbors (the maximum number of neighbors to include for a given motif) with a radius of 3x minimum distance using the regular matrix profile. The radius associates a neighbor by checking if the neighbor's distance is less than or equal to distance x radius.

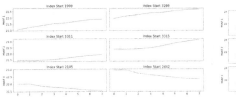

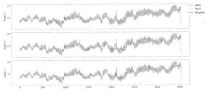

(a) The 3 motifs found, x-axis: window size found by the algorithm (7 hours), y-axis: variable values

(b) The 3 motifs shown in the entire series (6-month period), x-axis: hour index, y-axis: *temp1* average values

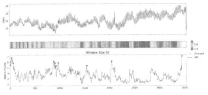

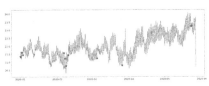

(c) Discords shown for a window slice

(d) Discords shown for the entire series

Fig. 3. MP results for variable temperature (*temp1*).

We can notice in Fig. 3a that the temperature has only common patterns of only ascending or only descending slope, which are found throughout the series. It is possible to see the same pattern, Motif 1, starting at index 1999 and starting later in the series at index 3269. Also, the presence of Motif 2 with the pattern starting at index 1011 and later at index 3315. Moreover, Motif 3 with the pattern starting at index 2105 and later at index 2442. These patterns are shown in red in Fig. 3b. Figure 3c shows the window size (92 hours) and the 10 discords found. Figure 3d shows the discords for variable *temp1* across the entire time series.

4.3 Forecasting Methods

Forecasting is divided in many steps. Before applying the forecasting model to find the best fit of a time series model, the Box-Jenkins method was used to study seasonality and stationarity. Differencing was needed to achieve stationarity and identify the order with the aid of autocorrelation and partial autocorrelation charts, Fig. 4.

Regarding our neural network model, we picked a simple architecture that consists of an LSTM layer stacked with a fully connected layer (LSTM-FC), trained using a mean squared loss and the Adam optimizer [11].

The results of forecasting using ARIMA, Holt-Winters and LSTM-FC for the CM data are presented. Holt-Winters is the method that presents smaller error rate. However ARIMA and LSTM-FC consistently manage to perform a good forecast for the variables and time period studied. Next examples show the model fit for the training set, the fit for the test set, and, for ARIMA and

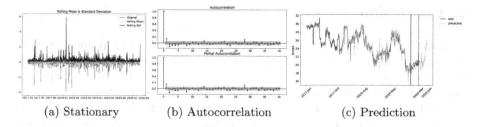

(a) Stationary   (b) Autocorrelation   (c) Prediction

Fig. 4. Forecasting steps for temperature *temp1*.

Holt-Winters, a small future projection. The example results for LSTM-FC take into account only one system of a single customer, which may favor forecasting, given that, in general, variable values of a system of a single customer do not present a large variation (Fig. 5).

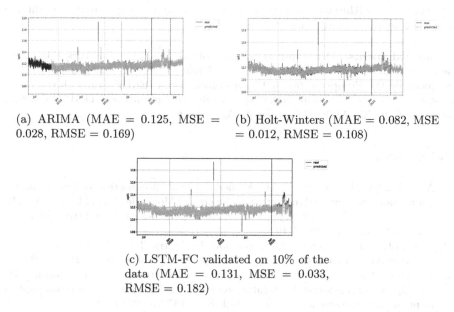

(a) ARIMA (MAE = 0.125, MSE = 0.028, RMSE = 0.169)

(b) Holt-Winters (MAE = 0.082, MSE = 0.012, RMSE = 0.108)

(c) LSTM-FC validated on 10% of the data (MAE = 0.131, MSE = 0.033, RMSE = 0.182)

Fig. 5. Forecasting results for the optical sensors type which belongs to one system of one customer.

For each time series we always take into account: (a) the direction of the trend, (b) the series seasonality that shows the presence of variations that occur at specific regular intervals, and (c) the residuals, which are the time series after removing trend and seasonal components. Finally, to evaluate the prediction error rates and model performance the most usual error metrics were used.

5 Conclusion

We managed to timely handle 400+ Gbytes of sensor data by properly resorting to state-of-the-art python libraries. Our models take into account the full data. We studied critical sensor attributes and concluded that when comparing a family of sensors with others in the same series, the average distance for some families can be very high, which may indicate some malfunctioning of those sensors. Regarding forecasting, our models produced reasonable results, even when handling all sensors of different systems and customers. We expected this modeling to cause problems, since variable values can differ significantly depending on the system's location (region). However DTW, ARIMA and Holt-Winters behaved well. LSTM-FC, which was applied to one system of a single customer also behaved well. The Matrix Profile analysis indicates the appropriate size of the sliding window for sensors maintenance. The ideal window size for the detection of motifs and discords, combined with forecasting methods such as ARIMA, Holt-Winters or LSTM-FC, can reinforce the prediction of the occurrence of events within the scope of industry 4.0, contributing to better predictive power for sensor maintenance.

Acknowledgments. This paper is a result of the project Safe Cities - Inovação para Construir Cidades Seguras, with the reference POCI-01-0247-FEDER-041435, co-funded by the European Regional Development Fund (ERDF), through the Operational Programme for Competitiveness and Internationalization (COMPETE 2020), under the PORTUGAL 2020 Partnership Agreement.

References

1. Abid, K., Sayed Mouchaweh, M., Cornez, L.: Fault prognostics for the predictive maintenance of wind turbines: state of the art. In: Monreale, A., et al. (eds.) ECML PKDD 2018. CCIS, vol. 967, pp. 113–125. Springer, Cham (2019). https://doi.org/10.1007/978-3-030-14880-5_10
2. Benschoten, A.H.V., Ouyang, A., Bischoff, F., Marrs, T.W.: MPA: a novel cross-language API for time series analysis. J. Open Sour. Softw. **5**(49), 2179 (2020)
3. Carvalho, T.P., Soares, F.A.A.M.N., Vita, R., Francisco, R.D.P., Basto, J.P., Alcalá, S.G.S.: A systematic literature review of machine learning methods applied to predictive maintenance. Comput. Ind. Eng. **137**, 106024 (2019)
4. Compare, M., Baraldi, P., Zio, E.: Challenges to IoT-enabled predictive maintenance for industry 4.0. IEEE Internet Things J. **7**(5), 4585–4597 (2020)
5. Cryer, J.D., Chan, K.S.: Time Series Analysis With Applications in R. Springer, Heidelberg (2008). https://doi.org/10.1007/978-0-387-75959-3
6. Goodfellow, I., Bengio, Y., Courville, A.: Deep Learning. The MIT Press, Cambridge (2017)
7. Hashemian, H.M., Bean, W.C.: State-of-the-art predictive maintenance techniques. IEEE Trans. Instrum. Meas. **60**(10), 3480–3492 (2011)
8. Him, L.C., Poh, Y.Y., Pheng, L.W.: IoT-based predictive maintenance for smart manufacturing systems. In: 2019 Asia-Pacific Signal and Information Processing Association Annual Summit and Conference (APSIPA ASC), pp. 1942–1944 (2019)

9. Hochreiter, S., Schmidhuber, J.: Long short-term memory. Neural Comput. **9**(8), 1735–1780 (1997)

10. Kanawaday, A., Sane, A.: Machine learning for predictive maintenance of industrial machines using IoT sensor data. In: 2017 8th IEEE International Conference on Software Engineering and Service Science (ICSESS), pp. 87–90 (2017)

11. Kingma, D.P., Ba, J.: Adam: a method for stochastic optimization. In: Bengio, Y., LeCun, Y. (eds.) 3rd International Conference on Learning Representations, ICLR 2015, San Diego, CA, USA, 7–9 May 2015, Conference Track Proceedings (2015)

12. Naskos, A., Kougka, G., Toliopoulos, T., Gounaris, A., Vamvalis, C., Caljouw, D.: Event-based predictive maintenance on top of sensor data in a real industry 4.0 case study. In: Cellier, P., Driessens, K. (eds.) ECML PKDD 2019. CCIS, vol. 1168, pp. 345–356. Springer, Cham (2020). https://doi.org/10.1007/978-3-030-43887-6_28

13. Paolanti, M., Romeo, L., Felicetti, A., Mancini, A., Frontoni, E., Loncarski, J.: Machine learning approach for predictive maintenance in industry 4.0. In: 2018 14th IEEE/ASME International Conference on Mechatronic and Embedded Systems and Applications (MESA), pp. 1–6 (2018)

14. Rakthanmanon, T., et al.: Searching and mining trillions of time series subsequences under dynamic time warping. In: Proceedings of the 18th ACM SIGKDD International Conference on Knowledge Discovery and Data Mining, KDD 2012, pp. 262–270. Association for Computing Machinery, New York (2012)

15. Raschka, S., Patterson, J., Nolet, C.: Machine learning in python: main developments and technology trends in data science, machine learning, and artificial intelligence. Information **11**(4), 193 (2020)

16. Rocklin, M.: Dask: parallel computation with blocked algorithms and task scheduling. In: Proceedings of the 14th Python in Science Conference, pp. 130–136 (2015)

17. Stetco, A., et al.: Machine learning methods for wind turbine condition monitoring: a review. Renew. Energy **133**, 620–635 (2019)

18. Zhu, Y., Zhao, Y., Zhang, J., Geng, N., Huang, D.: Spring onion seed demand forecasting using a hybrid holt-winters and support vector machine model. PLOS One **14**(7), 1–18 (2019)

Cloud Based Business Process Modeling Environment – The Systematic Literature Review

Anna Sołtysik-Piorunkiewicz$^{(\boxtimes)}$ ⓘ and Patryk Morawiec ⓘ

University of Economics in Katowice, 1 Maja 50 Street, 40-287 Katowice, Poland
anna.soltysik-piorunkiewicz@uekat.pl,
patryk.morawiec@edu.uekat.pl

Abstract. The article presents results of systematic literature review about business process modeling environment based on a cloud computing. The research methodology is evolved. In this paper the research set of 3901 articles is presented and examined. The sources were used for analyses came from Scopus, Web of Science, and EBSCOhost databases. The aim of the paper is to show the impact of cloud computing environment on business process modeling based on literature review. The research subject area is limited to *Computer science* category in cloud computing, business process modelling, and business process management subcategories. The research date set was limited to 313 items based on chosen literature sources indexed from year 2009 to 2020. The development of cloud based models in business process modelling and management, known as BPaaS, eBPM, and BPMaaS are described. The paper shows the state-of-art of business process lifecycle management and presents some classic methods of usage in modeling tools based on BPMN notation, UML, and Petri nets notations. The research methodology uses the different methods of data gathering and searching algorithms, with computer programming. The final findings based on the research questions (RQ1–RQ6) are described and presented in tables and on figures. The conclusion with future research ideas have been shown.

Keywords: BPMN · Cloud computing · Process modeling · BPaaS · Petri nets

1 Introduction

The cloud computing environment have some basic characteristics which are described in context of general essentials features, service models, and deployment models [16]. The impact of cloud computing is shown as a one of the trends in organizational strategies development of new business models [12, 13], and the methods of business process management and business process modelling are changing by cloud for improving business process lifecycle competitiveness in organization management, reducing costs and increasing flexibility of business processes. Cloud computing can offer many opportunities to improve business processes management in organization and use information

© Springer Nature Singapore Pte Ltd. 2021
T.-P. Hong et al. (Eds.): ACIIDS 2021, CCIS 1371, pp. 416–428, 2021.
https://doi.org/10.1007/978-981-16-1685-3_34

technology more efficiently. The aim of the paper is to show the impact of cloud computing environment on business process modeling and management based on literature review. The research dataset was based on Scopus, Web of Science, and EBSCOhost literature sources indexed during period of time 2009–2020. The structure of the paper is divided into five chapters. Chapter two presents the study of Business Process Management Systems background. There are shown some classic methods of business process management and tools based on BPMN notation, UML, and Petri nets notations as well. These methods are described in chapter three. The chapter four shows the cloud based models which have the impact on business modelling methods, and named in literature as BPaaS, eBPM, and BPMaaS. The research methodology was described in chapter five. According to the literature review two hypotheses were formulated, i.e.: H1: Cloud solutions influence the development of business processes in project lifecycle area in context of usage business process modeling tools; H2: Cloud solutions influence the development of business process management. The research methods were used with searching algorithms based on data collection, with computer aided tools. The findings have been presented in tables and on figures as well. The final chapter shows the conclusion and future research topics.

2   Business Process Lifecycle and Business Process Management

Business process is defined as *a series or network of value-added activities, performed by their relevant roles or collaborators, to purposefully achieve the common business goal* [15]. According to Bitkowska [4] a business processes are decomposable to elementary activities level, having clearly defined entry and exit limits integrated with organization, process owner, measurable goals, quantitative, qualitative and valuable efficacy measures, and provides added value. Business processes management in organization have to introduce an iterative procedure of process lifecycle [9]. Description of Oracle's business process lifecycle is shown in Table 1.

Table 1. Business process lifecycle according to Oracle

Phase	Activities	Activities area
Business process analysis	- Business process identification - Business processes maps development - Detailed modeling with visual notation use - Process efficacy and quality planning	Modeling, simulation
Business process execution	- Implementation and execution of implemented processes	Implementation, development
Business process monitoring	- Gathering business processes measures value - Reporting - Assessment of business processes impact on organization	Monitoring, improvement

Source: Own work based on [9]

Business processes can be divided into 3 groups [9]:

– Management processes – processes of planning, development goals setting.

- Operational processes – processes related to organization activity, leading to product/service creation.
- Supporting processes – simulation management and operational processes effectively.

IT solutions for business process management (BPMS - Business Process Management Systems) are designed to complex management of existing business processes in organization and also for its continuous optimization and improvement. BMPS process optimization functionality includes following features [14]: eliminate duplicated activities and tasks, removing discontinuities in process flow, eliminate downtimes and delays in process realization, and support for quality management models.

3   Business Processes Modeling – Classic Approach

Gawin & Marcinkowski presented 16 selected business processes modeling standards divided into 2 main categories adapted visual modeling standards and dedicated visual modeling standards [10]. Below are characterized some of the most popular of them with particular emphasis of BPMN notation.

3.1   BPMN

BPMN stands as Business Process Model and Notation, is graphic standard for representation of business process, created and developed by Object Management Group (OMG). Current version of BPMN standard (2.0) was introduced in 2011 and it's available on-line at www.omg.org as an open standard for everyone [8]. One of important aspects of using BPMN notation is the possibility to transform visual model into executable file based on XML tags. It's realized with use of Business Process Execution Language (BPEL) technique [19]. A BPMN core and layer structure is presented in [18]. BPMN 2.0 notation is organized into 4 diagrams: Process diagram, collaboration diagram, choreography diagram and conversation diagram. In Table 2 each of diagrams is briefly described.

Table 2. Business process model and notation diagram types

Diagram	Purpose	Description
Process diagram	Visualization	Built to illustrate advanced subprocesses, tasks and other objects aspects
Collaboration diagram	Communication	Communication exchange between business process participants
Choreography diagram	Interaction	Built to coordinate business process participants interaction
Conversation diagram	Interaction	Interaction aggregation between business partners

Source: Own work based on [10] and [18]

Table 3. Detail level of business processes model

Model type	Description
Illustrative model	Shows general business process course, without technical aspects like conditional flow parameters, tasks and activities types, undeveloped subprocesses, data objects are not presented on a model
Analytical model	Shows evaluation of executable process implementation works. Activities and tasks types are specified, developed subprocessed, data objects are shown without definition on a model
Executable model	Used for executable process precise description, all objects and parameters are shown on a model with their definitions

Source: Own work based on [8]

According to [8] in business process model can exists 3 levels of model detail described in Table 3.

BPMN standard gained popularity due to its simplicity, versatility and openness. Simplicity of this notation is not limiting it's possibilities to create complex models and additionally provides better clarity and legibility both for business owners, project management team and also technical staff.

3.2 Other Solutions

UML with Profile for Business Modeling

In software engineering a standard to model various phenomena e.g. data flow, sequence, use cases in system development is Unified Modeling Language – a notation consisting of 13 main diagrams and 4 abstract diagrams.

From all UML diagrams can be distinguished an activity diagram and state diagram. These types of diagrams can be implemented in business processes modeling. Activity diagram is a type of behavioral diagram consists of activities, actions, flows (control flow, object flow) and nodes. Similarly to BPMN it exists also swim lanes and partitions. State diagram (state machine diagram) is also example of behavioral diagram, it shows system states and events in system that cause transition from one state to another [25]. A state is defined by the object value of its attributes and relationship between another objects. Event is changing the value of an object state, it takes place at certain point in a time and the state of an object determines a response to event. A transition is a relationship representing changing state of an object to another state [7].

Petri Nets and Event-Driven Process Chain

The concept of communication between asynchronous components of computer system was introduced in 1962 by Carl Adam Petri in his PhD thesis titled *Komunikation mit Automaten* [20]. Nowadays Petri nets based on mathematical foundations are used commonly to model various phenomena, including also business processes. Petri net bases on 3 dependencies which are: Sequence, Alternative, Parallel sequence.

In business processes modeling Petri nets are mostly used for searching dependencies between flow objects and to indicate alternative courses [6]. According to Pasamonik use of Petri nets in business processes modeling is the most precise and deterministic due to formal mathematic model base and it should be used for detailed notation and process control in organization [19].

Event-driven Process Chain is simplified Petri net notation, more elastic and easier to understand than classic Petri net. EPC notation is based on events and activities elements and connections between them. Example of EPC process is used in ARIS notation [3]. Events in EPC notation are responsible for defining precondition and postcondition of a function. For decision making, there are used logical operators. Every EPC process starts with starting event, ends with ending event and have limitations as follows [2]: events can't make OR/XOR decisions; additional process only can connect to EPC function; events have to be linked with AND operator; and for decision making functions can be associated with all logical operators (AND, OR, XOR).

4   Cloud Environment for Business Processes Modeling

4.1   Business Process as a Service

BPaaS is relatively new concept of modeling business processes in a cloud environment. Solutions also known as BPMC (Business Process Management Cloud), provides functionality of business processes modeling, optimization, implementation, monitoring and reporting. There is no difference in functionality between classic on-premises BPMS solutions and BPMC, different is only the way of sharing, development, improvement and scalability for organization and end users [11].

According to Gzik [11], BPaaS is an overarching model over IaaS, PaaS and SaaS model of cloud computing integrating services derived from submodels [11].

BPMC software should follow 5 main features [5]: (1) User experience based - to simplify complex and challenging activity of business processes modeling; (2) Be document capable - due to critical character of documentation in business processes modeling. Modeled process will be better understandable with clear and simple documentation; (3) Provides teamwork opportunity - access to projects by team regardless of time and place; (4) Process library - possibility to archive work, versioning and quickly reengineering existing legacy processes; and (5) Support for BPMN 2.0 notation - avoid outdated notation use for better communication and create understandable processes.

Modeling business processes in cloud environment can have positive impact on project success [17].

4.2   Elastic Business Process in BPMaaS

Business Process Management requires high quality of software services, and the organizations are nowadays able to react rapidly to changing demands for computational resources. The BPM in the cloud gave the opportunity to grow up the flexibility of the business process management. The issues of cloud model environment development covers the architecture of an elastic Business Process Management System in context of existing work on scheduling, resource allocation, monitoring, decentralized coordination, and state management for elastic processes [24], i.e., processes which are carried out using elastic cloud resources. The elastic business process management (eBPM) approach evaluated in the cloud and the new model development has been discovered as BPMaaS (BPM as a Service) [22, 23], which have the opportunity to complete business processes in the cloud as well as an application's software, and hardware infrastructure, going through the idea of 'Everything as a Service'. There are a lot of benefits showed in case studies of the software solutions supporting BPM [20, 21], as an improvement factors of QoS in eBPM.

5   Research Methods and Findings

5.1   The Research Hypotheses

Based on the literature review of business process in cloud environment the following research hypotheses were formulated:

H1: Cloud solutions influence the development of business processes in project lifecycle area in context of usage business process modeling tools.
H2: Cloud solutions influence the development of business process management.

According to hypotheses the following research questions were formulated:

RQ1: What research areas/categories are linked with business processes modeling?
RQ2: What search phrases are linked with business processes modeling?
RQ3: What research areas/categories are linked with business processes management?
RQ4: What search phrases are linked with business processes management?

Additional research questions were formulated according to the character of study:

RQ5: What are the most common source publication types about related search queries?
RQ6: What is the current trend of literature publications related to research topic and when was the greatest interest of topic?

5.2   The Research Methodology - A Systematic Literature Review

The research methodology using in systematic literature review is evolved. A proposed research methodology based on systematic literature review study was conducted including scientific databases Scopus, Web of Science and EBSCOhost Academic Search Ultimate. The methodology [1] was adapted to perform this study and it's shown on a Fig. 1. There are some steps of the research methodology, including study design, selection of accessible full-text scientific database, formulate queries and search limitation, data collection and date set preparation, dataset evaluation, and summarization and conclusions.

The method of the study was based on literature review of business processes in cloud environment. The systematic literature review was designed and performed in

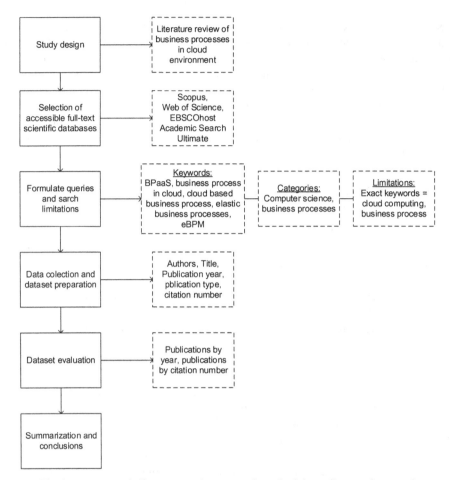

Fig. 1. A systematic literature review research methodology. Source: Own work

context of categories of research areas of cloud computing, business process modelling, and business process management.

The dataset was developed based on group of good known and the most cited sources from literature data bases for researchers and experts in studied area, i.e.: Scopus, Web of Science, and EBSCOhost.

Search keywords were proposed as follow: 'BPaaS', 'business processes in cloud', 'cloud based business processes', 'elastic business processes', 'eBPM'.

The next step was based on queries analysis with searching algorithm, and then the review of dataset was performed. The dataset has been limited, and then the research findings has been evaluated. The research findings were summarized and concluded.

5.3   Research Findings

The research findings were focused on queries results based on dataset from Scopus, Web of Science, and EBSCOhost published between 2009 and 2020. Search results are presented in Table 4. The amount of papers and literature sources were classified due to searching keywords. The chosen databases with sources, keywords, and number of articles were listed in Table 4. Research results were listed in 3901 items: 1041 items from Scopus, 2805 items from Web of Science, and 55 items from EBSCOhost.

Table 4. Databases search results

Database	Keyword	Number of articles
Scopus	"BPaaS"	68
	"business process in cloud"	475
	"cloud based business processes"	286
	"elastic business processes"	207
	"eBPM"	5
		$\sum 1041$
Web of science	"BPaaS"	31
	"business process in cloud"	1662
	"cloud based business processes"	1030
	"elastic business processes"	79
	"eBPM"	3
		$\sum 2805$
EBSCOhost	"BPaaS"	3
	"business process in cloud"	46
	"cloud based business processes"	6
	"elastic business processes"	0
	"eBPM"	0
		$\sum 55$
Summary		$\sum 3901$

Source: Own work

The H1 hypothesis can be confirmed in results of the query related to 'BPaaS', 'business processes in cloud', and 'cloud based business processes' phrases. Answers to research questions (RQ 1, RQ 2) related to H1 hypothesis indicate it exist related papers indexed in all searched databases. Hypothesis H2 can also be confirmed due to results of query (RQ 3, RQ 4) related to 'elastic business processes' and 'eBPM' phrases.

Limitations in search was subject area limited to *Computer science* category, and keywords were limited to cloud computing, business process modelling and business process management. Exact search queries were as follows.

1. For Scopus database:

   - TITLE-ABS-KEY(BPaaS) AND (LIMIT-TO(SUBJAREA,"COMP"))
   - TITLE-ABS-KEY(business processes in cloud) AND (LIMIT-TO(SUBJAREA,"COMP")) AND (LIMIT-TO(EXACTKEYWORD,"Business Process"))
   - TITLE-ABS-KEY(cloud based business processes) AND (LIMIT-TO (SUBJAREA,"COMP")) AND (LIMIT-TO(EXACTKEYWORD,"Business Process"))
   - TITLE-ABS-KEY(elastic business processes) AND (LIMIT-TO(SUBJAREA,"COMP"))
   - TITLE-ABS-KEY(eBPM) AND (LIMIT-TO(SUBJAREA,"COMP"))

2. For Web of Science database:

   - TOPIC: (BPaaS): Refined by: WEB OF SCIENCE CATEGORIES: (COMPUTER SCIENCE INFORMATION SYSTEMS OR COMPUTER SCIENCE THEORY METHODS OR COMPUTER SCIENCE SOFTWARE ENGINEERING OR COMPUTER SCIENCE INTERDISCIPLINARY APPLICATIONS) Indexes = SCI-EXPANDED, SSCI, A&HCI, CPCI-S, CPCI-SSH, BKCI-S, BKCI-SSH, ESCI, CCR-EXPANDED, IC Timespan = All years
   - TOPIC: (business process in cloud) Refined by: WEB OF SCIENCE CATEGORIES: (COMPUTER SCIENCE THEORY METHODS OR COMPUTER SCIENCE INFORMATION SYSTEMS OR COMPUTER SCIENCE SOFTWARE ENGINEERING OR COMPUTER SCIENCE INTERDISCIPLINARY APPLICATIONS) Indexes = SCI-EXPANDED, SSCI, A&HCI, CPCI-S, CPCI-SSH, BKCI-S, BKCI-SSH, ESCI, CCR-EXPANDED, IC Timespan = All years
   - TOPIC: (cloud based business processes) Refined by: WEB OF SCIENCE CATEGORIES: (COMPUTER SCIENCE INFORMATION SYSTEMS OR COMPUTER SCIENCE THEORY METHODS OR COMPUTER SCIENCE SOFTWARE ENGINEERING OR COMPUTER SCIENCE INTERDISCIPLINARY APPLICATIONS) Indexes = SCI-EXPANDED, SSCI, A&HCI, CPCI-S, CPCI-SSH, BKCI-S, BKCI-SSH, ESCI, CCR-EXPANDED, IC Timespan = All years
   - TOPIC: (elastic business processes) Refined by: WEB OF SCIENCE CATEGORIES: (COMPUTER SCIENCE THEORY METHODS OR COMPUTER SCIENCE INFORMATION SYSTEMS OR COMPUTER SCIENCE SOFTWARE ENGINEERING OR COMPUTER SCIENCE INTERDISCIPLINARY APPLICATIONS) Indexes = SCI-EXPANDED, SSCI, A&HCI, CPCI-S, CPCI-SSH, BKCI-S, BKCI-SSH, ESCI, CCR-EXPANDED, IC Timespan = All years
   - TOPIC: (eBPM) Refined by: WEB OF SCIENCE CATEGORIES: (COMPUTER SCIENCE INFORMATION SYSTEMS OR COMPUTER SCIENCE INTERDISCIPLINARY APPLICATIONS) Indexes = SCI-EXPANDED, SSCI,

A&HCI, CPCI-S, CPCI-SSH, BKCI-S, BKCI-SSH, ESCI, CCR-EXPANDED, IC Timespan = All years.

Search queries for EBSCOhost database were not able to export. Search query in Web of Science database for 'business process in cloud' and 'cloud based business processes' keywords gave to many various results, so it wasn't taken into account. After removing 445 duplicate records between databases and between keywords, considered number of articles to final selection was 763. After manual selection of left articles, total number of selected articles is 313. Schema for article acceptance is presented on Fig. 2.

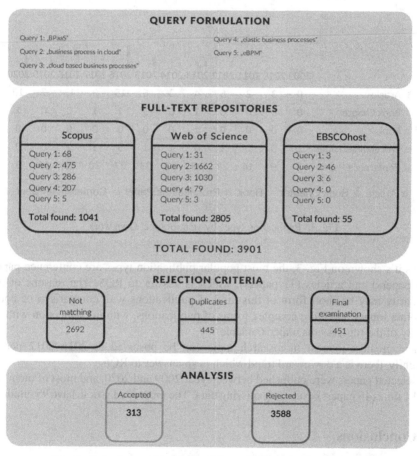

Fig. 2. Selection papers schema. Source: Own work based on [2]

The selection process was based on three rules as follows:

(1) studies highly relevant to business process management in cloud environment. Researches only partially related to the both topics e.g. cloud security were rejected;

(2) articles about general topic were rejected e.g. "32nd International Conference on Information System 2011";
(3) final decision about article relevancy acceptance was based on a title.

The chart of selected papers over time with specification of publication types is presented on Fig. 3.

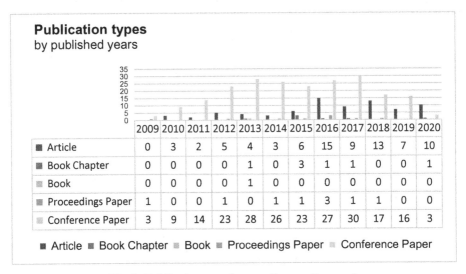

Publication types
by published years

	2009	2010	2011	2012	2013	2014	2015	2016	2017	2018	2019	2020
■ Article	0	3	2	5	4	3	6	15	9	13	7	10
■ Book Chapter	0	0	0	0	1	0	3	1	1	0	0	1
▨ Book	0	0	0	0	1	0	0	0	0	0	0	0
■ Proceedings Paper	1	0	0	1	0	1	1	3	1	1	0	0
▨ Conference Paper	3	9	14	23	28	26	23	27	30	17	16	3

■ Article ■ Book Chapter ▨ Book ■ Proceedings Paper ▨ Conference Paper

Fig. 3. Publication types by year. Source: Own work

As it's shown on Fig. 3, the most popular publication types are conference papers (219 papers) and articles (77 papers), which is answer to RQ5. The reasons of this popularity may be short form of this kind of publications with comparison of books and other longer and more complex forms of publications, with in connection with the novelty of the topic seems understandable.

The greatest interest in research topic can be observed in 2016–2017 period. Currently, there is a downward trend which is an answer to RQ6.

Selected papers were published between year 2009 and 2020, and most of them was cited 1 time (46 papers) excluding missing data. The most cited article have 95 citations.

6   Conclusions

Business processes modeling is an important issue in organizations. Eliminating bottle-necks and malfunctioning processes can influence on organization performance and it's helpful in ensuring proper functioning of organization at all and individual departments.

Cloud computing as a technology based on availability upon requests seems to be good environment for business processes modeling and simulations.

The aim of the paper was to investigate the current trends in scientific literature about cloud based business processes modeling environment. To achieve the chosen goal, a systematic literature review method were performed. During the research it was formulated hypotheses H1 and H2 about the impact of cloud solutions on business processes modeling and business processes management, and accuracy of research study. Hypotheses were verified with the help of research questions. Both hypotheses H1 and H2 was verified positively due to literature review results and showed in Table 4 and on Fig. 3. The described study have some limitation, so the further research need to be focused more on showing the state-of-art in the field, and the impact of cloud computing on modeling and simulation tools development in IT project management.

References

1. Abideen, A.Z., et al.: Mitigation strategies to fight the COVID-19 pandemic—present, future and beyond. J. Health Res. **34**(6), 1–17 (2020)
2. Ajmad, A., et al.: Event-driven process chain (EPC) for modeling and verification of business requirements – a systematic literature review. IEEE Access **6**, 9027–9048 (2018)
3. ARIS Community. https://www.ariscommunity.com/event-driven-process-chain. Accessed 15 Nov 2020
4. Bitkowska, A.: Zarządzanie procesowe we współczesnych organizacjach. Difin, Warsaw (2013)
5. Business Process as a Service: the features of a good app. Heflo blog. https://www.heflo.com/blog/bpm/business-process-as-a-service/. Accessed 10 Nov 2020
6. Chomiak-Orsa, I., Kołtonowska, A.: Modelowanie procesów biznesowych z wykorzystaniem sieci Petriego i BPMN. Próba oceny metod. Informatyka Ekonomiczna **2**(48), 9–18 (2018)
7. Dennis, A., Wixom, B.H., Tegarden, D.: System Analysis & Design with UML Version 2.0. An Object-Oriented Approach, 4th edn., p. 253. Wiley, Hoboken (2012)
8. Drejewicz, Sz.: Zrozumieć BPMN. Modelowanie procesów biznesowych, Helion, Gliwice (2012)
9. Gawin, B.: Modelowanie procesów biznesowych. In: Wrycza, S., Maślankowski, J. (eds.) Informatyka Ekonomiczna. Teoria i zastosowania. Polish Scientific Publishers, Warsaw (2019)
10. Gawin, B., Marcinkowski, B.: Symulacja procesów biznesowych. Standardy BPMS i BPMN w praktyce, Helion, Gliwice (2013)
11. Gzik, T.: Procesy biznesowe w chmurze obliczeniowej, Zeszyty Naukowe Politechniki Częstochowskiej Zarządzanie No. 31 (2018)
12. Jelonek, D., Stępniak, C., Turek, T., Ziora, L.: Identification of mental barriers in the implementation of cloud computing in the SMEs in Poland. In: 2014 Federated Conference on Computer Science and Information Systems, Warsaw, pp. 1251–1258 (2014)
13. Jelonek, D., Wysłocka, E.: Barriers to the development of cloud computing adoption and usage in SMEs in Poland. Adv. Inf. Sci. Appl. **1**, 128–133 (2014)
14. Jurek, J.: Wdrożenia informatycznych systemów zarządzania. Polish Scientific Publishers, Warsaw (2016)
15. Ko, R.K.L.: A computer scientist's introductory guide to business process management (BPM). Crossroads **15**(4), 11–18 (2009)
16. Mell, P., Grance, T.: The NIST definition of cloud computing, recommendations of the National Institute of Standards and Technology, NIST Special Publication 800-145 (2011). https://faculty.winthrop.edu/domanm/csci411/Handouts/NIST.pdf. Accessed 30 Nov 2020
17. Morawiec, P., Sołtysik-Piorunkiewicz, A.: The new role of cloud technologies in management information systems implementation methodology. In: Arai, K., Kapoor, S., Bhatia, R. (eds.) FTC 2020. AISC, vol. 1290, pp. 423–441. Springer, Cham (2021). https://doi.org/10.1007/978-3-030-63092-8_29

18. OMG BPMN v2.0.1 Specification. https://www.omg.org/spec/BPMN/2.0.1/PDF. Accessed 13 Nov 2020

19. Pasamonik, P.: Modelowanie procesów biznesowych zorientowane na czynności. Zesz. Nauk. WSInf **9**(2), 102–116 (2010)

20. Peterson, J.L.: Petri Net Theory and the Modeling of Systems. Prentice Hall, Englewood Cliffs (1981)

21. Rosinosky, G., Labba, C., Ferme, V., Youcef, S., Charoy, F., Pautasso, C.: Evaluating multitenant live migrations effects on performance. In: Panetto, H., Debruyne, C., Proper, H.A., Ardagna, C.A., Roman, D., Meersman, R. (eds.) OTM 2018. LNCS, vol. 11229, pp. 61–77. Springer, Cham (2018). https://doi.org/10.1007/978-3-030-02610-3_4

22. Rosinosky, G., Youcef, S., Charoy, F.: An efficient approach for multitenant elastic business processes management in cloud computing environment. In: IEEE, pp. 311–318, June 2016

23. Rosinosky, G., Youcef, S., Charoy, F.: A genetic algorithm for cost-aware business processes execution in the cloud. In: Pahl, C., Vukovic, M., Yin, J., Yu, Qi. (eds.) ICSOC 2018. LNCS, vol. 11236, pp. 198–212. Springer, Cham (2018). https://doi.org/10.1007/978-3-030-03596-9_13

24. Schulte, S., Janiesch, Ch., Venugopal, S., Weber, I., Hoenisch, P.: Elastic business process management: state of the art and open challenges for BPM in the cloud. Future Gen. Comput. Syst. 46, 36–50 (2015)

25. Sommerville, I.: Software Engineering, 9th edn., pp. 135–136. Addison-Wesley, Boston (2010)

Terrain Classification Using Neural Network Based on Inertial Sensors for Wheeled Robot

Artur Skoczylas[✉], Maria Stachowiak, Paweł Stefaniak, and Bartosz Jachnik

KGHM CUPRUM Research and Development Centre Ltd., gen. W. Sikorskiego Street 2-8, 53-659 Wroclaw, Poland

{askoczylas,mstachowiak,pkstefaniak,bjachnik}@cuprum.wroc.pl

Abstract. In the article, a method of terrain recognition for robotic application has been described. The main goal of the research is to support the robot's motor system in recognizing the environment, adjusting the motion parameters to it, and supporting the location system in critical situations. The proposed procedure uses differences between calculated statistics to detect the diverse type and quality of ground on which wheeled robot moves. In the research IMU (Inertial Measurement Unit) has been used as a main source of data, especially 3-axis accelerometer and gyroscope. The experiment involved collecting data with a sensor mounted on a remotely controlled wheeled robot. This data was collected from 4 hand-made platforms that simulated different types of terrain. For terrain recognition, a neural network-based analytical model has been proposed. In this paper authors present results obtained from the application model to experimental data. The paper describes the structure of NN and the whole analytical process in detail. Then, based on a comparison of the obtained results with the results from other methods, the value of the proposed method was shown.

Keywords: Neural network · Terrain classification · Pattern recognition

1 Introduction

Currently, terrain recognition techniques are a very popular topic in the literature. Over the years, one can observe a lot of different ideas to solve this problem. Many works focus on different ways of finding differences in soil characteristics. The majority of the terrain recognition approaches are focused primarily on classification using vision data. In [1] the method of rural roads' segmentation using laser images (with laser range-finder) and camera images is described. [2] also uses laser data, but the emphasis is taken on distinguishing between thin objects such as wires or tree branches, and solid objects such as ground surfaces, rocks, or tree trunks. The experiment was carried out with the vehicle ground using a variety of stationary laser sensors. Classification algorithms that are based on visual characteristics (such as color or texture) are often sensitive to changes in lighting, which in closed spaces or underground conditions very often can lead to poor quality predictions [3]. In [4], however, a legged robot that collected vibration data from an on-board inertial measurement unit and magnetic encoders was used. The results was

© Springer Nature Singapore Pte Ltd. 2021
T.-P. Hong et al. (Eds.): ACIIDS 2021, CCIS 1371, pp. 429–440, 2021.
https://doi.org/10.1007/978-981-16-1685-3_35

accurate and robust to variations in stride frequency and low-lag, encouraging to further observation.

One of the first works on vibration-based terrain classification has been proposed in 2002 by Iagnemm and Dubowsky [5] as an innovative detection mode for determining the terrain class for hazard detection. For the investigated case study, using vibration data has proven to be a good choice. The differences in the substrate are visible in various statistics. Using Fast Fourier Transform (FFT) and on the Power Spectral Density (PSD), it was possible to train a model distinguishing 7 different types of a substrate [6]. The model was trained and classified the data with a Support Vector Machine (SVM). Similar relationships were investigated in [7], however, a neural network was used there along with the low-pass filter and the FFT. Other noteworthy studies are [8] and [9], which also served as an inspiration for this article. The first work involved a walking robot that tried to classify a terrain based on force/torque data obtained from the robot's cube sensors and a signal created by a discrete wavelet transform. In the second paper, the authors investigated the difference in surface type using both cameras and a force-torque (FT) sensor. The FT sensor collected vibration and force data. A different approach was used here and a feature vector from vibration data, among other things, was used to teach the model. A similar feature vector will be used in this work. The advantage of choosing the neural network method is its versatility. The proposed layers in the model are intended to give greater accuracy than could be achieved by using other methods.

2 Experiment

The task of this article is to describe a method based on AI that identifies the type of surface and assigns it an appropriate quality. The samples on which the neural network was built and tested were obtained from an experiment that used a wheeled robot and built modules of the substrate. During the experiment, a wheeled robot with mounted sensors was traveling successively over various types of models simulating various types of ground. Then the data from these passes were collected and processed. In this way, a set of samples from each category was obtained, which were then randomly divided into two subsets, training and testing. As the name suggests, they were used to training and checking the prediction effectiveness of the described network.

2.1 Wheeled Robot

A simple, remote-controlled wheeled robot was used in experimental works in order to obtain data for further analytical model development. The robot is presented in Fig. 1. The robot is equipped with 4 engines, which in addition to propelling it allows him to turn in place (obtained using the engine speed changes). The device is controlled using Bluetooth.

The robot has been equipped with a number of sensors to measure values that could convey information about the type of ground on which it moves. Data recordings are stocked on the micro SD card.

Fig. 1. Wheeled robot

2.2 Modules of the Terrain

In order to simulate various types of substrate, special, modular mock-ups were created. Each module was designed to simulate a different type of terrain that can be briefly divided from easiest to hardest to drive. For the purposes of the experiment, 4 modules were created:

- **Plain cardboard** - plain surface, no-frills. The task of this module was to simulate *the ideal quality of the road*. Illustrated in Fig. 2(a).
- **Corrugated Cardboard** - the module is covered with corrugated cardboard. This cardboard includes some kind of resistance while driving, without generating huge vibrations. The task of this module was to simulate *the good quality of the road*. The module presented in Fig. 2(b).
- **Plasticine** - the ground on which the hills were built, made of hardened plasticine. These hills, going pointwise, caused considerable difficulties while driving. It can be seen in Fig. 2(c), the hills occur irregularly, which was supposed to simulate *the average quality of the road* with point deterioration.
- **Caps** - a module whose entire running surface has been irregularly covered with PET bottle caps inverted. The module prepared in this way offers significant difficulties in the passage over its entire surface, which affects the measurements of the robot, therefore the module presents *the bad quality of the road*. The module is shown in Fig. 2(d).

As can be seen in the figures above, each of the mock-ups has a designated area to move around (black lines). This area has the same dimensions for all modules, which allows them to be easily connected. Also, the ArUco code has been placed in each corner of the modules. The codes were used to identify individual modules as well as cut out

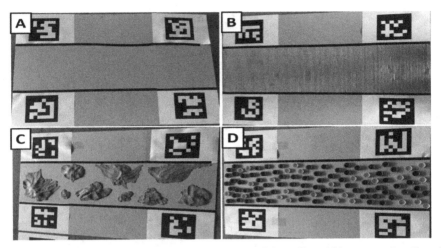

Fig. 2. Modules simulating various types of substrate: a) plain cardboard, b) corrugated cardboard, c) plasticine, d) caps

the module's area in the video recording. This facilitated signal segmentation, however, it was only used in cases where the robot was traveling on several consecutive modules.

2.3 The Course of the Experiment

The experiment consisted of repeatedly driving over four modules with a wheeled robot (plain cardboard, corrugated cardboard, plasticine and caps). Between each pass, the robot was paused for a few seconds to facilitate the subsequent segmentation of signals from its sensors. During the ride, efforts were made to make the ride as stable as possible and in a straight line. Attention was also drawn to the passage outside the pattern to be as short as possible so that the route outside the mock-up would not interfere with the results. The described procedure is presented in Fig. 3.

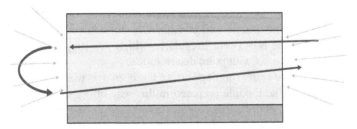

Fig. 3. The procedure of driving along the mock-up module

During the entire experiment, each module was driven around 70 times. Single trips were not categorized into directions (up - down, down - up). Between successive passes, the robot was manually turned over (again to facilitate segmentation) by lifting it and rotating it by 180°.

2.4 Data Preparation

The data series from the experiment containing multiple runs of one module was saved in a separate file for each module. A sample fragment of the variable on which the pattern is the most visible is presented in Fig. 4. The figure shows the signal from the X axis of the accelerometer, which occurs greater and lesser excitations at intervals. This is the result of the steps described earlier to facilitate segmentation. Large excitations are the result of the robot passing through the module, while the smaller ones correspond to the procedure of robot rotation by lifting. If the travel vibrations are too similar to those related to rotation, segment the signal concerning other axes, where the lift will be much more visible.

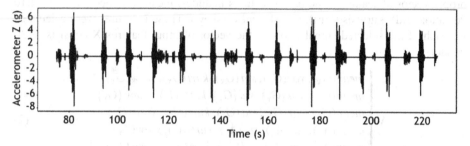

Fig. 4. Raw signal from Z-axis accelerometer after completing a group of passes

The procedure of segmentation and extraction of rides from the above signal started by dividing it into smaller fragments concerning places with very low values (stops). Then, each of the fragments had their values of the accelerometer and gyroscope in selected axes (depending on the location of the sensor) averaged and on this basis it was decided whether the given fragment was a ride or a rotation.

The next step in data processing was their further segmentation. Each run of the robot was divided into 3 shorter fragments by overlapping windows. The position of these windows was random, however, they had to meet two conditions. Firstly, all 3 windows in total had to cover the entire signal, this meant that no signal fragment could be omitted. The second condition specifies the minimum size of such a window. The window cannot be too small due to a significant reduction in information and the possibility of causing interference. The segmentation of a single ride is shown in Fig. 5.

2.5 Vector of Features

By properly preparing the data, there were prepared about 840 samples, which were divided into 2 subsets: training and testing, in a 4:1 ratio. This division took place at random. Then, to create the vector of features 2 signals that were considered the most informative were selected. They are:

$$Gyroscope \begin{cases} G_x - in\ axis\ X \\ G_y - in\ axis\ Y \\ G_z - in\ axis\ Z \end{cases} \quad Accelerometer \begin{cases} A_x - in\ axis\ X \\ A_y - in\ axis\ Y \\ A_z - in\ axis\ Z \end{cases} \quad (1)$$

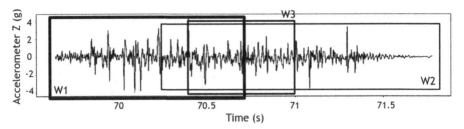

Fig. 5. Single-trip segmentation procedure

Next, from these signals, statistical measures were calculated and then used to the proper vector of features. The list of features includes: mean, variance (var), standard deviation (std), kurtosis (kurt), and skewness (skew). These measures were calculated for each of the selected signals, so that the vector of input features X consists of 30 variables:

$$X \begin{cases} mean(G_x), var(G_x), std(G_x), kurt(G_x), skew(G_x) \\ mean(G_y), var(G_y), std(G_y), kurt(G_y), skew(G_y) \\ mean(G_z), var(G_z), std(G_z), kurt(G_z), skew(G_z) \\ mean(A_x), var(A_x), std(A_x), kurt(A_x), skew(A_x) \\ mean(A_y), var(A_y), std(A_y), kurt(A_y), skew(A_y) \\ mean(A_z), var(A_z), std(A_z), kurt(A_z), skew(A_z) \end{cases} \qquad (2)$$

3 The Neural Network

The neural network (NN) model was used to classify the data for a given type of terrain. Neural networks are multi-layer networks of neurons (nodes) connected. Each connection can then transmit a signal to the other. A node, which receives that signal then processes it and can send the signal further to their connected nodes. Different sections perform different transformations on their inputs. Neural network approaches have strong power to classify various patterns similarly to a human's brain. Neural networks have been exploited in a wide variety of applications, the majority of which are concerned with pattern recognition in one form or another. From a pattern recognition point of view, neural networks can be considered as an extension of many standard techniques that have been developed over many decades [10]. In [11] its great possibilities for image processing are shown compared to the conventional methods like Bayesian and minimum distance method. For terrain classification, neural networks have been used before. In [12] authors using a skid steer mobile with some optional on-board sensors and neural networks could classify terrain in real-time and during the robot's actual mission. Moreover, in [13] artificial neural network correctly recognized almost 99% of terrain classes by using data from RGB-D sensor, which provides vision data.

Selection of the network structure for the task is one of the most important tasks in projects related to neural networks. Invalid choice of network parameters lead to the unfavourable events like underfitting or overfitting [14]. An example of the network structure selection method for a physical phenomenon was presented in [15], where an algorithm was used to create different network structures (in specified boundaries). These structures were then subjected to an efficiency assessment on the basis of which the best one for a given problem was selected. The neural network model was created using the PyTorch library [16] for Python. The presented network consists of 5 layers as shown in Fig. 6.

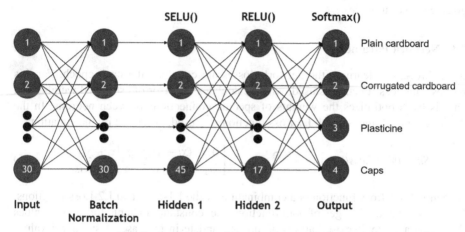

Fig. 6. Architecture of the neural network model

The first layer of the network is the input layer. It has 30 neurons, one for each feature vector element. Each neuron from this layer is connected to each neuron from the second layer, which is the batch normalization layer. This layer is responsible for recentering and scaling the signals received from the input layer, thanks to which the network is faster and more stable [17]. The 30 neurons of this layer are connected to the linearly corresponding neurons of the first hidden layer. It performs the activation function for this layer SELU (Scaled Exponential Linear Unit) described by the formula:

$$SELU\,(x) = \lambda \begin{cases} x, x > 0 \\ \alpha e^x - \alpha, x \le 0 \end{cases} \tag{3}$$

SELU was proposed to have self-normalizing properties and makes the network more robust to noise and permutations [18]. Each neuron of the first hidden layer (45 neurons) is then connected to each neuron of the second hidden layer. The second layer has been reduced to only 17 neurons, and the activation function of this layer is ReLU (Rectified Linear Unit) which is described as follows:

$$ReLU\,(x) = \max(0, x) \tag{4}$$

ReLU was proposed to reduce the gradient vanishing problem induced by the sigmoid or tahn functions. The advantage of using this function is that the neurons are not activated at the same time and because of that ReLU is more efficient than other functions [19].

The final layer is the output layer with the function Softmax(). Softmax() takes as input a vector of K real numbers which it normalizes to a probability distribution consisting of K probabilities proportional to the exponents of the input numbers. Softmax() compared to some sigmoid functions that can be used for binary classification, can be used to solve multi-class classification problems [19]. The final layer consists of 4 neurons, one for each possible substrate category. The net effect is a table containing the probability estimated by the network that a given vector of input features describes a given group of modules.

3.1 Network Parameters

In the process of learning the network, losses are counted after each epoch. For this purpose, the loss function is used, which calculates the error of estimated weights and on this basis optimizes the weights of specific connections between neurons. In the described model, the Smooth L1 Loss function was used, described by the equation:

$$Smooth\,L1\,Loss(x, y) = \frac{1}{n}\sum_i z_i, \quad z_i \begin{cases} 0.5 \cdot (x_i - y_i)^2, & if \ |x_i - y_i| < 1 \\ |x_i - y_i| - 0.5, & otherwise \end{cases} \quad (5)$$

Smooth L1 Loss function is a combination of the L1-Loss and L2-Loss functions. It combines the advantages of both functions, i.e. constant gradients for large x values (L1-Loss) and smaller oscillations during the update in the case of smaller x values (L2-Loss). This function is less sensitive to outliers and it eliminates the sensitivity of adjusting the learning, which otherwise when using L2-Loss function alone could lead to exploding gradients [20].

Another important parameter is the pace of teaching the network. Neural networks, as mentioned before, use an optimization algorithm during the training process that estimates the error gradient for the current state of the model, and updates the model weights using back propagation. The amount by which these weights are updated during training is called the Learning Rate. In the case of the described model, the rate was equal 0.0001.

The last parameter discussed is the number of epochs on which a given network has been trained. The number of epochs determines how many times we optimize the weights of our model. However, it should be noted that this parameter should be selected appropriately because too small will cause under fitting, and otherwise, if it is too high, overfitting will occur. In the model described, the number of epochs was 5000.

3.2 Network Performance

Once the network is ready, its accuracy in categorization should be checked. This is done by using completely new, yet unknown data for classification. After the network has calculated the resulting probability for this data, it is checked against the markings and the accuracy of the network is calculated from this. However, due to the relatively small data set, it was decided to slightly extend the described algorithm. Namely, the network was launched, i.e. trained and tested, a greater number of times (approximately 50). Before each such run, the dataset was randomly mixed before subsets were selected, so that each run provided a different training and test set. For each such run, the percentage of correctly estimated labels was calculated. At Fig. 7 the results are presented broken down into the type of the classified substrate.

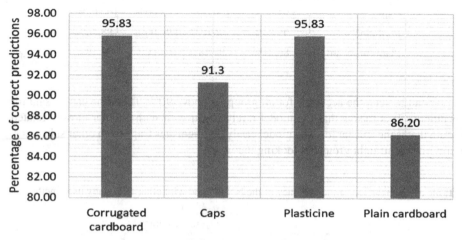

Fig. 7. Percentage of correct predictions per group

It can be seen that the results achieved for all groups are above 85%. The average correctness of the network was calculated at 89.64%. The worst result is for plain cardboard. It is possible that this is due to the fewest characteristics of a given substrate. Plain cardboard is just a straight path with no bulges. In Table 1 the confusion matrix for the data is presented.

Table 1. Confusion matrix for the data

Confusion matrix				
Pred	True			
	Corrugated cardboard	Caps	Plasticine	Plain cardboard
Corrugated cardboard	23	0	0	3
Caps	0	21	0	0
Plasticine	0	2	23	1
Plain cardboard	1	0	1	25

Plain cardboard and corrugated cardboard, due to their similarity, are most often confused with each other. It is similar with the caps in relation to plasticine. However, these are only individual cases. Further analysis of the results are shown in Table 2.

Table 2. Measures of the diagnostic value

	Sensitivity	Specificity	Accuracy	Precision
Corrugated cardboard	95.83%	96.05%	96.00%	88.46%
Caps	91.30%	100.00%	98.00%	100.00%
Plasticine	95.83%	96.05%	96.00%	88.46%
Plain cardboard	86.21%	97.18%	94.00%	92.59%

As can be seen the accuracy for all groups is above 90%. The same can be said for specificity. This means that the model rarely confuses it with another closely related terrain that is the true target. The model easily distinguishes Caps from other substrates. Furthermore, straight ground is looking the worst.

Table 3. Comparison of the accuracy of the NN method with other frequently used methods

	Overall accuracy
NN method	95.5%
Random forest method	84.0%
kNN method	87.58%
SVM method	91.83%

The quality of the prepared model was also compared with other methods, including the SVM method, which is commonly used for such problems. The results are presented in Table 3. Each of the methods gave satisfactory results, which proves well-chosen statistics. Only two of the methods have the accuracy above 90%, of which the method used in the article achieved the highest value.

4   Conclusions

The article presents the preparation and course of an experiment aimed at creating a neural network model, which would define the type of road from given data. Various types of road were simulated by preparing 4 mock-up modules. The proposed model of the neural network was trained using data collected from the IMU sensor. The model uses basic statistics calculated on signals from the accelerometer and the gyroscope.

Looking at the quality of the results obtained, the work can be expanded. This type of signal and the statistics obtained from it seem to distinguish the types of substrate very well. Moreover, the prepared model gives the best results among the most popular methods used in this type of problem. The next step would be to validate the model on real data, for example on different road conditions, where changes in signals may be smaller.

The possibility of developing and using the model for the classification and quality monitoring of mine roads is an interesting development direction to take in the future. In typical mining conditions, the nature of the surface substrate, which is a very important factor, often varies greatly. Large-scale mining machinery, which often transports large amounts of excavated material, is much more prone to failure if speed is not matched to the quality of the road.

Acknowledgment. This work is a part of the project which has received funding from the European Union's Horizon 2020 research and innovation programme under grant agreement No 780883.

References

1. Rasmussen, C.: Combining laser range, color, and texture cues for autonomous road following. In: Proceedings 2002 IEEE International Conference on Robotics and Automation (Cat. No. 02CH37292), vol. 4, pp. 4320–4325. IEEE, May 2002
2. Vandapel, N., Huber, D.F., Kapuria, A., Hebert, M.: Natural terrain classification using 3-d ladar data. In: Proceedings of the IEEE International Conference on Robotics and Automation, 2004. Proceedings. ICRA 2004, vol. 5, pp. 5117–5122. IEEE April, 2004
3. Brooks, C.A., Iagnemma, K.: Vibration-based terrain classification for planetary exploration rovers. IEEE Trans. Rob. **21**(6), 1185–1191 (2005)
4. Bermudez, F.L.G., Julian, R.C., Haldane, D.W., Abbeel, P., Fearing, R.S.: Performance analysis and terrain classification for a legged robot over rough terrain. In: 2012 IEEE/RSJ International Conference on Intelligent Robots and Systems, pp. 513–519. IEEE, October 2012
5. Iagnemma, K.D., Dubowsky, S.: Terrain estimation for high-speed rough-terrain autonomous vehicle navigation. In: Unmanned Ground Vehicle Technology IV, vol. 4715, pp. 256–266. International Society for Optics and Photonics, July 2002
6. Weiss, C., Frohlich, H., Zell, A.: Vibration-based terrain classification using support vector machines. In: 2006 IEEE/RSJ International Conference on Intelligent Robots and Systems, pp. 4429–4434. IEEE, October 2006
7. DuPont, E.M., Roberts, R.G., Selekwa, M.F., Moore, C.A., Collins, E.G.: Online terrain classification for mobile robots. In: ASME International Mechanical Engineering Congress and Exposition, vol. 42169, pp. 1643–1648, January 2005
8. Walas, K., Kanoulas, D., Kryczka, P.: Terrain classification and locomotion parameters adaptation for humanoid robots using force/torque sensing. In: 2016 IEEE-RAS 16th International Conference on Humanoid Robots (Humanoids), pp. 133–140. IEEE, November 2016
9. Walas, K.: Terrain classification and negotiation with a walking robot. J. Intell. Rob. Syst. **78**(3–4), 401–423 (2015)
10. Bishop, C.M.: Neural networks: a pattern recognition perspective. Aston University (1996)
11. Yoshida, T., Omatu, S.: Pattern recognition with neural networks. In: Proceedings of the IEEE 2000 International Geoscience and Remote Sensing Symposium. Taking the Pulse of the Planet: The Role of Remote Sensing in Managing the Environment, IGARSS 2000. (Cat. No. 00CH37120), vol. 2, pp. 699–701. IEEE, July 2000
12. Ojeda, L., Borenstein, J., Witus, G., Karlsen, R.: Terrain characterization and classification with a mobile robot. J. Field Robot. **23**(2), 103–122 (2006)
13. Kozlowski, P., Walas, K.: Deep neural networks for terrain recognition task. In: 2018 Baltic URSI Symposium (URSI), pp. 283–286. IEEE May, 2018

14. Dudzik, M., Stręk, A.M.: ANN architecture specifications for modelling of open-cell aluminum under compression. Math. Probl. Eng. **2020**, 1–26 (2020)
15. Dudzik, M.: Towards characterization of indoor environment in smart buildings: modelling PMV index using neural network with one hidden layer. Sustainability **12**(17), 6749 (2020)
16. Paszke, A., et al.: PyTorch: an imperative style, high-performance deep learning library. In: Wallach, H., Larochelle, H., Beygelzimer, A., d'Alché-Buc, F., Fox, E., Garnett, R. (eds.) Advances in Neural Information Processing Systems 32, pp. 8024–8035. Curran Associates, Inc. (2019). https://papers.neurips.cc/paper/9015-pytorch-an-imperative-style-high-performance-deep-learning-library.pdf
17. Ioffe, S., Szegedy, C.: Batch normalization: accelerating deep network training by reducing internal covariate shift. arXiv preprint arXiv:1502.03167 (2015)
18. Zhang, J., Yan, C., Gong, X.: Deep convolutional neural network for decoding motor imagery based brain computer interface. In: 2017 IEEE International Conference on Signal Processing, Communications and Computing (ICSPCC), pp. 1–5. IEEE October, 2017
19. Sharma, S.: Activation functions in neural networks. Towards Data Science, 6 (2017)
20. Girshick, R.: Fast R-CNN. In: Proceedings of the IEEE International Conference on Computer Vision, pp. 1440–1448 (2015)

Author Index

Printed in the United States
by Baker & Taylor Publisher Services